International Conference on Inverse Problems 2010

Journal of Physics: Conference Series Volume 290

Hong Kong, China
13-17 December 2010

ISBN: 978-1-61782-911-6
ISSN: 1742-6588

Printed from e-media with permission by:

Curran Associates, Inc.
57 Morehouse Lane
Red Hook, NY 12571

Some format issues inherent in the e-media version may also appear in this print version.

Printed by Curran Associates, Inc. (2011)

For permission requests, please contact the Institute of Physics
at the address below.

Institute of Physics
Dirac House, Temple Back
Bristol BS1 6BE UK

Phone: 44 1 17 929 7481
Fax: 44 1 17 920 0979

techtracking@iop.org

Additional copies of this publication are available from:

Curran Associates, Inc.
57 Morehouse Lane
Red Hook, NY 12571 USA
Phone: 845-758-0400
Fax: 845-758-2634
Email: curran@proceedings.com
Web: www.proceedings.com

TABLE OF CONTENTS

PREFACE

Following the first *International Conference on Inverse Problems – Recent Theoretical Development and Numerical Approaches* held at the City University of Hong Kong in 2002, the fifth International Conference was held again at the City University during December 13–17, 2010.

This fifth conference was jointly organized by Professor Yiu-Chung Hon (Co-Chair, City University of Hong Kong, HKSAR), Dr Leevan Ling (Co-Chair, Hong Kong Baptist University, HKSAR), Professor Jin Cheng (Fudan University, China), Professor June-Yub Lee (Ewha Womans University, South Korea), Professor Gui-Rong Liu (University of Cincinnati, USA), Professor Jenn-Nan Wang (National Taiwan University, Taiwan), and Professor Masahiro Yamamoto (The University of Tokyo, Japan). It was agreed to alternate holding the conference among the above places (China, Japan, Korea, Taiwan, and Hong Kong) once every two years. The next conference has been scheduled to be held at the Southeast University (Nanjing, China) in 2012.

The purpose of this series of conferences is to establish a strong collaborative link among the universities of the Asian-Pacific regions and worldwide leading researchers in inverse problems. The conference addressed both theoretical (mathematics), applied (engineering) and developmental aspects of inverse problems. The conference was intended to nurture Asian-American-European collaborations in the evolving interdisciplinary areas and it was envisioned that the conference would lead to long-term commitments and collaborations among the participating countries and researchers.

There was a total of more than 100 participants. A call for the submission of papers was sent out after the conference, and a total of 19 papers were finally accepted for publication in this proceedings. The papers included in the proceedings cover a wide scope, which reflects the current flourishing theoretical and numerical research into inverse problems.

Finally, as the co-chairs of the Inverse Problems Conference 2010, we wish to express our cordial thanks to all the keynote, plenary, and invited speakers and members of the Advisory Board. We would like to thank the City University of Hong Kong, the Hong Kong Mathematical Society, the K C Wong Education Foundation, the Lee Hysan Foundation, and the Wei Lun Foundation for their generous financial support. Particularly, we wish to thank Miss Lonn Chan for her excellent administrative work in making all the arrangements for the conference.

March 2011
Yiu-Chung Hon, City University of Hong Kong, HKSAR
Leevan Ling, Hong Kong Baptist University, HKSAR

Invited Keynote Speaker
H Thomas Banks, North Carolina State University, USA

Invited Plenary Speakers
Ryuichi Ashino, Osaka Kyoiku University, Japan
Andrea Caponnetto, City University of Hong Kong, HKSAR
Jeng-Tzong Chen, National Taiwan Ocean University, Taiwan
Hiromichi Itou, Gunma University, Japan
Hyeonbae Kang, Inha University, South Korea
Michael Klibanov, University of North Carolina at Charlotte, USA
Kim Knudsen, Technical University of Denmark, Denmark
Rainer Kress, Georg-August-Universität Göttingen, Germany
Jérôme Le Rousseau, University of Orleans, France
Chang-Ock Lee, Korea Advanced Institute of Science and Technology (KAIST), South Korea
Gui-Rong Liu, University of Cincinnati, USA
Shuai Lu, Fudan University, China
Fadil Santosa, University of Minnesota, USA
Tomoya Takeuchi, North Carolina State University, USA
Xiang Xu, Zhejiang University, China
Jun Zou, The Chinese University of Hong Kong, HKSAR

Invited Speakers
Mohamed Abdel-Mooty, The American University in Cairo, Egypt
Krishna Agarwal, National University of Singapore, Singapore
Hui Cao, Sun Yat-sen University, China
Wen Chen, Hohai University, China
Jin Cheng, Fudan University, China
Lian Duan, University of Oxford, UK
Meibao Ge, Zhejiang Sci-Tech University, China
Ralf Hielscher, TU Chemnitz, Germany
Guanghui Hu, Weierstrass Institute for Applied Analysis and Stochastics (WIAS), Germany
Kiwan Jeon, National Institute for Mathematical Sciences, South Korea
Yu Jiang, Hokkaido University, Japan
De-Xing Kong, Zhejiang University, China
Olga Krivorotko, Novosibirsk State University, Russia
Philipp Küegler, Austrian Academy of Sciences, Austria
Kiwoon Kwon, Dongguk University, South Korea
Chun-Kong Law, National Sun Yat-sen University, Taiwan
June-Yub Lee, Ewha Womans University, South Korea
Ming Li, City University of Hong Kong, HKSAR
Mikyoung Lim, Korea Advanced Institute of Science and Technology (KAIST), South Korea
Jijun Liu, Southeast University, China
Xiaodong Liu, Chinese Academy of Sciences, China

Fuming Ma, Jilin University, China
Hang Ma, Shanghai University, China
Valdemar Melicher, Ghent University, Belgium
Gen Nakamura, Hokkaido University, Japan
Long Tuan Nguyen, Ruhr-Universitat Bochum, Germany
Etibar Panakhov, Firat University, Turkey
Robert Plato, University of Siegen, Germany
Jean-Pierre Puel, The University of Tokyo, Japan
Hans-Jürgen Reinhardt, University of Siegen, Germany
Maxim Shishlenin, Sobolev Institute of Mathematics, Russia
Ilya Silvestrov, Institute of Petroleum Geology and Geophysics SB RAS, Russia
Mourad Sini, Austrian Academy of Sciences, Austria
Ján Sládek, Slovak Academy of Sciences, Slovakia
Zhijun Tan, Sun Yat-sen University, China
Quan-Fang Wang, The Chinese University of Hong Kong, HKSAR
Wenyan Wang, The University of Tokyo, Japan
Pihua Wen, Queen Mary, University of London, UK
Zhihai Xiang, Tsinghua University, China
Xiang-Tuan Xiong, Northwest Normal University, China
Dinghua Xu, Zhejiang Sci-Tech University, China
Yongzhi Steve Xu, University of Louisville, USA
Jiaqing Yang, Chinese Academy of Sciences, China
Jian Zhang, Sichuan Normal University, China
Ying Zhang, Fudan University, China
Ting Zhou, University of Washington, USA
Jianxin Zhu, Zhejiang University, China

Advisory Board
H Thomas Banks, North Carolina State University, USA
Tony F Chan, The Hong Kong University of Science and Technology, HKSAR
Heinz W Engl, Johannes Kepler Universitát Linz, Austria
Rainer Kress, Georg-August-Universitát Góttingen, Germany
Tatsien (Daqian) Li, Fudan University, China
Gunther Uhlmann, University of Washington, USA

Organizing Committee
Jin Cheng, Fudan University, China
Yiu-Chung Hon (Co-Chair), City University of Hong Kong, HKSAR
June-Yub Lee, Ewha Womans University, South Korea
Leevan Ling (Co-Chair), Hong Kong Baptist University, HKSAR
Gui-Rong Liu, University of Cincinnati, USA
Jenn-Nan Wang, National Taiwan University, Taiwan
Masahiro Yamamoto, The University of Tokyo, Japan

PEER REVIEW STATEMENT

All papers published in this volume of *Journal of Physics: Conference Series* have been peer reviewed through processes administered by the proceedings Editors. Reviews were conducted by expert referees to the professional and scientific standards expected of a proceedings journal published by IOP Publishing.

Multipoles-based linear sampling method: impact of using multipole expansion

K Agarwal[1] and X Chen
Department of Electrical and Computer Engineering,
National University of Singapore, Singapore, 117576

E-mail: krishna.agarwal@nus.edu.sg, elechenx@nus.edu.sg

Abstract. In this article, we present a physical interpretation of the linear sampling method (LSM). The proposed interpretation is based on the multipole expansion of the induced currents at the sampling point, thus called multipoles-based linear sampling method (MLSM). MLSM does not interpret LSM as the focusing of fields at a sampling point. It rathers considers LSM as creating a current distribution that effectively radiates like a monopole at the sampling point. In this sense, MLSM brings LSM closer to the original scattering problem. We show that if infinite multipoles are considered, MLSM performs similar to LSM. This fact can be used to understand the poor performance of LSM in some particular cases, for example, circular and annular scatterers with specific permittivities. We show that it is not necessary to use infinite multipoles in MLSM. The truncation of higher order multipoles in MLSM serves as a regularization technique. As compared to the conventionally used Tikhonov regularization in LSM, the regularization in MLSM is not based on mathematical theory. It is rather a physics-based truncation technique as we truncate the multipoles which anyways do not contribute significantly to the scattered field. Such regularization is simple to understand and implement for practical engineering purposes.

1. Introduction

Linear sampling method (LSM) is a qualitative imaging method used in electromagnetic inverse scattering problems for estimating the scatterer support [1-6]. In LSM, we try to choose the amplitudes of various sources, such that the scattered field due to the chosen source amplitudes is similar to an isotropic monopole radiation pattern centered at the sampling point. If for a particular sampling point, finite source amplitudes can result into a monopole radiation pattern, the sampling point is concluded as a scattering point. Though LSM has shown good potential in practical applications, its development and understanding is based on the mathematical theory. Its relation to the forward scattering problem and its physical interpretation is very less understood.

In one physical interpretation, the basic mathematical model of LSM is interpreted as a focusing problem [7, 8], where LSM is understood as finding the source amplitudes that can focus the fields to a sampling point. However this interpretation fails in specific examples, where the so called focusing is seen when not expected or vice versa. This prompts us to take another look into the relation of LSM with the forward scattering problem and find another, more consistent physical explanation of LSM.

[1] To whom any correspondence should be addressed.

We therefore propose an alternative explanation of LSM in terms of the multipole expansion of the currents induced on the scatterers at a sampling point, first introduced in [9]. Our approach is called multipoles-based linear sampling (MLSM) for ease of reference. In MLSM, we try to choose the amplitude of various sources, such that when the current induced on the scatterers is represented as effective multipoles on a sampling point, the monopole is the single most prominent multipole at the sampling point. If for a particular sampling point, finite source amplitudes can result into such induced currents, the sampling point is concluded as a scattering point.

The focus of the article is to discuss the fundamental idea behind MLSM and the impact of using multipole expansion in MLSM.

2. Introduction to the linear sampling method and the focusing interpretation

Linear sampling method (LSM) is a qualitative method used to reconstruct the support of extended scatterers. Below, we introduce the basic formulation of linear sampling method for the TM case.

Let the scatterer support be denoted by Σ. Let Γ denote the circle on which the detectors and sources are arranged. The radius of Γ is very large as compared to the wavelength, such that far field approximation is applicable. Accordingly, let $E^{\infty}(\vec{r}_{\varphi},\vec{r}_{\theta})$ be the $\hat{z}$ directed scattered far-field pattern measured on Γ in the direction of φ, when a unit amplitude plane wave impinges from the direction θ. Let the far field expression of the Green's function be denoted by $\mathrm{g}^{\infty}(\vec{r},\vec{r}')$.

For a generic point $\vec{r}_p$ in Ω, LSM consists of solving the far-field integral equation for the unknown $h(\vec{r}_{\theta},\vec{r}_p)$ [2, 4, 5, 7]:

$$\int_{\Gamma} E^{\infty}(\vec{r}_{\varphi},\vec{r}_{\theta})h(\vec{r}_{\theta},\vec{r}_p)d\theta = \mathrm{g}^{\infty}(\vec{r}_{\varphi},\vec{r}_p) \ . \tag{1}$$

It is obvious that the right hand side represents a circularly symmetric scattered field with respect to $\vec{r}_p$, which, for later convenience, is referred to as the monopole radiation pattern. The value of $h(\vec{r}_{\theta},\vec{r}_p)$ becomes unbounded if the sampling point $\vec{r}_p$ does not belong to the scatterer support. While reconstructing, (1) is cast into a matrix equation as follows:

$$\mathbf{K}\cdot\overline{h} = \overline{\mathrm{g}}^{\infty}, \tag{2}$$

where, $\mathbf{K}$ is the far field multistatic matrix (an operator that maps the currents at the sources to the scattered fields at the detectors), $\overline{h}$ contains the elements $h(\vec{r}_{\theta},\vec{r}_p)$, and $\overline{\mathrm{g}}^{\infty}$ contains the elements $\mathrm{g}^{\infty}(\vec{r}_{\varphi},\vec{r}_p)$. For the sake of convenience of reference, we shall refer to the vector $\overline{h}$ computed using LSM as $\overline{h}_{\mathrm{LSM}}$, which is computed using the following equation:

$$\overline{h}_{\mathrm{LSM}} = \mathbf{K}^{\dagger}\cdot\overline{\mathrm{g}}_{\infty}, \tag{3}$$

This above equation is solved for $\overline{h}$ typically using Tikhonov regularization. The sampling points $\vec{r}_p$ for which $\left|\overline{h}\right|$ (Euclidean norm) are significantly large are concluded to belong to the background. The remaining points are identified as the scatterer support.

The model in (2) can effectively be understood as a focusing problem [7, 8]. Considering that $h(\vec{r}_{\theta},\vec{r}_p)$ denotes the source amplitude at an incidence angle θ, the fields scattered by such current distribution is given by $\mathbf{K}^{\infty}\cdot\overline{h}$ using the definition of the multistatic matrix. Then solving (2) implies finding such source distribution $h(\vec{r}_{\theta},\vec{r}_p)$ that results into an isotropic monopole radiation pattern centered at $\vec{r}_p$. If $\left|\overline{h}\right|$ is finite and small, it implies that we can use practically finite energy sources to focus the waves on the point $\vec{r}_p$. On the other hand, if $|\overline{h}|$ is very large (theoretically infinite), then it implies that there is no finite energy source distribution that can be used to focus on the point $\vec{r}_p$.

3. Multipoles-based linear sampling

Another interpretation of the LSM model, closer to the scattering model, is that the solution of (2) is a source distribution that results into an induced current distribution, which in turn scatters the electric field similar to a monopole radiation from $\vec{r}_p$. Thus, in solving (2), we are trying to generate an induced current distribution, such that it can be effectively represented as a monopole at $\vec{r}_p$. To this end, we study the multipole expansion of the scattered field and the connection of the multipole expansion with the induced current distribution. In fact, to solve (2) is to find a superposition $h(\vec{r}_\theta,\vec{r}_p)$ of $E^\infty(\vec{r}_\varphi,\vec{r}_\theta)$ such that among all multipoles, the monopole radiation is the only dominant contributor in the resultant total radiation.

The scattered field received at a point $\vec{r}_\varphi$, $E_z(\vec{r}_\varphi,\vec{r}_\theta)$, whose far field notation is $E^\infty(\vec{r}_\varphi,\vec{r}_\theta)$, is radiated by the induced currents on the scatterer, i.e.,

$$E_z(\vec{r}_\varphi,\vec{r}_\theta)=\int_\Sigma I_z^{\text{ind}}(\vec{r},\vec{r}_\theta)\,\mathrm{g}(\vec{r}_\varphi,\vec{r})d\vec{r}, \tag{4}$$

where, $I_z^{\text{ind}}(\vec{r},\vec{r}_\theta)$ is the current induced on $\vec{r}$. The scattered fields received at the detectors can be decomposed into various independent terms corresponding to the multipole radiation from a sampling point $\vec{r}_p$. Using addition theorem [10] on $\mathrm{g}(\vec{r}_\varphi,\vec{r})$, the expression of $E_z(\vec{r}_\varphi,\vec{r}_\theta)$ can be rewritten in terms of various multipoles corresponding to a sampling point $\vec{r}_p$ as:

$$E_z(\vec{r}_\varphi,\vec{r}_\theta)=\sum_{n=-\infty}^{\infty}\alpha_n(\vec{r}_p,\vec{r}_\theta)\mathrm{g}_n(\vec{r}_\varphi,\vec{r}_p), \tag{5}$$

where,

$$\alpha_n(\vec{r}_p,\vec{r}_\theta)=\int_\Sigma I_z^{\text{ind}}(\vec{r},\vec{r}_\theta)\mathrm{J}_n\left(k\left|\vec{r}_p-\vec{r}\right|\right)\mathrm{e}^{-in\arg(\vec{r}-\vec{r}_p)}d\vec{r}, \tag{6}$$

$$\mathrm{g}_n(\vec{r}_\varphi,\vec{r}_p)=\frac{i}{4}\mathrm{H}_n^{(1)}\left(k\left|\vec{r}_\varphi-\vec{r}_p\right|\right)\mathrm{e}^{in\arg(\vec{r}_\varphi-\vec{r}_p)}, \tag{7}$$

and $\mathrm{J}_n(\bullet)$ is the Bessel function of n th order. It is evident that $\alpha_n(\vec{r}_p,\vec{r}_\theta)$ represents the n th effective multipole current at $\vec{r}_p$, such that the sum of the radiated fields from all such multipoles is equal to the measured scattered field.

The above formulation suggests that in MLSM, the key physical quantity of interest is not the scattered field measured at the detectors (as in LSM), but the induced current distribution that leads to the detected scattered fields. We highlight that this is the central idea behind MLSM and enables the alternative interpretation in terms of the induced multipoles.

Using (5), the fundamental equation of LSM (1) is equivalent to:

$$\int_\Gamma \alpha_n(\vec{r}_p,\vec{r}_\theta)h(\vec{r}_\theta,\vec{r}_p)d\theta=\begin{cases}1 & n=0\\ 0 & \text{otherwise}\end{cases}. \tag{8}$$

For convenience of further use, we define:

Journal of Physics: Conference Series **290** (2011) 012001 doi:10.1088/1742-6596/290/1/012001

$$\upsilon_n = \int_\Gamma \alpha_n(\vec{r}_p, \vec{r}_\theta) h(\vec{r}_\theta, \vec{r}_p) d\theta \ . \tag{9}$$

Physically, υ_n can be understood as the multipole currents induced by the source distribution $h(\vec{r}_\theta, \vec{r}_p)$. MLSM therefore implies that as long as the multipole expansion of the induced current distribution at a sampling point is such that the monopole is the only prominent contributor, the sampling point will be detected as a scatterer.

4. Reconstruction using MLSM and the physical regularization

Now, we present a reconstruction approach based on the above model. Although one needs to consider infinite number of multipoles to fully account for the scattered field in (5), usually, a sufficiently large finite number of multipoles is enough to approximate $E_z(\vec{r}_\varphi, \vec{r}_\theta)$, especially so in the presence of noise. Considering $(2N+1)$ number of multipoles, the expression for the far field (5) can be rewritten as:

$$E_z(\vec{r}_\varphi, \vec{r}_\theta) \approx \sum_{n=-N}^{N} \alpha_n(\vec{r}_p, \vec{r}_\theta) \, g_n(\vec{r}_\varphi, \vec{r}_p) \, . \tag{10}$$

The equation (10) suggests that the multipole radiation functions $g_n(\vec{r}_\varphi, \vec{r}_p)$ can be understood as a mapping from the effective multipole current at a sampling point $\vec{r}_p$ to the measured scattered electric field. Thus, by approximating (5) as (10), we have shrunk the mathematically infinite dimensional domain of this mapping to a finite $(2N+1)$ dimensional domain. Thus, the truncation of higher order multipoles serves as the regularization scheme in MLSM.

For computing $\bar{h}_{\mathrm{MLSM}}$, we perform two stage inversion:

$$\mathbf{A} = \left(\mathbf{G}^{\mathrm{mul}}\right)^{\dagger} \cdot \mathbf{K}; \quad \bar{h}_{\mathrm{MLSM}} = \mathbf{A}^{\dagger} \cdot \bar{\mathrm{D}} \, . \tag{11}$$

Here, the matrix $\mathbf{A}$ contains $\mathrm{A}(n, \theta;\ n = -N \text{ to } N) = \alpha_n(\vec{r}_p, \vec{r}_\theta)$, $\mathbf{G}^{\mathrm{mul}}$ contains $\mathrm{G}^{\mathrm{mul}}(\varphi, n;\ n = -N \text{ to } N) = g_n(\vec{r}_\varphi, \vec{r}_p)$, and $\bar{\mathrm{D}}$ contains $(2N+1)$ elements, all of which are zero, except the middle element, which corresponds to the monopole and has a value 1. In the proposed method, we have used least squares based pseudoinverse. Thus effectively, $\bar{h}_{\mathrm{MLSM}}$ is given as follows:

$$\bar{h}_{\mathrm{MLSM}} = \left(\left(\mathbf{G}^{\mathrm{mul}}\right)^{\dagger} \cdot \mathbf{K}\right)^{\dagger} \cdot \bar{\mathrm{D}} \, . \tag{12}$$

5. Comparison of LSM and MLSM with regard to $h(\vec{r}_s, \vec{r}_p)$

Generally in LSM, the Tikhonov regularization parameter α is chosen using the general discrepancy principle and is evaluated for each sampling point individually. Effectively, Tikhonov regularization makes the solution $h(\vec{r}_s, \vec{r}_p)$ of (3) stable and smooth by assigning lower weights to the vectors in the signal subspace and relatively higher weights to the vectors in the noise subspace. Evidently, this regularization scheme is purely mathematical and does not have any physical background. Tikhonov regularization serves to approximate the monopole radiation in terms of the strengths given to the various spectral vectors that span the space of the scattered fields. Thus, it is a mathematical approximation which can be tuned or changed by changing the value of the Tikhonov regularization parameter α .

However, the nature of approximation in MLSM is entirely different. MLSM employs the approximation in terms of the truncation of higher order multipoles. This approximation is justified from the perspective of physics because the contribution of higher order multipoles to the scattered field is indeed small. When the value of N is very large, in the noise-free case, the solution of $\overline{h}_{\text{LSM}}$ is close to $\overline{h}_{\text{MLSM}}$.

Physically, it is well known that the monopole and dipoles are the most prominent sources induced on the scatterers. Not only the strengths of these multipoles is higher than the other multipoles in general, the strength of radiation from the higher order multipoles also successively diminishes. Thus, it is reasonable to use only the monopole and the dipoles. Thus, out of the infinite multipoles, our scheme uses only the monopole and dipole terms (i.e. $N=1$) for reconstruction and truncates all the higher order multipoles.

As compared to a large value of N, the use of $N=1$ implies the following. Now, solving (11) would mean that we seek an optimal combination of the monopole and dipole currents such that the resultant radiation fields match the scattered fields as closely as possible. The solution $h(\vec{r}_s,\vec{r}_p)$ is such that the contribution from the determined dipole current is very small. Thus, effectively the requirement on $h(\vec{r}_s,\vec{r}_p)$ has reduced, which would otherwise require to suppress the contribution from all higher multipoles. In terms of the subspace, considering the contribution from each multipole as an independent dimension, using $N=1$ implies that we use a very small subspace from the infinite-dimensional space of induced multipoles, which serves as the data space for determining $\overline{h}_{\text{MLSM}}$ as well as the domain of the operator $\mathbf{G}^{\text{mul}}$.

6. Numerical examples

For all the examples presented in this paper, the number of sources and detectors is 13. The sources and detectors are placed uniformly along a circle of radius 10λ, where λ is the wavelength of the incident wave. The region of investigation is a square region of dimension 2λ.

We consider two examples in which the conventional LSM behaves in an anomalous manner and the focusing interpretation cannot explain the anomalous behaviour. In the first example, the scatterer is a circular scatterer of radius 0.3λ and relative permittivity 7, as shown in Figure 1(a). The result of LSM is plotted in Figure 1(b). It is clearly seen that LSM does not detect the origin as the scatterer support, though the origin actually belongs to the scatterer support. According to the focusing interpretation, since the origin belongs to a scatterer support, using LSM, we should be able to choose $\overline{h}_{\text{LSM}}$ such that the fields can be focused at the origin.

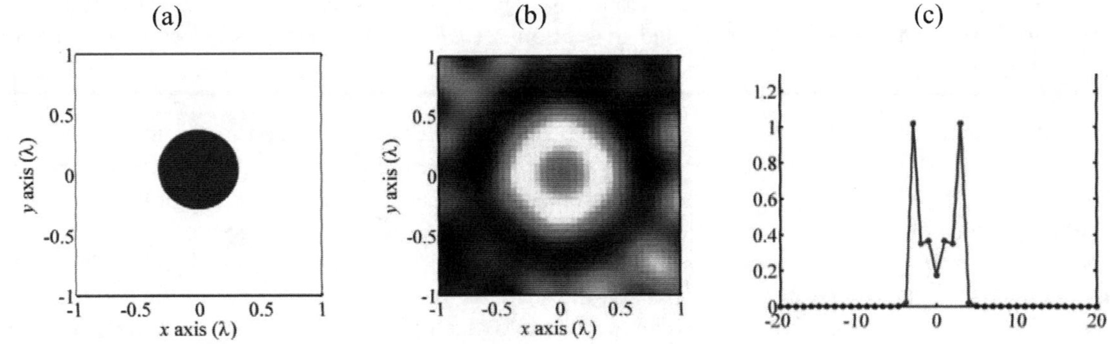

Figure 1: Numerical example 1. (a) the scatterer profile: circular scatterer of radius 0.3λ and relative permittivity 7. (b) reconstruction using LSM, the origin is not detected as scatterer support (c) The multipole expansion of the induced current distribution, υ_n, at the origin.

While the focusing approach fails in explaining this, MLSM can explain this behaviour of LSM. MLSM behaves similar to LSM if sufficiently high number of multipoles are considered. Thus, we use $N = 20$ and plot the multipole expansion of the induced currents (i.e. υ_n of (9) using $\bar{h}_{\text{LSM}}$) at the origin. It is seen that the monopole is not the prominent multipole at the origin. Due to this, the LSM formulation fails to detect the origin as the scatterer support.

In the second example, we consider an annular scatterer with inner and outer radii of 0.2λ and 0.4λ, and relative permittivity of 2. The scatterer is shown in Figure 2(a). The LSM result is plotted in Figure 2(b). It is seen that though the origin does not belong to the scatterer support, LSM detects origin as belonging to the scatterer support. According to the focusing interpretation, since origin does not belong to the scatterer support, finite valued $\bar{h}_{\text{LSM}}$ should not be able to focus the fields at the origin. Thus the focusing interpretation cannot explain this anomalous behaviour. Following the basic principle of MLSM, we plot the multipole expansion of the induced current at the origin in Figure 2(c). We see that the monopole is the most prominent multipole at the origin. Due to this, LSM detects the origin as belonging to the scatterer support.

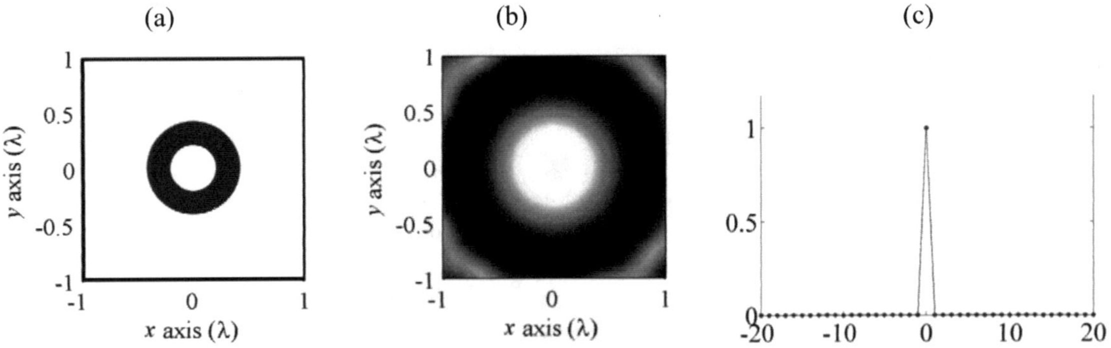

Figure 2: Numerical example 2. (a) the scatterer profile: circular scatterer of radius 0.3λ and relative permittivity 7. (b) reconstruction using LSM, the origin is not detected as scatterer support (c) The multipole expansion of the induced current distribution, υ_n, at the origin.

From the above examples, we can clearly see that MLSM (with large value of *N*) can indeed be used to explain the behaviour of LSM for various cases. Now, we consider the truncation of multipoles as the physical regularization. As discussed in section 4.1, using $N = 1$ in MLSM should work for most cases. Detailed examples are presented in [9]. Here we consider the two examples presented above.

The MLSM reconstruction using $N = 1$ for both the examples are presented in Figure 3. It is seen that while using $N = 1$ is sufficient for example 2 (Figure 3(b)), it is not sufficient for example 1 (Figure 3(a)). The reason is again evident in the multipole expansion. Since the octopole is the most prominent induced multipole at the origin for example 1 (see Figure 1(c)), we need to consider at least $N = 3$ for example 1. The MLSM reconstruction result using $N = 3$ is shown in Figure 4. It is seen that the origin is now correctly identified as the scatterer support. This example shows how the physical nature of the induced current can be used to determine the regularization parameter in MLSM.

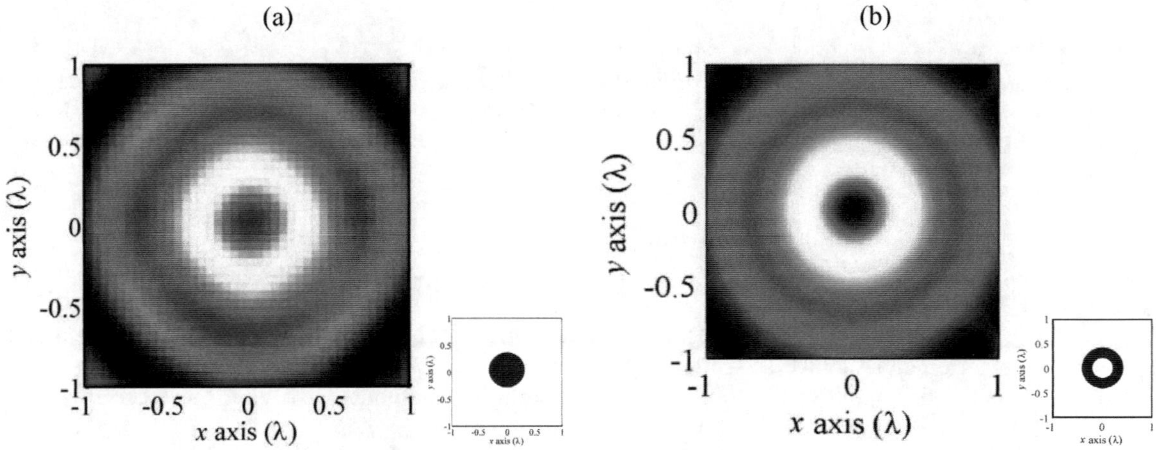

Figure 3: MLSM reconstruction using $N = 1$ for (a) Example 1 and (b) Example 2. The actual scatterers are shown in the inset.

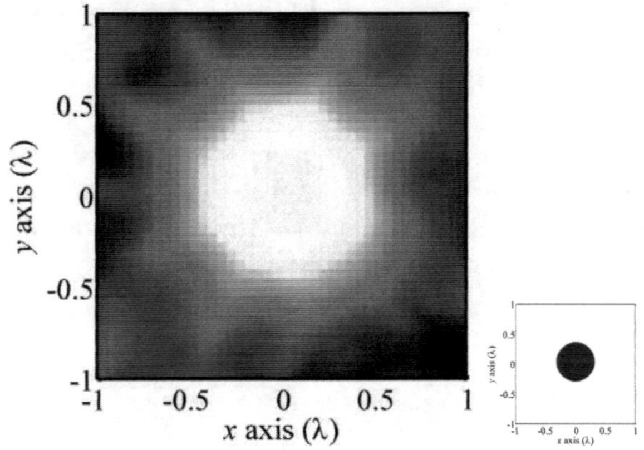

Figure 4: MLSM reconstruction using $N = 1$ for example 1.

7. Conclusion

Multipole based linear sampling method is discussed in the context of the popular linear sampling method. The discussion and formulation highlight that as opposed to LSM, MLSM does not consider the scattered fields at the detectors as the key physical quantity. Instead, it uses the induced current distribution as the key physical quantity that explains the linear sampling method correctly in the context of the scattering phenomenon.

To highlight the impact, we show that when very large number of multipoles is considered in MLSM, MLSM formulation is equivalent to LSM. This fact can be used to explain the reconstruction results of LSM. This becomes more important for the cases where LSM produced anomalous or incorrect reconstruction.

In addition to the above, MLSM incorporates a physical regularization scheme that involves truncation of higher order multipoles. This scheme is based on the nature of induced currents, thus strongly related to the scattering phenomenon itself. It is not strictly mathematical and can be chosen easily for most practical applications. Based on the fact that monopoles and dipoles are the most prominent induced multipoles for most scatterers, MLSM with $N = 1$ works well for most scatterers. For other scatterers, we may consider using higher values of N.

Acknowledgments
The authors acknowledge the fruitful discussion with Dr. Zhong Yu from National University of Singapore. This work was supported by the Singapore Ministry of Education under grant no. R-263-000-575-112.

[1] Arens 2004 Why linear sampling works. *Inverse Problems* **20** 1 pp. 163-173.
[2] Cakoni and Colton 2003 The linear sampling method for cracks. *Inverse Problems* **19** 2 pp. 279-295.
[3] Collino, Fares and Haddar 2003 Numerical and analytical studies of the linear sampling method in electromagnetic inverse scattering problems. *Inverse Problems* **19** 6 pp. 1279-1298.
[4] Colton, Haddar and Monk 2002 The linear sampling method for solving the electromagnetic inverse scattering problem. *SIAM Journal on Scientific Computing* **24** 3 pp. 719-731.
[5] Colton, Haddar and Piana 2003 The linear sampling method in inverse electromagnetic scattering theory. *Inverse Problems* **19** 6 pp. S105-S137.
[6] Colton and Kress 2006 Using fundamental solutions in inverse scattering. *Inverse Problems* **22** 3 pp. R49-R66.
[7] Catapano, Crocco and Isernia 2007 On simple methods for shape reconstruction of unknown scatterers. *IEEE Transactions on Antennas and Propagation* **55** 5 pp. 1431-1436.
[8] Shelton and Warnick 2003 Behavior of the regularized sampling inverse scattering method at internal resonance frequencies. *Progress In Electromagnetics Research* **38** pp. 29-45.
[9] Agarwal, Chen and Zhong 2010 A multipole-expansion based linear sampling method for solving inverse scattering problems. *Optics Express* **18** 6 pp. 6366-6381.
[10] Abramowitz and Stegun 1972 Handbook of mathematical functions : with formulas, graphs, and mathematical tables (New York: ^Dover Publications).

International Conference on Inverse Problems 2010 IOP Publishing
Journal of Physics: Conference Series **290** (2011) 012002 doi:10.1088/1742-6596/290/1/012002

Signal and image source separations by wavelet analysis

Ryuichi Ashino[1], Shusuke Kataoka[2], Takeshi Mandai[3] and Akira Morimoto[2]

[1] Mathematical Sciences, Osaka Kyoiku University, Japan
[2] Information Science, Osaka Kyoiku University, Japan
[3] Department of Engineering Science, Osaka Electro-Communication University, Japan

E-mail: ashino@cc.osaka-kyoiku.ac.jp, j109604@ex.osaka-kyoiku.ac.jp, mandai@isc.osakac.ac.jp and morimoto@cc.osaka-kyoiku.ac.jp

Abstract. Blind Source Separation (BSS) is one of the hottest emerging areas in signal processing. The purpose of BSS is to separate and to estimate the original sources from the sensor array, without knowing the transmission channel characteristics. Besides methods based on independent component analysis which is one of the most powerful tools for BSS, several methods based on time-frequency analysis have been proposed. One of them is the quotient signal estimation method which can estimate the unknown number of sources. We present a new method using wavelet analysis and apply it to signal and image source separations.

1. Introduction

The human visual and auditory systems are sensors to measure the frequency of light and sound, respectively. By this reason, we humans understand the nature using time and frequency. Let us consider a music. By recording the music, we have a function $f(x)$, $x \in \mathbb{R}$. The music can be reproduced in full if we use all the values of $f(x)$. On the other hand, we can also determine completely all properties of $f(x)$ from its Fourier transform, because the Fourier transform is one-to-one and we have the inverse Fourier transform. Either one of $f(x)$ and $\widehat{f}(\xi)$ is sufficient as information representing the music. When we play a music with an instrument, say piano, we hit a certain key in a certain time according to a musical score, which is called a *time-frequency information.* Thus the human recognizes the nature based on time-frequency information.

Figure 1. Time-frequency information of a music — a musical score.

The *cocktail party effect* [5, 8] is a challenging problem in auditory perception. One of basic questions for the cocktail party problem is how to build a machine to solve the cocktail party problem in a satisfactory manner. This question corresponds to *Blind Source Separation* (BSS) [8], which is one of the hottest emerging areas in signal processing [6, Chapter 1]. The purpose of

BSS is to separate and to estimate the original sources from the sensor array, without knowing the transmission channel characteristics. Certain *a priori* knowledge on the original sources is needed for this separation, and the original sources cannot be uniquely estimated. Besides methods based on independent component analysis [9] which is one of the most powerful tools for BSS, several methods based on time-frequency analysis have been proposed. One of them is the quotient signal estimation method [4, 10, 14, 15], which was refined in [1, 3] by using the analytic wavelet transform based on the analytic signal.

For image source separation, we combined the edge analysis by the wavelet transform and the quotient signal estimation method in [13]. As the monogenic signal is a two-dimensional generalization of the analytic signal [7], we proposed to use the monogenic wavelet transform for image source separation in [12]. In this lecture, we improve this method by introducing a new algorithm.

2. Blind separation problem of images

For the sake of simplicity, we will deal with only gray scale images having $P \times Q$ pixels, which can be regarded as matrices in $\mathbb{R}^{P\times Q}$. We will use the following notation:

$$a_{\downarrow} = (a_1, \dots, a_n)^T.$$

The full-rank matrix $A = (a_{j,\ell}) \in \mathbb{R}^{M\times N}$ with non-negative components is called the *mixing matrix*. Assume that there are N-tuple of unknown original images:

$$s_\ell = \big(s_\ell[p,q]\big) \in \mathbb{R}^{P\times Q}, \qquad \ell = 1, \dots, N,$$

where the number N is also unknown. The mathematical model of observed images are

$$x_j = \sum_{\ell=1}^{N} a_{j,\ell}\, s_\ell, \qquad j = 1, \dots, M, \tag{1}$$

which can be represented as

$$x_{\downarrow}[p,q] = A s_{\downarrow}[p,q].$$

Estimating the original images from the observed images is called *blind separation*. Once we know N and A, we can reconstruct original images s_ℓ, $\ell = 1, \dots, N$ by solving the equation (1). Hence, in our case, the blind separation problem is reduced to the estimation of the unknown quantities N and A.

3. Riesz transform and monogenic signal

The n-dimensional Fourier transform of $f(x) \in L^1(\mathbb{R}^n)$ is defined by

$$\mathcal{F}[f](\xi) = \widehat{f}(\xi) = \int_{\mathbb{R}^n} f(x)\, e^{-ix\cdot\xi}\, dx,$$

where $i = \sqrt{-1}$. The inverse Fourier transform of $g(\xi) \in L^1(\mathbb{R}^n)$ is defined by

$$\mathcal{F}^{-1}[g](x) = (2\pi)^{-n} \int_{\mathbb{R}^n} g(\xi)\, e^{ix\cdot\xi}\, d\xi.$$

The *Riesz transforms*, denoted by $\mathcal{R}_j$, $j = 1, \dots, n$, are the Fourier multiplier operators $-i\, D_j/|D|$, that is,

$$\mathcal{R}_j f(x) = \mathcal{F}^{-1}\left[-i\, \frac{\xi_j}{|\xi|}\, \widehat{f}(\xi)\right].$$

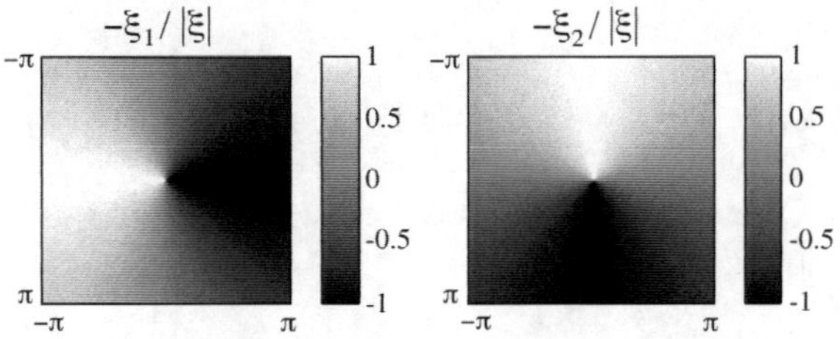

Figure 2. The imaginary parts $-\xi_1/|\xi|$, $-\xi_2/|\xi|$ of the masks.

The Riesz transforms $\mathcal{R}_j f$, $j = 1, \ldots, n$ of real-valued function f are also real-valued. The masks $-i\xi_1/|\xi|$, $-i\xi_2/|\xi|$ of the two-dimensional Riesz transforms $\mathcal{R}_1$, $\mathcal{R}_2$ are pure imaginary. The imaginary parts $-\xi_1/|\xi|$, $-\xi_2/|\xi|$ of the masks are illustrated in Figure 2.

Let us define three fundamental unitary operators for time-frequency analysis. Denote by $\mathcal{T}_b$, the *translation* operator by $b \in \mathbb{R}^n$, that is,

$$(\mathcal{T}_b f)(x) = f(x - b),$$

by $\mathcal{D}_a$, the *dilation* operator by $a > 0$, that is,

$$(\mathcal{D}_a f)(x) = a^{-n/2} f(x/a),$$

and by $\mathcal{M}_\omega$, the *modulation* operator by $\omega \in \mathbb{R}^n$, that is,

$$(\mathcal{M}_\omega f)(x) = e^{i\omega \cdot x} f(x).$$

Definition 1 *The monogenic signal $\mathcal{M}f$ of $f \in L^2(\mathbb{R}^n)$ is defined by*

$$\mathcal{M}f = (f, \mathcal{R}_1 f, \mathcal{R}_2 f, \ldots, \mathcal{R}_n f).$$

4. Monogenic wavelet transform

The *continuous wavelet transform* $W_\psi f(b, a)$ of $f \in L^2(\mathbb{R}^n)$ with respect to a window function ψ, called *wavelet function*, is defined by

$$\begin{aligned} W_\psi f(b,a) &= \langle f, \mathcal{T}_b \mathcal{D}_a \psi \rangle \\ &= a^{-n/2} \int_{\mathbb{R}^n} f(x) \overline{\psi\left(\frac{x-b}{a}\right)}\, dx, \end{aligned}$$

where $b \in \mathbb{R}^n$ and $a \in \mathbb{R}_+ = (0, \infty)$. The inversion formula, called the *inverse continuous wavelet transform*,

$$f(x) = \frac{1}{C_\psi} \int_{b \in \mathbb{R}^n} \int_{a \in \mathbb{R}_+} (W_\psi f)(b, a) \mathcal{T}_b \mathcal{D}_a \psi \frac{da}{a^{n+1}}\, db$$

is obtained under the hypothesis that

$$\int_0^\infty \frac{|\widehat{\psi}(\omega t)|^2}{t}\, dt = C_\psi < \infty, \quad \text{independent of } \omega,$$

which is called the *admissibility condition*. We have the following theorem:

Theorem 2

$$\mathcal{R}_j W_\psi = W_\psi \mathcal{R}_j = W_{-\mathcal{R}_j \psi}, \qquad j = 1, \ldots, n.$$

The monogenic signal

$$\mathcal{M}(W_\psi f) = ((W_\psi f), \mathcal{R}_1(W_\psi f), \ldots, \mathcal{R}_n(W_\psi f))$$

of the continuous wavelet transform $W_\psi f$ is called the *monogenic wavelet transform* of f. By Theorem 2, we have the following Lemma 3.

Lemma 3

$$\mathcal{M}(W_\psi f) = ((W_\psi f), (W_{-\mathcal{R}_1\psi} f), \ldots, (W_{-\mathcal{R}_n\psi} f)).$$

By Lemma 3, the monogenic wavelet transform of f can be regarded as the $(n+1)$-tuple of continuous wavelet transforms of f with $n+1$ different wavelet functions ψ and $-\mathcal{R}_j\psi$, $j = 1, \ldots, n$. The vector valued function $(\psi, -\mathcal{R}_1\psi, \ldots, -\mathcal{R}_n\psi)$ is called the *monogenic wavelet function* generated from ψ. Therefore, we can use the framework of the method for BSS using time-frequency information matrices made by several different wavelet functions proposed in [2].

5. Algorithms for image separation

To deal with images, we discretize the monogenic wavelet transform. The parameters b and a of the continuous wavelet transform $W_\psi f(b, a)$ are called *position* and *scale*, respectively. Substitute b with $(p, q) \in \mathbb{N}^2$ and a with α^r, where $\alpha > 1$ is a properly chosen constant and $r \in \mathbb{Z}$. Then, (p, q) and r are discrete version of position and scale parameters instead of b and a, respectively. For an image $x = (x[p, q]) \in \mathbb{R}^{P \times Q}$, the *spatial-scale information* X of x is defined by

$$\begin{aligned} X[p, q, r] &= \mathcal{M}(W_\psi x)(p, q, \alpha^r) \\ &= ((W_\psi x)(p, q, \alpha^r), \mathcal{R}_1(W_\psi x)(p, q, \alpha^r), \mathcal{R}_2(W_\psi x)(p, q, \alpha^r)). \end{aligned}$$

Denote by S_ℓ and X_j, the spatial-scale informations of the original images s_ℓ and the observed images x_j, respectively. By the linearity of the continuous wavelet transform and the Riesz transforms, (1) can be represented as

$$X_j[p, q, r] = \sum_{\ell=1}^{N} a_{j,\ell}\, S_\ell[p, q, r], \quad j = 1, \ldots, M,$$

or

$$X_\downarrow[p, q, r] = AS_\downarrow[p, q, r]. \tag{2}$$

The following Algorithm 4 estimates the number N of sources and the mixing matrix A.

Algorithm 4

(i) *For each point (p, q, r), compute the singular value decomposition (SVD) of $X_\downarrow[p, q, r] = U\Sigma V^T \in \mathbb{R}^{M \times 3}$, where s_1, s_2, s_3 are singular values with $s_1 \geq s_2 \geq s_3 \geq 0$. When $s_1 \gg s_2$, record the first column u_1 of U if every element of u_1 is positive, record $-u_1$ else if every element of u_1 is negative.*

(ii) *Plot the histogram of recorded u_1 on the sphere $\mathbb{S}^{M-1}$ by the self-organizing map [11]. Note that $\|u_1\| = 1$, because u_1 is the first column of the orthogonal matrix U.*

(iii) *Estimate the number N of sources by the number of peaks by plotting the histogram.*

(iv) *Estimate the normalized columns of the mixing matrix A by the spherical coordinate of the peaks.*

6. Conclusion

The monogenic wavelet transform of an n-dimensional function f can be regarded as the $(n+1)$-tuple of continuous wavelet transforms of f with $n+1$ different wavelet functions ψ and $-\mathcal{R}_j\psi$, $j = 1, \ldots, n$. By generalizing the time-frequency information matrix method to the n-dimensional case, we have proposed a novel method for blind image separation using the monogenic wavelet transform.

Acknowledgment

This work was partially supported by JSPS.KAKENHI (C)20540168, (C)20540193, (C)22540130.

References

[1] R. Ashino, C. A. Berenstein, K. Fujita, A. Morimoto, M. Morimoto, D. Napoletani, and Y. Takei, *Mathematical background for a method on quotient signal decomposition*, Appli. Anal., **86** (5), 577–609, 2007.

[2] R. Ashino, K. Fujita, T. Mandai, A. Morimoto, and K. Nishihara, *Blind source separation using time-frequency information matrix given by several wavelet transforms*, Information, **10** (5), 555–568, 2007.

[3] R. Ashino, T. Mandai, A. Morimoto, and F. Sasaki, *Blind source separation of spatio-temporal mixed signals using time-frequency analysis*, Appli. Anal., **88** (3), 425–456, 2009.

[4] R. Balan and J. Rosca, *Statistical properties of STFT ratios for two channel systems and applications to blind source separation*, in Proceedings of the 2nd International Workshop on Independent Component Analysis and Blind Signal Separation, Helsinki, Finland, 19–22, 2000.

[5] C. Cherry, *Some experiments on the recognition of speech, with one and with two ears*, Journal of Acoustical Society of America, **25** (5), 975–979, 1953.

[6] A. Cichocki and S. Amari, *Adaptive blind signal and image processing: Learning algorithms and applications*, John Wiley & Sons, Chichester, West Sussex, England, 2002.

[7] M. Felsberg and G. Sommer, *The monogenic signal*, IEEE Trans. Signal Processing, **49** (12), 3136–3144, 2001.

[8] S. Haykin and Z. Chen, *The cocktail party problem*, Neural Computation, **17** (9), 1875–1902, 2005.

[9] A. Hyvärinen, J. Karhunen, E. Oja, *Independent component analysis*, John Wiley & Sons, New York, NY, 2001.

[10] A. Jourjine, S. Rickard, and O. Yilmaz, *Blind separation of disjoint orthogonal signals: Demixing N sources from 2 mixtures*, in Proceedings of the 2000 IEEE International Conference on Acoustics, Speech, and Signal Processing, **5**, 2985–2988, 2000.

[11] T. Kohonen, *Self-Organizing Maps*, Second Edition, Springer-Verlag, Berlin, 1997.

[12] A. Morimoto, R. Ashino, T. Mandai, *Image separation using the monogenic wavelet transform*, in Proceedings of the 10th International Symposium on Communications and Information Technologies 2010 (ISCIT 2010), 707–712, 2010.

[13] A. Morimoto, H. Koyama, D. Inoue, A. Oomichi, H. Nishimura, R. Ashino, and T. Mandai, *Image Separation using Wavelet Analysis*, Transactions of the Japan Society for Industrial and Applied Mathematics, **19** (3), 257–278, 2009, in Japanese.

[14] D. Napoletani, C. A. Berenstein, P. S. Krishnaprasad, and D. C. Struppa, *Quotient signal estimation*, in Progress in Mathematics, **238**, Birkhäuser, Boston, MA, 151–162, 2005.

[15] O. Yilmaz and S. Rickard *Blind source separation of speech mixtures via time-frequency masking*, IEEE Transactions on Signal Processing, **52**, 1830–1847, 2004.

A direct numerical reconstruction algorithm for the 3D Calderón problem

Fabrice Delbary[1], **Per Christian Hansen**[1], **Kim Knudsen**[2]

[1]Department of Informatics and Mathematical Modelling, Technical University of Denmark, 2800 Kgs. Lyngby, Denmark

[2]Department Mathematics, Technical University of Denmark, 2800 Kgs. Lyngby, Denmark

E-mail: fdel@imm.dtu.dk, pch@imm.dtu.dk, k.knudsen@mat.dtu.dk

Abstract. In three dimensions Calderón's problem was addressed and solved in theory in the 1980s in a series of papers, but only recently the numerical implementation of the algorithm was initiated. The main ingredients in the solution of the problem are complex geometrical optics solutions to the conductivity equation and a (non-physical) scattering transform. The resulting reconstruction algorithm is in principle direct and addresses the full non-linear problem immediately. In this paper we will outline the theoretical reconstruction method and describe how the method can be implemented numerically. We will give three different implementations, and compare their performance on a numerical phantom.

1. Introduction

The Calderón problem (or the so-called inverse conductivity problem) was formulated in Calderón's seminal paper [4]. The problem concerns the unique determination and reconstruction of an electric conductivity distribution in a bounded domain from knowledge of the Dirichlet-to-Neumann (or Voltage-to-Current) map on the boundary of the domain. Among the numerous applications of Calderón's problem we single out the emerging technology for medical imaging called Electrical Impedance Tomography [3, 8], where the Dirichlet-to-Neumann map can be obtained (in principle) from electrostatic boundary measurements.

In this paper we consider the particular domain $\Omega = B(0,1) \subset \mathbb{R}^3$, the unit ball in $\mathbb{R}^3$, and denote by $\sigma \in C^\infty(\overline{\Omega})$ the conductivity distribution in Ω. Suppose $\sigma \geq c > 0$, where c is a constant, and suppose further that $\sigma = 1$ near $\partial\Omega$. Imposing a voltage potential $f \in H^{1/2}(\partial\Omega)$ on $\partial\Omega$ induces a voltage potential $u \in H^1(\Omega)$ given as the unique solution to

$$\begin{aligned} \nabla \cdot \sigma \nabla u = 0 \text{ in } \Omega, \\ u = f \text{ on } \partial\Omega. \end{aligned} \tag{1}$$

The resulting current flux through the boundary is given by $g = \nu \cdot \nabla u|_{\partial\Omega} = \partial_\nu u|_{\partial\Omega}$, where $\nu(x) = x$ is the outward unit normal to $\partial\Omega$ at x. This defines the Dirichlet-to-Neumann map

$$\Lambda_\sigma : f \mapsto g.$$

The injectivity of $\sigma \mapsto \Lambda_\sigma$, i.e. the uniqueness question for the inverse problems, was settled affirmatively in [15] using the so-called complex geometrical optics solutions to the governing equation, and the theoretical framework for a reconstruction algorithm was developed in [11, 12]. The algorithm

Published under licence by IOP Publishing Ltd

essentially divides the non-linear problem into three steps involving the solution of linear equations: first a boundary integral equation for the boundary trace of the complex geometrical optics solutions is solved and a scattering transform is computed, then the inverse Fourier transform is applied, and finally a boundary value problem is solved for the conductivity. The first numerical reconstructions following this approach were obtained in [1, 2, 9], but in these papers only the zeroth order approximation of the solution to the boundary integral equation was employed.

The reconstruction algorithm we consider here solves the full non-linear problem directly. In contrast to the reconstruction methods based on the formulation of the inverse problem as an optimization problem solved by iterative schemes, we do not face the potential problem of convergence to local minima. We emphasize, that even though our implementation focus on the sphere geometry and take several assumptions on the conductivity, the algorithm is valid in general domains and general C^2 conductivities.

In this paper we will outline the theoretical reconstruction method and discuss three different approximations and implementations. The first reconstruction method is based on the solution of the boundary integral equation, however we suggest to replace the Faddeev Green's function by the usual fundamental solution for the Laplacian in $\mathbb{R}^3$. The second reconstruction method is based on the zero order approximation in the boundary integral equation, while the third reconstruction is a linearization method tantamount to Calderón's linear reconstruction method.

2. The reconstruction algorithms

For the solution u to (1) define $v = \sigma^{1/2}u$. The v satisfies

$$\begin{aligned}(-\Delta + q)v &= 0 \text{ in } \Omega,\\ v &= f \text{ on } \partial\Omega.\end{aligned} \tag{2}$$

with $q = \Delta\sigma^{1/2}/\sigma^{1/2}$. Extend q by $q = 0$ outside Ω and consider

$$(-\Delta + q)v = 0 \text{ in } \mathbb{R}^3. \tag{3}$$

The reconstruction algorithm is now based on the scattering transform of q defined by

$$\mathbf{t}(\xi,\zeta) = \int_\Omega e^{-ix\cdot(\xi+\zeta)}\psi(x,\zeta)q(x)dx, \tag{4}$$

where $\psi(x,\zeta)$ denotes the complex geometrical optics solutions to (3) satisfying the asymptotic condition $e^{-ix\cdot\zeta}\psi(x,\zeta) \approx 1$ for large x. The parameter $\zeta \in \mathbb{C}^3$ is assumed to satisfy $\zeta^2 = (\xi+\zeta)^2 = 0$. In a sense $\mathbf{t}$ can be understood as a non-linear Fourier transform of q. Integration by parts yields

$$\mathbf{t}(\xi,\zeta) = \int_{\partial\Omega} e^{-ix\cdot(\xi+\zeta)}(\Lambda_\sigma - \Lambda_1)\psi(x,\zeta)dx. \tag{5}$$

In order to formulate an equation for $\psi|_{\partial\Omega}$ we define the single-layer potential

$$(S_\zeta\phi)(x) = \int_{\partial\Omega} G_\zeta(x-y)\phi(y,\zeta)ds(y),$$

with the Faddeev Green's function

$$G_\zeta(x) = \frac{e^{ix\cdot\zeta}}{(2\pi)^3}\int_{\mathbb{R}^3}\frac{e^{ix\cdot\xi}}{|\xi|^2 + 2\xi\cdot\zeta}d\xi.$$

The reconstruction algorithm consists of three steps:

$$\Lambda_\sigma \xrightarrow{\text{(i)}} \mathbf{t}(\xi,\zeta) \xrightarrow{\text{(ii)}} q(x) \xrightarrow{\text{(iii)}} \sigma(x).$$

(i) $\psi|_{\partial\Omega}$ can be computed from boundary measurements by solving

$$\psi + S_\zeta(\Lambda_\sigma - \Lambda_1)\psi = e^{ix\cdot\zeta}, \qquad x \in \partial\Omega \tag{6}$$

and $\mathbf{t}$ can be computed by (5).

(ii) q can be computed from $\mathbf{t}$ using

$$\lim_{\zeta\to\infty} \mathbf{t}(\xi,\zeta) = \hat{q}(\xi).$$

(iii) σ can be computed from q by solving

$$(\Delta + q)\sigma^{1/2} = 0 \text{ in } \Omega, \quad \sigma^{1/2}|_{\partial\Omega} = 1. \tag{7}$$

It can be shown that ill-posedness of the problem is present only in step (i). Steps (ii)-(iii) give rise to well-posed problems.

2.1. *The* $\mathbf{t}_0$ *approximation*

To avoid working with the exponentially growing Faddeev Green's function in the boundary integral equation (6) we suggest to consider

$$\psi_0 + S_0(\Lambda_\sigma - \Lambda_1)\psi_0 = e^{ix\cdot\zeta}, \qquad x \in \partial\Omega, \tag{8}$$

where the kernel of S_0 is the usual fundamental solution for the Laplacian in $\mathbb{R}^3$, $G_0(x) = 1/(4\pi|x|)$. It can be shown [6] that this equation is uniquely solvable for any ζ. We then compute $\mathbf{t}_0$ by inserting ψ_0 for ψ in (5), that is

$$\mathbf{t}_0(\xi,\zeta) = \int_{\partial\Omega} e^{-ix\cdot(\xi+\zeta)}(\Lambda_\sigma - \Lambda_1)\psi_0(x,\zeta)dx, \tag{9}$$

and carry out steps (ii) and (iii) of the algorithm as before. We will call the obtained reconstruction σ_0. For the 2D problem a similar approximation was used in [7, 10].

2.2. *The* $\mathbf{t}^{\text{exp}}$ *approximation*

A crude approximation of $\mathbf{t}$ is obtained by substituting the asymptotic value of $\psi \approx e^{ix\cdot\zeta}$ for ψ in (5). This was suggested for the 2D problem in [14]. The resulting approximation of $\mathbf{t}$ is denoted by $\mathbf{t}^{\text{exp}}$, i.e.

$$\mathbf{t}^{\text{exp}}(\xi,\zeta) = \int_{\partial\Omega} e^{-ix\cdot(\xi+\zeta)}(\Lambda_\sigma - \Lambda_1)e^{ix\cdot\zeta}dx. \tag{10}$$

We then propose to implement steps (ii) and (iii) with $\mathbf{t}^{\text{exp}}$ for $\mathbf{t}$, thus obtaining an approximation of σ denoted by σ^{exp}. Effectively this approach gives a linearization of step (i) in the algorithm.

2.3. *The Calderón inversion*

Step (ii) and (iii) in the algorithm can be linearized as well. This gives the reconstruction formula

$$\sigma^{\text{app}}(x) = 1 - \frac{1}{2(2\pi)^n}\int \frac{\mathbf{t}^{\text{exp}}(\xi,\zeta)}{|\xi|^2}e^{ix\cdot\xi}d\xi. \tag{11}$$

3. Details on the numerical implementation

To test the different reconstruction schemes we chose a numerical phantom composed of three objects with constant conductivity different from the conductivity of the background medium (conductivity 1).

- A ball with center in the (Oxy) plane and with a conductivity larger than 1.
- Two prolate ellipsoids with axis in the (Oxy) plane, with slightly different sizes and with a same conductivity smaller than 1.

The phantom is illustrated in figure 1.

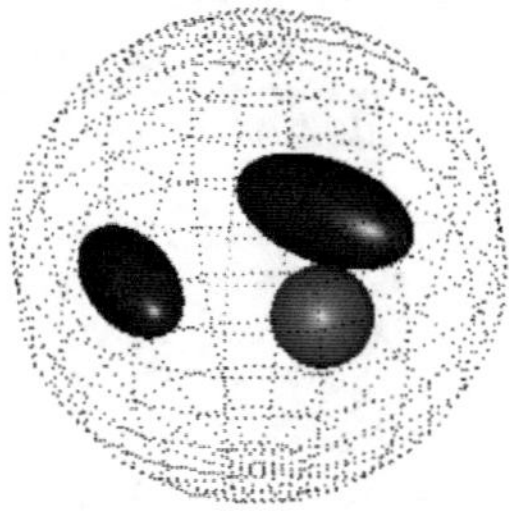

Figure 1. Phantom

In order to be in the framework of the theory, the conductivity distribution must be smooth. Hence we have smoothed out the discontinuities in the phantom. A cross sectional plot of the resulting conductivity distribution and its support can be seen in figure 2.

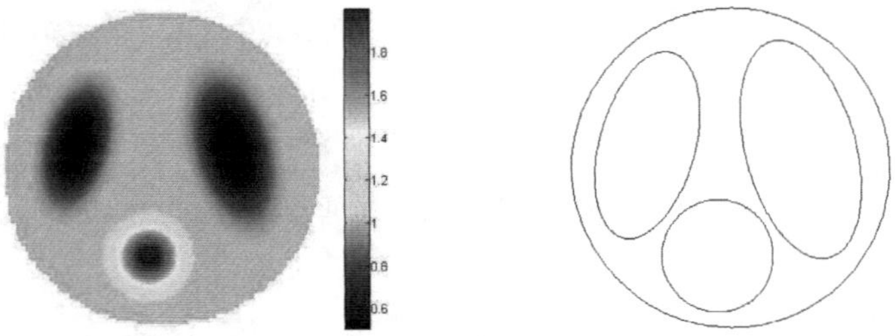

Figure 2. Left: profile of the conductivity σ in the (Oxy) plane. Right: support of σ.

The Dirichlet-to-Neumann map Λ_σ is computed using the numerical phantom and represented in the basis of spherical harmonics Y_n^m at the boundary of Ω. We fix $f = Y_n^m$ for n, m up to a fixed order and solve numerically the Schrödinger equation (2) for v using the moment method. Then we expand $\Lambda_\sigma Y_n^m = \partial_\nu v|_{\partial\Omega}$ in the same basis of spherical harmonics at the boundary and store the coefficients up to the fixed order.

Quadrature points are chosen on the unit sphere, as defined in [16] and [5] for solving boundary integral equations of the second kind involved in acoustic scattering. All computations involved in the reconstruction methods we propose are done using pointwise values of functions at these quadrature points.

To briefly summarize, we consider a positive integer N and the N^2 points x_{pq}, $1 \leq p \leq N, 0 \leq q \leq 2N-1$ on the unit sphere $\partial\Omega$ given by the spherical coordinates

$$x_{pq} = (\sin\theta_p \cos\varphi_q, \sin\theta_p \sin\varphi_q, \cos\theta_p),$$

where $\theta_p = \arccos t_p$ and $\varphi_q = \pi q/N$, the real numbers t_p being the N zeros of the Legendre polynomial P_N. These quadrature points allow us to construct an approximation formula for functions $\psi \in C(\partial\Omega)$ by linear combinations of spherical harmonics, using pointwise values of ψ at the quadrature points, more precisely

$$\psi \simeq \frac{\pi}{N}\sum_{p=1}^{N}\sum_{q=0}^{2N-1}\alpha_p\psi(x_{pq})\sum_{n=0}^{N-1}\sum_{m=-n}^{n}Y_n^{-m}(x_{pq})Y_n^m, \tag{12}$$

where α_p are the weights of the Gauss-Legendre quadrature rule on $[-1,1]$ and Y_n^m are the usual spherical harmonics. For analytic functions ψ, the approximation (12) converges exponentially in the maximum norm as $N \to \infty$. From this approximation formula ensues the quadrature formula

$$\int_{\partial\Omega}\psi\, ds \simeq \frac{\pi}{N}\sum_{p=1}^{N}\sum_{q=0}^{2N-1}\alpha_p\psi(x_{pq}), \tag{13}$$

and the approximation formula

$$\int_{\partial\Omega}\frac{\psi(y)}{|x-y|}\,ds(y) \simeq \frac{\pi}{N}\sum_{p=1}^{N}\sum_{q=0}^{2N-1}\alpha_p\psi(x_{pq})\sum_{n=0}^{N-1}P_n(x_{pq}\cdot x)\quad \text{for } x \in \partial\Omega. \tag{14}$$

Equation (13) is used in the different schemes for numerical integrations involved in (9) and (10) and the formulas (12), (14) are used to build the matrix of the discretized operator $I + S_0(\Lambda_\sigma - \Lambda_1)$ for the $\mathbf{t}_0$ method. The computation of $1-\sigma^{\text{app}}$ from $\mathbf{t}^{\text{exp}}(\xi,\zeta)/|\xi|^2$ (Calderón inversion) or the computation of the approximation of q ($\mathbf{t}^{\text{exp}}$ and $\mathbf{t}_0$ approximations) is simply done by a fast Fourier transform. For the $\mathbf{t}^{\text{exp}}$ and $\mathbf{t}_0$ approximations, the solution of the Schrödinger equation (7) is done with a finite elements method of order 1.

3.1. The $\mathbf{t}_0$ approximation

The reconstruction scheme for the $\mathbf{t}_0$ approximation requires steps to compute the solution of (8), thus, we proceed as follows

(i) Build the matrix of the operator $I + S_0(\Lambda_\sigma - \Lambda_1)$ using (12), (14).

(ii) Solve the discretized system of (8), using a Tikhonov regularization with Morozov discrepancy principle because of the ill-posedness, to get values of ψ at the quadrature points.

(iii) Compute (9) using (13) for $|\xi| < \xi_M$ where $\xi_M > 0$ and where, despite the fact that in the theory $|\zeta| \to \infty$, $|\zeta|$ is chosen of minimal norm to avoid numerical instabilities.

(iv) Compute the fast inverse Fourier transform of $\mathbf{t}_0(\xi,\zeta)$ to get the approximation of q_0 of q.

(v) Solve the Schrödinger equation (7) to get σ_0 using a finite elements method.

3.2. The $\mathbf{t}^{\text{exp}}$ approximation

The reconstruction scheme for the $\mathbf{t}^{\text{exp}}$ method does not require to solve the first boundary integral equation, hence, the steps are the following

(i) Compute (10) using (13) for $|\xi| < \xi_M$ where $\xi_M > 0$ and where $|\zeta|$ is chosen of minimal norm to avoid numerical instabilities.

(ii) Compute the fast inverse Fourier transform of $\mathbf{t}^{\text{exp}}(\xi,\zeta)$ to get an approximation q^{exp} of q.

(iii) Solve the Schrödinger equation (7) to get σ^{exp} using a finite elements method.

3.3. The Calderón inversion

The Calderón inversion does not require to solve the Schrödinger equation since the approximation of the conductivity is directly computed from $\mathbf{t}^{\text{exp}}$, so that the steps are limited to

(i) Compute (10) using (13) for $|\xi| < \xi_M$ where $\xi_M > 0$ and where $|\zeta|$ is chosen of minimal norm to avoid numerical instabilities.

(ii) Compute the fast inverse Fourier transform of $\mathbf{t}^{\text{exp}}(\xi, \zeta)/|\xi|^2$ to get $1 - \sigma^{\text{app}}$.

4. Numerical results

We consider reconstructions with $\mathbf{t}_0$ and $\mathbf{t}^{\text{exp}}$ approximations and with the Calderón inversion, first with $|\xi| \leq 6$ and then with $|\xi| \leq 8$. We take $N = 16$, that is 512 points on the unit sphere. The cross sectional profile of the reconstructions for the cases $|\xi| \leq 6$ and $|\xi| \leq 8$ can be seen in figure 3 and 4 respectively.

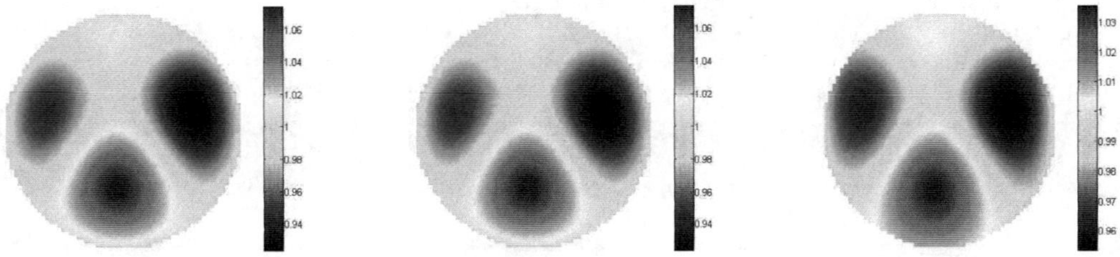

Figure 3. Reconstructions with truncation $\xi_M = 6$: left σ_0, middle σ^{exp}, and right σ^{app}.

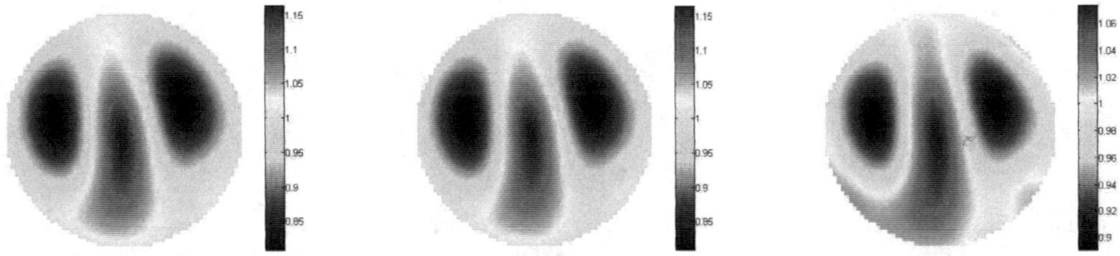

Figure 4. Reconstructions with truncation $\xi_M = 8$: left σ_0, middle σ^{exp}, and right σ^{app}.

From figure 3 and 4 we note that the shape of the inclusions are determined better with the lower value of ξ_M, whereas the contrast is better achieved with the larger value of ξ_M. This is perhaps due to the exponential growth of the involved functions with respect to ξ. Also, we see that even though the reconstructions are quite similar the reconstructions σ_0 and σ^{exp} seem to be slightly better than σ^{app}. This holds true with both truncations.

5. Conclusions

In this article we have outlined three different reconstruction methods for the Calderón problem and given their numerical implementations. The performance of the methods was illustrated with a numerical phantom consisting of three inclusions. All three methods produce reconstructions that separate and localize the inclusions, however the contrast is not reliable. A comparison of the three methods shows that the mere linearization performs the poorest, whereas the most involved method based on solving the perturbed boundary integral equation gives the best reconstructions. However, there is still room for improvement, and this may be achieved by solving the correct boundary integral equation as a first step in the algorithm. The example shows that there is a delicate balance between the noise in the problem, the errors the computation, and the truncation of the scattering transform, but more analysis must be done to reveal the connections and understand the regularization effect.

Acknowledgments
The work of Fabrice Delbary was supported by a post doctoral fellowship from the Villum Kann Rasmussen Foundation.

References

[1] J. Bikowski *Electrical Impedance Tomography Reconstructions in Two and Three Dimensions; from Calderón to Direct Methods* PhD thesis, Colorado State University, Fort Collins, CO, 2008.
[2] J. Bikowski, K. Knudsen, and J. L. Mueller, *Direct numerical reconstruction of conductivities in three dimensions using scattering transforms*, Inverse Problems **27** (2011), 015002 (19pp).
[3] L. Borcea, *Electrical impedance tomography*, Inverse Problems **18** (2002), pp. 99–136.
[4] A-P. Calderón, *On an inverse boundary value problem*, Seminar on Numerical Analysis and its Applications to Continuum Physics (Rio de Janeiro, 1980), Soc. Brasil. Mat., Rio de Janeiro, 1980, pp. 65–73.
[5] D. Colton and R. Kress, *Inverse Acoustic and Electromagnetic Scattering Theory*, vol. 93 of Applied Mathematical Sciences, Springer-Verlag, Berlin, second ed., 1998.
[6] H. Cornean, K. Knudsen, and S. Siltanen, *Towards a d-bar reconstruction method for three-dimensional EIT*, J. Inverse Ill-Posed Probl. **14** (2006), no. 2, pp. 111–134.
[7] M. DeAngelo and J. L. Mueller *D-bar reconstructions of human chest and tank data using an improved approximation to the scattering transform,* Physiological Measurement **31** (2010), no. 2, pp. 221–232.
[8] D. Holder, *Electrical impedance tomography*, Institute of Physics Publishing, Bristol and Philadelphia, 2005.
[9] K. Knudsen, J. L. Mueller, The Born approximation and Calderón's method for reconstruction of conductivities in 3-D, Proceedings of the Eighth AIMS International Conference on Dynamical Systems, Differential Equations and Applications, Dresden 2010. To appear.
[10] J. Mueller and S. Siltanen, *Direct reconstruction of conductivities from boundary measurements*, Siam J. Sci. Comp. **24** (2003), no. 4, 1232–1266.
[11] A. I. Nachman, *Reconstructions from boundary measurements*, Ann. of Math. (2) **128** (1988), no. 3, pp. 531–576.
[12] R. G. Novikov, *A multidimensional inverse spectral problem for the equation* $-\Delta\psi+(v(x)-Eu(x))\psi=0$, Funktsional. Anal. i Prilozhen. **22** (1988), no. 4, pp. 11–22, 96, translation in Funct. Anal. Appl., **22** (1988), no. 4, pp. 263–272.
[13] A. I. Nachman, J. Sylvester, and G. A. Uhlmann, *An* n*-dimensional Borg-Levinson theorem*, Comm. Math. Phys. **115** (1988), no. 4, pp. 595–605.
[14] S. Siltanen, J. Mueller and D. Isaacson *An implementation of the reconstruction algorithm of A. Nachman for the 2-D inverse conductivity problem,* Inverse Problems **16** (2000), pp. 681-699 (Erratum: *Inverse problems* **17**, 1561–1563).
[15] J. Sylvester and G. A. Uhlmann, *A global uniqueness theorem for an inverse boundary value problem*, Ann. of Math. (2) **125** (1987), no. 1, pp. 153–169.
[16] L. Wienert, *Die numerische Approximation von Randintegraloperatoren für die Helmholtzgleichung im* $\mathbb{R}^3$, Dissertation, Göttingen (1990).

Data Assimilation using Bayesian Filters and B-spline Geological Models

Lian Duan[1], **Chris Farmer**[1], **Ibrahim Hoteit**[2], **Xiaodong Luo**[2], **Irene Moroz**[3]

[1] Oxford Centre for Collaborative Applied Mathematics, Mathematical Institute, University of Oxford, 24 - 29 St Giles, Oxford, OX1 3LB

[2] King Abdullah University of Science and Technology, Thuwal, Saudi Arabia

[3] Oxford Centre for Industrial and Applied Mathematics, Mathematical Institute, University of Oxford, 24 - 29 St Giles, Oxford, OX1 3LB

E-mail: `lian.duan@maths.ox.ac.uk`

Abstract. This paper proposes a new approach to problems of data assimilation, also known as history matching, of oilfield production data by adjustment of the location and sharpness of patterns of geological facies. Traditionally, this problem has been addressed using gradient based approaches with a level set parameterization of the geology. Gradient-based methods are robust, but computationally demanding with real-world reservoir problems and insufficient for reservoir management uncertainty assessment. Recently, the ensemble filter approach has been used to tackle this problem because of its high efficiency from the standpoint of implementation, computational cost, and performance. Incorporation of level set parameterization in this approach could further deal with the lack of differentiability with respect to facies type, but its practical implementation is based on some assumptions that are not easily satisfied in real problems.

In this work, we propose to describe the geometry of the permeability field using B-spline curves. This transforms history matching of the discrete facies type to the estimation of continuous B-spline control points. As filtering scheme, we use the ensemble square-root filter (EnSRF). The efficacy of the EnSRF with the B-spline parameterization is investigated through three numerical experiments, in which the reservoir contains a curved channel, a disconnected channel or a 2-dimensional closed feature. It is found that the application of the proposed method to the problem of adjusting facies edges to match production data is relatively straightforward and provides statistical estimates of the distribution of geological facies and of the state of the reservoir.

1. Introduction

History matching is the art of adjusting reservoir models until they reproduce the past reservoir behaviour, consisting of historical production and pressure data. Once a model has been history matched, it is often assumed that it can then be used to simulate future reservoir behaviour with a higher degree of confidence, particularly if the adjustments are constrained by known geological properties in the reservoir [1]. However, it can be shown within the Bayesian framework that such history matched models are modes of the posterior density and by using an over-smoothed model that may only be a local maximum of the posterior density there are no reasons whatever for assuming that history matched models have any particular predictive power. Only by adopting a fully Bayesian approach one can hope to quantify the uncertainty in the history matched models and then only if the equations used to model the reservoir are a faithful representation of the reservoir physics.

Before automatic history matching of realistic reservoir models became feasible, reservoir models were history matched manually using engineering judgement and a work flow that has developed through years of experience (such as in [2]). However, such a procedure can be efficient only when a small number of parameters needs to be retrieved from the observed production data. It took Agarwal *et al.* [3] one year of intensive work to match 25 years worth of production and performance data for a complex fractured chalk reservoir in the Norwegian North Sea, with 76 production wells, 13 gas injectors and 37 water injectors.

Automatic strategies for history matching were initially based on the perturbation method. This method computes the single grid block sensitivity coefficients in order to evaluate the change of an objective function to small perturbations of the permeability field [4, 5]. For every iteration, the direct method of computing these sensitivity coefficients requires integrating the reservoir model as many times as the number of parameters to be determined. In consequence, this approach is computationally quite demanding for real reservoir history matching problems. Later methods used adjoint techniques which can be very efficient, but are very difficult to code. Nowadays, advanced automatic strategies are statistical and are based on the Ensemble Kalman filter (EnKF) approach [6, 7] and sometimes the method of randomised maximum likelihood [8, 9, 10] which does use adjoint methods [11]. EnKF methods build the solution sequentially every time a new observation is available by updating the mean and the posterior correlation matrix, while deterministic variational methods produce maximum-likelihood solutions through minimization of an objective function.

Given the limited amount of data and the complexity of the problem, many engineers prefer to adjust relatively small numbers of parameters ("global parameters"), even if the history matching problem is formulated as an automated process. For instance, large scale permeability trends, like barriers and channels, have a large impact on how the fluid flows in a porous medium. Reliable knowledge of these geological structures is therefore important for the reservoir engineers controlling the production in the reservoir. One can then turn the history matching problem into an estimation problem of the geological facies pattern [12].

In geology, facies are a body of rock with specified characteristics such as permeability, porosity, etc[13]. In the existing literature, the history matching problem of the facies pattern has been addressed using a level set parameterization of the facies pattern with the parameters estimated as the solution of a gradient-based variational problem. The piecewise constant level set method [14] has been applied to many areas since the original method was introduced by Osher and Sethian [15, 16], e.g. for tracing interfaces between different phases [17], for image segmentation [18], or for optimal shape design problems [19].

Nielsen *et al.* [20] and Dorn *et al.* [21] applied the level set method to model the facies bodies and history matched the observed data of several synthetic reservoir models by minimising a misfit function between the simulated and observed data. This approach was found to be robust. However, it can be more computationally expensive for realistic problems than, say an EnKF method, because it requires integrating the adjoint of the model to determine the gradient descent minimization directions. More importantly, this method only provides a deterministic estimation without any information about the associated uncertainty, which might be insufficient for reservoir management.

Sequential ensemble-based data assimilation techniques, based on the work of the Ensemble Kalman Filter (EnKF) of [6, 7], appear to offer attractive features and powerful capabilities for the automatic history matching problem of facies pattern. The EnKF actually provides an efficient approach to estimate a large number of variables and is based on a simple conceptual formulation and relative ease of implementation (it does not need the linearization of the reservoir model nor the development of its adjoint and only requires forward integrations of the forward reservoir model). It furthermore computes estimates of the solution uncertainties as part of its algorithm at reasonable computational cost.

The EnKF originated as a variant of the Kalman filter designed for large dimensional nonlinear systems and is now one of the most used data assimilation schemes. The filter propagates an ensemble of prior realizations with the nonlinear model and uses the linear Kalman filter correction step to update the ensemble every time new observations are available. Despite the underlying nature of its linear analysis step, the efficiency of EnKF method was demonstrated on different applications and in many situations achieved accurate estimation much faster than fully nonlinear Bayesian methods such as the particle filter [22]. The original formulation of the EnKF requires perturbing the observations so that its analysis covariance stochastically matches that of the Kalman filter. This might however introduce noise in the estimation process especially when the size of the ensemble is not much larger than the number of observations. The Ensemble Square Root Filter (EnSRF) was introduced as an alternative ensemble implementation that exactly matches the analysis covariance of the Kalman filter [23].

Because ensemble and other Kalman-based filters are only designed for the estimation of continuous variables, the main challenge in applying an EnKF to the history matching problem of the geological facies pattern is to transform the discrete facies variables into continuous variables through an appropriate parameterization of the facies.

Liu *et al.* [24, 25] used a threshold truncation map to describe the geostatistical model of the problem of interest, transforming the problem into a position estimation problem in each cell. The initial ensemble was generated using truncated pluri-Gaussian models. Very recently, Chang [26] combined the level set method and the traditional EnKF, naturally transforming the discrete facies variable of -1 and 1 to a continuous variable of real numbers less than or greater than 0. Their initial ensemble was generated using multipoint geostatistics methods, such that the facies in each cell of the initial ensemble are centered and with a standard deviation of 0.5. After the history matching process, the mean of the artificial facies values are corrected towards one side of zero with reduced uncertainty. Both schemes are however difficult to apply to a real reservoir problem because their assumptions for generating the initial ensemble are not often satisfied in real history matching problems.

In this paper, we introduce a new method to incorporate the level set method in an EnKF. The idea is based on a B-spline parameterization of the facies pattern. B-splines were introduced more than 50 years ago. The method is mainly used for curve fitting [27] and image registration [28]. It has been also applied in many other areas, such as neural networks [29] and cardiac motion estimation [30]. As in [20], our method used a binary level set function to represent the permeability field. The discontinuous curves of the level set functions are then parameterized by either two one-dimensional unclamped cubic B-splines for the channel features, or one two-dimensional clamped cubic B-spline for the two-dimensional closed features. Using this continuous parameterization of the edges of the facies, we will demonstrate that the history matching problem of the facies pattern can be solved using the EnSRF.

The paper is organised as follows. Section 2 presents the reservoir model and describes the history matching problem and its state-space formulation. The EnSRF which is used to solve the history matching problem is summarised in Section 3. The general framework for level set modelling is presented in Section 4. An introduction of the B-spline parameterizations for the discontinuous curve of the level set functions is presented in Section 5. Section 6 discusses the key B-spline parameterizations and the typical form of filter realizations. Section 7 presents and discusses the results of three different numerical experiments of facies pattern history matching problems. A summary and a discussion conclude the work in Section 8.

Figure 1. The facies of the reservoir model, blue cells (shale), red cells (sandstone), blue pluses (injectors) and black pluses (producers).

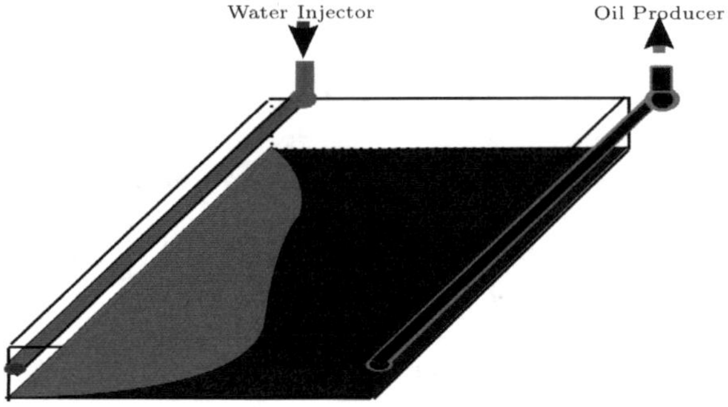

Figure 2. Three-dimensional view of the oil enhanced recovery using water flooding experiment as presented in [31].

2. Reservoir Model and History Matching Problem

We consider a two dimensional horizontal rectangular reservoir model with isotropic permeability and assume that it is fully saturated with water and oil initially (as shown in Fig. 1). The reservoir is formed by two types of rock, the large shale area with low permeability κ_l and a small amount of sandstone with high permeability κ_h. Using the water flooding experiment in [31], secondary oil recovery is performed by injecting water on the left boundary of the reservoir with a constant injection rate and producing oil from the right boundary of the reservoir with a constant bottom-hole pressure as shown in Fig. 2. Assuming incompressibility and immiscibility of the fluids, the model equations are the conservation of mass equations:

$$\phi(\underline{x})\frac{\partial S_o}{\partial t} - \nabla\cdot\left(\frac{\kappa(\underline{x})\kappa_{ro}(S_o)}{\mu_o}\nabla p_o\right) = q_o(\underline{x}), \tag{1}$$

$$\phi(\underline{x})\frac{\partial S_w}{\partial t} - \nabla\cdot\left(\frac{\kappa(\underline{x})\kappa_{rw}(S_w)}{\mu_w}\nabla p_w\right) = q_w(\underline{x}), \tag{2}$$

where $(\underline{x}, t) \in \Omega \times [0, T]$, $\Omega \subset \mathcal{R}^2$ is the bounded reservoir domain. Here, the subscripts o and w refer to the phases, oil and water, respectively. Also S_f denotes the saturation, μ_f the viscosity, p_f the pressure, q_f the external volumetric flow rate and κ_{rf} is the relative permeability, where f is the fluid phase. The porosity and the absolute permeability are given by $\phi(\underline{x})$ and $\kappa(\underline{x})$, respectively. Closure of the system is obtained through the assumption of a completely saturated medium, such that

$$S_o + S_w = 1, \tag{3}$$

and a function P_c defining the capillary pressure,

$$p_o - p_w = P_c. \tag{4}$$

The quantities $\phi, \kappa, \kappa_{rf}$ and P_c are all properties of the porous medium and are not accessible through direct measurements except at the wells. In the forward problem, they are assumed to be known such that the reservoir behaviour (ie. pressure and saturation of each fluid phrase) can be predicted using a finite volume simulator. In the history matching problem, these rock properties are the unknowns. More specifically, with some prior knowledge about the reservoir ϕ, κ_{rf}, κ values, the existence of features (ie. a channel or a close feature) and the rock properties along the left and right boundaries, our goal, in this paper, is to estimate the geological facies pattern (hence the absolute permeability $\kappa(\underline{x})$) to predict the reservoir future behaviour conditioned upon pressure and water saturation measurements in the production well.

In the forward problem (or the reservoir simulator), the finite volume method is employed to discretize the reservoir domain Ω into $N = N_y \times N_x$ uniform spatial cells, where the variables within each cell are assumed constant. Equations (1, 2, 3, 4) are solved using a custom finite volume simulator, in which cell residuals are minimised implicitly in time by applying Newton iterations and the generalized minimal residual method (GMRES) as the linear solver. The cell index from top to bottom and then from left to right is denoted by superscript j from 1 to N as shown in Fig. 1. A typical (input) state vector x consists of state variables ($\mathbf{P} \in \mathcal{R}^{N\times 1}, \mathbf{S} \in \mathcal{R}^{N\times 1}$) and model parameters ($\mathbf{m} \in \mathcal{R}^{n_m\times 1}$), as follows:

$$x = \{\mathbf{P}, \mathbf{S}, \mathbf{m}\}^T = \{P^1, \ldots, P^N, S^1, \ldots, S^N, \mathbf{m}\}^T \in \mathcal{R}^{(2N+n_m)\times 1}. \tag{5}$$

with the superscript T denoting a matrix transpose. In each cell j, S^j is the water saturation and assuming a zero capillary pressure $P_c = 0$, the pressure for each fluid phase is

$$P^j = p_o^j = p_w^j. \tag{6}$$

The model parameters $\mathbf{m}$ are time-independent variables and include unknown rock properties (such as the permeability and porosity fields). The choice of $\mathbf{m}$ will be discussed in detail in Sections 4, 5 and 6.

In the history matching problem, the simulation of the pressure and saturation in every cell and the production data in the wells can be described using the state-space equations as follows:

$$x_k = f(x_{k-1}), \tag{7a}$$

$$s_k = Hx_k + v_k. \tag{7b}$$

Here, the subscript k is the time index; the function $f(\cdot)$ denotes the reservoir simulator and H is a $(2N_y) \times (2N + n_m)$ measurement matrix relating the state vector to the observations. In the present study it is of the form

$$H = \begin{pmatrix} \underbrace{\begin{matrix} I^{N_y} \\ 0 \end{matrix}}_{N_y} & \underbrace{\begin{matrix} 0 \\ 0 \end{matrix}}_{2N-2N_y} & \underbrace{\begin{matrix} 0 \\ I^{N_y} \end{matrix}}_{N_y} & \underbrace{\begin{matrix} 0 \\ 0 \end{matrix}}_{n_m} \end{pmatrix}, \tag{8}$$

where the matrix I^{N_y} is an $N_y \times N_y$ identity matrix. The first and second matrices I^{N_y} take the measurements on the pressure in the injection wells and water saturation in the production wells, respectively. The measurement noise v_k is modelled here as an independent multivariate Gaussian distribution with mean 0 and covariance matrix

$$R = \begin{pmatrix} \sigma_P^2 I^{N_y} & 0 \\ 0 & \sigma_S^2 I^{N_y} \end{pmatrix}, \tag{9}$$

where σ_P and σ_S are respectively the standard deviation of the pressure measurement noise in the injection wells and the water saturation measurement in the production wells.

3. Ensemble Square Root Filter (EnSRF)

The Ensemble Square Root Filter (EnSRF) is an ensemble Kalman filter suitable for large-scale nonlinear models. It was introduced by [23] to avoid the noise introduced by the perturbed observation needed in the original EnKF. At time t_k, it provides estimates of the state vector

Journal of Physics: Conference Series **290** (2011) 012004 doi:10.1088/1742-6596/290/1/012004

X_k and the associated uncertainty based on an ensemble of r realizations from the distribution of the state denoted by

$$X_k = \{x_{k,1}, x_{k,2}, \dots, x_{k,i}, \dots, x_{k,r}\} \in \mathcal{R}^{(2N+n_m)\times r}. \tag{10}$$

Here, the subscript $i \in [0, r]$ denote the realization index and each realization $x_{k,i} \in \mathcal{R}^{(2N+n_m)\times 1}$ consists of state variables and model parameters, as in equation (5):

$$x_{k,i} = \{P^1_{k,i}, \dots, P^N_{k,i}, S^1_{k,i}, \dots, S^N_{k,i}, \mathbf{m}_{k,i}\}^T \in \mathcal{R}^{(2N+n_m)\times 1}, \tag{11}$$

Following these definitions and notations, the algorithm of the EnSRF can be summarized as an initialization step, a sequence of forecast steps (with the model to update the ensemble forward in time) followed each time by an analysis step to update the forecast ensemble conditioned on the new observations. After a prediction or an analysis step, the state estimate is taken as the mean and the associated uncertainty is represented by the ensemble sample covariance matrix.

Initialization step:

(i) The probability density function $\pi(\mathbf{P}_0, \mathbf{S}_0, \mathbf{m})$ of the initial state and model parameters is assumed to be given and is represented statistically by an ensemble $X^a_0 = \{x^a_{0,1}, \dots, x^a_{0,r}\}$.

(ii) For every filtering step $k = 1, 2, \cdots$, proceed as follows;

Forecasting step

(a) Integrate the available analysis ensemble $X^a_k = \{x^a_{k,1}, \dots, x^a_{k,r}\}$ forward with the nonlinear reservoir model (7a) to the time of the next available observation to compute the forecast ensemble $X^f_k = \{x^f_{k,1}, \dots, x^f_{k,r}\}$ as

$$x^f_{k,i} = f(x^a_{k-1,i}) \qquad \text{for} \quad i = 1, \dots, r, \tag{12}$$

where the superscripts f and a denote forecast and analysis, respectively.

(b) Compute the forecast state $\hat{x}^f_k$, the forecast ensemble anomalies $A^f_k = [A^f_{k,1}, ..., A^f_{k,r}]$:

$$\hat{x}^f_k = \frac{1}{r}\sum_{i=1}^{r} x^f_{k,i}, \tag{13}$$

$$A^f_{k,i} = \frac{1}{\sqrt{r-1}}(x^f_{k,i} - \hat{x}^f_k). \tag{14}$$

The forecast error covariance matrix is then

$$P^f_k = A^f_k (A^f_k)^T, \tag{15}$$

but does not need to be explicitly computed for the filter algorithm.

Analysis step

(c) Compute the Kalman Gain Matrix K_k:

$$K_k = P^f_k H^T (H P^f_k H^T + R)^{-1}. \tag{16}$$

(d) Calculate the analysis $\hat{x}^a_k$ by using the Kalman analysis equation:

$$\hat{x}^a_k = \hat{x}^f_k + K_k(s_k - H_k \hat{x}^f_k). \tag{17}$$

(e) Perform an eigenvalue decomposition on $(I - K_kH)$, such that $(I - K_kH) = UDU^{-1}$ with positive eigenvalues only, $D_{ii} > 0$.

(f) Generate a new analysis ensemble X_k^a using a transform matrix T and the previous ensemble anomalies A_k^a as:

$$T = UD^{1/2}U, \tag{18a}$$

$$A_k^a = TA_k^f, \tag{18b}$$

$$X_k^a = \hat{x}_k^a \cdot \mathcal{U}^{1\times r} + \sqrt{r-1}A_k^a, \tag{18c}$$

where $\mathcal{U}^{1\times r}$ denotes a $1 \times r$ unit matrix (matrix of 1s).

4. Level Set Modelling of the Permeability Field

With the aim of estimating the facies pattern, hence the permeability field, it is natural to consider the vector of facies as the parameter vector $\mathbf{m}$ to be estimated with the ensemble Kalman filter in the history matching problem [32]. Using prior knowledge about the existence of features (ie. a channel or a closed feature) and two types of rocks with absolute permeability κ_h and κ_l in the reservoir, one can model the facies pattern and absolute permeability field using level set functions.

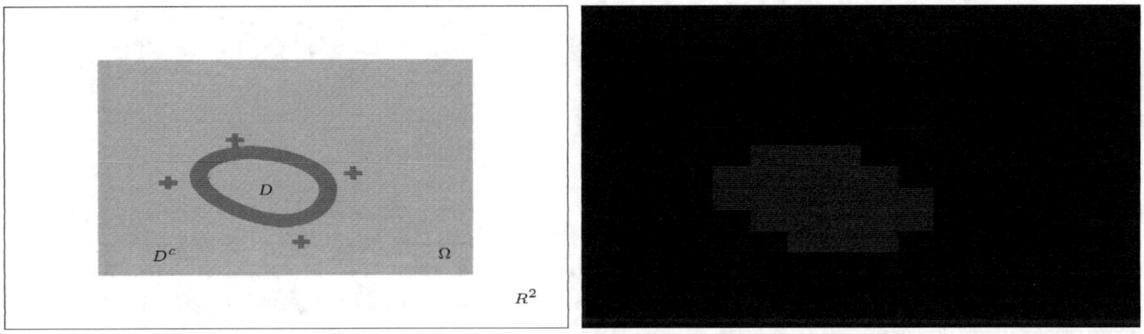

Figure 3. Permeability field with a two-dimensional closed feature modelled by a level set function via B-spline parameterization of the discontinuous curve. Plots show the reservoir domain Ω (gray rectangle), the B-spline control point domain (outer white rectangle bounded by black edges), discontinuous curve of the level set function (left figure in red), B-spline control points (left figure red pluses) and the corresponding permeability field (right figure) using equations (19) and (20).

Level set functions are discontinuous functions which at convergence take the values 1 and -1 inside and outside two subregions. If a domain Ω is divided into two subregions, D and D^c ($D^c = \Omega \backslash D$ is the complement of D), a level set function Φ is defined as:

$$\Phi(\underline{x}) = 1, \ \forall \underline{x} \ in \ D, \qquad \Phi(\underline{x}) = -1, \ \forall \underline{x} \ in \ D^C. \tag{19}$$

The curve separating D and D^c representing the edges of the sandstone is implicitly given as the discontinuous curve ς of Φ. The properties of Φ can be used to construct the facies pattern and the permeability field $\kappa(\underline{x})$ with distinct constant values inside the two different subdomains. The facies variable in each cell is modelled as a discrete variable taking value of -1 or 1. As in Fig. 3, if we assume that the value of $\kappa(\underline{x})$ is equal to κ_h in D and equal to κ_l in D^c, the $\kappa(\underline{x})$ can be written as

$$\kappa(\underline{x}) = \frac{1}{2}[\kappa_h(\Phi(\underline{x}) + 1) - \kappa_l(\Phi(\underline{x}) - 1)], \tag{20}$$

with constraints

$$\Phi(\underline{x})^2 - 1 = 0. \tag{21}$$

5. B-spline Parameterization

Estimating the facies type of each cell by including these as parameters in the EnSRF state vector is not compatible with the formulation of the filter. Indeed, Kalman type ensemble filters are designed to work with continuous variables while the facies type is a discrete variable taking one of two values -1 or 1 (or κ_h or κ_l for permeability field) only. Even if the filter starts with a forecast ensemble consisting of two types of rocks, the updated ensemble will not necessarily consist of the same type of rocks after the analysis step. This is the biggest challenge of using an ensemble-based Kalman filter for the history matching of facies in a reservoir.

To go around this problem and use an ensemble filter for the estimation of facies patterns, one needs to introduce a parameterization that transforms the discrete variables representing the facies (or permeability) into continuous variables [24, 25, 26] and then use the latter in the estimation process. In this paper, this is achieved by parameterizing the discontinuous curve ς of the level set function using B-spline functions. These are designed around local control points with global smoothness constraints making them particularly useful for our problem [33].

The idea of the B-spline method is briefly recalled hereafter. The reader is referred to [33] for a detailed description of the method. A B-spline is a spline function that has minimal support with respect to a given degree, smoothness and domain partition. It is a generalisation of a Bézier curve. The definition of a two-dimensional uniform unclamped B-spline approximation that can be summarized as follows:

$$\varsigma(\chi) = \sum_{c=0}^{n} N_{c,d}(\tau)\underline{b}^c. \tag{22a}$$

The subscript $c \in [0, n]$ is the index of the control point; the subscript d is the degree of the curve; $\chi \in [0, 1]$ is an evaluation point; $\underline{b}^c = (a^c, b^c) \in \mathcal{R}^2$ are the $n+1$ control points, where a^c is along the direction perpendicular to the left boundary of the reservoir with 0 at the left boundary, and b^c is along the direction horizontal to the left boundary of the reservoir with 0 at the top boundary. The $N_{c,d}(\chi)$ are the B-spline basis functions and are recursively constructed as a function of $n+d+1$ knots χ_m. The knots χ_m are constructed in the standard open uniform fashion, as follows:

$$\chi_m = \begin{cases} 0, & 0 \le m \le d \\ \frac{m-d}{n+1-d}, & d+1 \le m \le n \\ 1, & n+1 \le m \le n+d+1 \end{cases}. \tag{22b}$$

The recursive definition of the B-spline functions starts from the basis function

$$N_{c,d}(\chi) = \begin{cases} 1, & \chi_c \le \chi \le \chi_{c+1} \\ 0, & \text{otherwise} \end{cases} \tag{22c}$$

for $0 \le c \le n+d$. The functions are then computed recursively as

$$N_{c,l}(\chi) = \frac{\chi - \chi_c}{\chi_{c+l} - \chi_c} N_{c,l-1}(\chi) + \frac{\chi_{c+l+1} - \chi}{\chi_{c+l+1} - \chi_{c+1}} N_{c+1,l-1}(\chi) \tag{23}$$

for $1 \le l \le d$ and $0 \le c \le n+d-l$. The support of any B-spline function $N_{c,l}(\chi)$ is then $[\chi_c, \chi_{c+l+1}]$.

For parameterizing closed curves, it is advisable to use the clamped B-splines. These are computed using wrapping control points (or joining the ends) and a degree d B-spline with $n+1$ control points which can be generated in the following way:

(i) Generate d fictitious control points $\{\underline{b}^{n+1}, \ldots, \underline{b}^{n+d}\}$ such that:

$$\underline{b}^{n+1} = \underline{b}^{1}, \quad \underline{b}^{n+2} = \underline{b}^{2}, \quad \ldots, \quad \underline{b}^{n+d} = \underline{b}^{d}. \tag{24}$$

The number of control points is artificially increased from $n+1$ to $n+d+1$. The index of the control point is now between 0 to $\tilde{n}$, where $\tilde{n} = n + d$.

(ii) By replacing n by $\tilde{n}$, the clampled B-splines is generated using equation (22a) with basis function in equations (22c, d) and uniform knots (22b).

6. Updating Facies Patterns with EnSRF

Let us first consider a reservoir with a channel feature. The discontinuous curve ς of the level set function is then parameterized by two one-dimensional unclamped B-splines. Although the control points $\underline{b}^c = (a^c, b^c)$ are two-dimensional, they are located uniformly along the direction perpendicular to the left boundary of the reservoir with fixed a^cs. Assuming prior knowledge of the existence of a facies feature and the rock properties along the left and right boundaries, the upper edge of the discontinuous curve ς (or the channel) is parameterized by a one-dimensional unclamped B-spline function with fixed starting and ending control points and $n+1$ intermediate control points $\{b^0, b^1, \ldots, b^n\}$, as shown in Fig. 4. Similarly, we consider $n+1$ intermediate control points $\{b^{n+1}, \ldots, b^{2n+1}\}$ to parameterize the lower edge of the discontinuous curve. It is noted that the control points are continuous variables in R^2, which contains the reservoir domain Ω and the region where the upper edge of the discontinuous curve ς below the lower edge is assumed to be contained in D^c. The model parameters to be included as part of the the filter state vector will therefore consist of the positions of the intermediate control points. A typical realization of the ensemble of the EnSRF will therefore be of the form:

$$x_{k,i} = \{P^1_{k,i}, \ldots, P^N_{k,i}, S^1_{k,i}, \ldots, S^N_{k,i}, b^0_{k,i}, \ldots, b^{2n+1}_{k,i}\}^T. \tag{25}$$

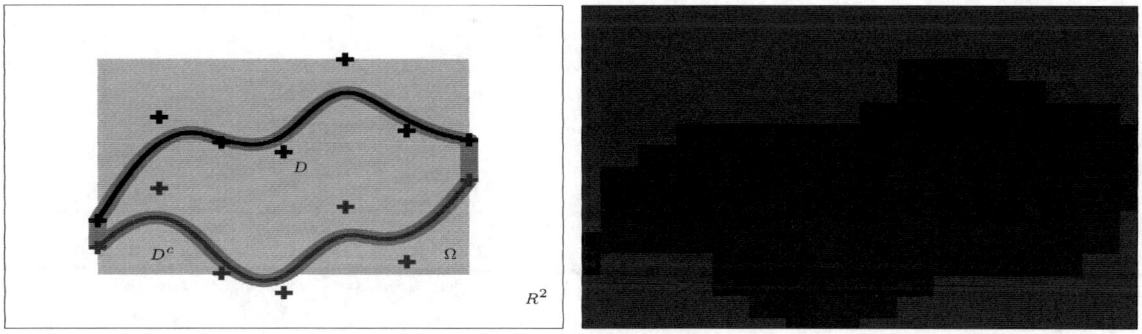

Figure 4. Permeability field featuring a channel modelled by a level set function via B-spline parameterization of the discontinuous curve. Plots show the reservoir domain Ω (gray rectangle), the B-spline control point domain (outer white rectangle bounded by black edges), discontinuous curve of the level set function (left figure in red), B-spline control points (left figure black pluses) for the upper edge of the discontinuous curve (left figure in black), B-spline control points (left figure blue pluses) for the lower edge of the discontinuous curve (left figure in blue), and the discontinuous curve of the level set function (left figure in red) and the corresponding permeability field (right figure) computed using equations (19) and (20).

For a reservoir with two-dimensional closed features, as in Fig. 3, the curve ς of the level set function is parameterized by a two-dimensional clamped B-spline. The model parameters to be included in the state vector will be the values of the $n+1$ control points. In this case, typical realization of the ensemble of the EnSRF will be of the form:

$$x_{k,i} = \{P_{k,i}^1, \ldots, P_{k,i}^N, S_{k,i}^1, \ldots, S_{k,i}^N, a_{k,i}^0, \ldots, a_{k,i}^n, b_{k,i}^0, \ldots, b_{k,i}^n\}^T. \quad (26)$$

The estimation process we are dealing with is greatly complicated by the nature of the geological heterogeneity, which is not amenable to simple parameter descriptions. One may always consider a large enough number of control points to allow for more flexibility to reproduce the complicated spatial configuration. This might however make the estimation problem of the facies pattern more difficult because it increases the chances of having different parameter sets producing comparable fits to the available measurements [11, 34, 35]. In this case, it would be more appropriate to constrain the number of control points which could lead to a lower-dimensional estimation problem but at the price of missing some features of the the true reservoir geometry. In this work, we choose $n = 4$ for reservoirs with channel features and $n = 3$ for reservoirs with two-dimensional closed features.

7. Numerical Experiments

Three numerical experiments were performed to test the proposed method and to evaluate its performance. More specifically, the EnSRF was used to estimate the facies patterns of three different permeability fields using synthetic pressure and saturation measurements. Facies patterns were parameterized using B-splines as described in section 6. In these experiments, the true permeability field consists of two distinct values and the test cases are a curved channel (Fig. 5), a disconnected channel (Fig. 7), and a two-dimensional closed feature (Fig. 10). The reservoir is square and horizontal with constant thickness and non-flow outer boundary. The relative permeability functions are defined by the Corey models:

$$\kappa_{rw} = \hat{\kappa}_{rw}\left(\frac{S_w - S_{wr}}{1 - S_{wr} - S_{or}}\right)^{e_w},$$
$$\kappa_{ro} = \hat{\kappa}_{ro}\left(\frac{S_o - S_{or}}{1 - S_{or} - S_{wr}}\right)^{e_o}.$$

where the Corey exponents, e_w and e_o, the residual saturations, S_{wr} and S_{wo}, and the endpoint permeabilities, $\hat{\kappa}_{rw}$ and $\hat{\kappa}_{ro}$, are assumed to be known. The numerical values of the reservoir properties are listed in table 1.

For each reference permeability field we simulate the 'true' values of saturation (S_w) and pressure (p_w) for the applied time steps on the given grid by integrating the reservoir model forward in time. Synthetic measurements are then extracted from the true states and perturbed with random noise to test the system in the Prefect Model Scenario (PMS). The measurement noise is assumed to be uncorrelated Gaussian noise with zero mean. The standard deviation of the pressure measurement noise in the injector wells and the water saturation measurement in the projection wells are $\sigma_P = 2$ bar and $\sigma_S = 0.002$, respectively. The automatic history matching of the facies pattern is then performed by assimilating these measurements every 16 days and over an 800 day period. The estimated permeability field is then used to predict the reservoir behaviour from day 801 to day 1600. To mitigate the problem of ensemble collapse, large ensembles with $r = 600$ members were used in the experiments. The mean of the ensemble pressure and saturation are taken as the estimated pressure and saturation in the cells, respectively. In all experiments cubic B-splines were used.

We use two different approaches for the estimation of log permeability fields:

(i) *Calculate the log permeability field from the mean of the B-spline control points (LPMB)*: The discontinuous curve ς of the level set function is first computed from the mean of the

Reservoir dimensions:	$1000m \times 1000m \times 40m$	
Simulation grid:	$16 \times 16 \times 1$ cells	
Porosity:	0.2	
Viscosity:	$\mu_w = 0.5 \cdot 10^{-3}$ Pa·s	$\mu_o = 0.5 \cdot 10^{-3}$ Pa·s
Endpoint relative permeability:	$\hat{\kappa}_{rw} = 0.1$	$\hat{\kappa}_{ro} = 1$
Residual saturations:	$S_{wr} = 0.2$	$S_{or} = 0.2$
Corey exponents:	$e_w = 2$	$e_o = 3$
Initial saturations:	$S_w = 0.2$	$S_o = 0.8$
Capillary pressure function:	$P_c(S_w) \equiv 0$ kPa	
Injector rate:	8% of total pore volume per year	
Production rate:	constant BHP= 200.0 bar	
Number of timesteps:	100	
Total production time:	1600 days	

Table 1. Key properties in the reservoir simulator

analysed ensemble. The log permeability field is then obtained using equations (19 and 20). The LPMB figures consist of the same number of rock types as the reference permeability field and will be used for the comparison against the reference field.

(ii) *Take the mean of the log permeability fields computed from each set of the B-spline control points (MLP)*: The log permeability fields associated with the B-spline parameterization of the discontinuous curve ς in each realization of the analysed ensemble are first calculated. The estimated log permeability field is then found by averaging the log permeability fields cell-by-cell. During this process, artificial rock types might be generated. The MLP figures will be used to evaluate the quality of the estimated permeability field during the history matching period. The standard deviation of the log permeability field (SDLP) estimated from the realizations of the B-spline control points of the analysed ensemble is also used as a measure of the uncertainty about the estimated permeability.

To evaluate the estimates of the facies pattern, we present figures plotting the discontinuous curve of the reference permeability field, the estimated discontinuous curve, and the cells with large facies uncertainty. The estimated discontinuous curve is further plotted along the boundary of the cells where a sign change occurs in the mean level set function. We define the cells with facies standard deviation greater than 0.8 as the cells with large facies uncertainty, and label these in black dots in their corresponding cell centers.

7.1. Curved and Disconnected Channels

In the case the permeability field has a curved channel or a disconnected channel, a typical realization of the state vector in the EnSRF will be of the same form as equation (25).

Fig. 5 presents the evolutions of the log permeability field and the facies patterns estimates with the corresponding uncertainties over the history matching period for the curved channel case. The true log permeability field is plotted in the top panel, where the location and index of the injection and production wells are labelled by blue and black pluses, respectively. The time-evolution of the log permeability fields as estimated by the mean of the B-spline control points (LPMB) is plotted in the second row, the mean (MLP) and standard deviation (SDLP) of the log permeability field are plotted in the third and fourth row, respectively; and figures comparing the true and estimated discontinuous curves, and indicating the cells with large facies uncertainty are plotted in the fifth row.

Because of the lack of complete knowledge of the initial state of the reservoir (on day 0) in realistic applications, the initial realizations of the EnSRF are considered to be quite different from the true reservoir permeability field and facies pattern. One can see from Fig. 5 that the convergence toward the true permeability field is rather slow and after 192 assimilation days (12 assimilation cycles) the facies estimates are still quite different from the true facies. This is due to the limited amount of information (observation) that have been used in the estimation process. The filter however preserves standard deviations indicating large uncertainties on the estimates during that period. As more information become available over time, the estimation of the permeability field and facies pattern gradually improves. At the end of the history matching period, the estimated permeability field and facies pattern are almost identical to the true fields. Only one cell is showing a significant mismatch between the estimated and the true permeability, but the truth lies within the estimated uncertainty.

Fig. 6 shows typical data assimilation processes in an injection well (no. 7), and three production wells (no. 3, 14 and 16) over time. The true states are plotted in red, the measurements are plotted in blue, the simulations of the initial ensembles are plotted in magenta, and the EnSRF ensembles at the end of each analysis steps are plotted in green. One can see that the history matching of the pressure in the injection well and saturation in the production wells have been very well reproduced. The uncertainty level (the vertical width of the green lines) provided by the EnSRF which is much smaller than the initial uncertainty level (the vertical width of the magenta lines). The uncertainty of the estimated pressure field slightly increases during the prediction step, before it decreases after an analysis step. This is a typical behaviour of a sequential assimilation system. More specifically, water breakthroughs are not observed in all production wells during the history matching period. The true reservoir pressure and saturation behaviour are well predicted and within the uncertainty level. Similar good results are obtained for the other injection and production wells.

Similar conclusions can be made from the results in the disconnected channel case as can be seen from Figs. 7 and 8 which are plotted in the same spirit of Figs. 5 and 6. During the first 192 days, the estimated log permeability field and facies pattern indicate a disconnected channel, but with a quite large uncertainty. The filter efficiently recovers the true fields in the last 608 days. The corresponding uncertainties further decreases over time and become much smaller than the initial uncertainties. Again, the history matched permeability field and facies pattern are almost identical to the truth at the end of the history matching period, and the truth lies within the uncertainties estimated by the filter. Fig. 7 further demonstrates that the estimated state of the disconnected channel reservoir are also quite good.

7.2. Two-Dimensional Closed Feature

For a two-dimensional closed feature, a typical realization of the state vector for the EnSRF will be of the same form as equation (26).

In practical applications, very often some prior information about the existence of a closed feature reservoir are available but not about the shape and location. We therefore generate an initial ensemble that reflects the reality of these problems, mainly having a closed features quite different than the true permeability field in term of shape and location. The discontinuous curve of the level set functions and the corresponding permeability fields for four realizations of the initial ensemble are presented in Fig. 9. The choice of the initial ensemble is also reflected in the log permeability and facies pattern plots on day 0 of Fig. 10.

As in the channel cases, Fig. 10 shows that the proposed history matching system is capable of recovering the basic structures of the true permeability field after 576 assimilation despite starting the estimation process from a very poor initial prior. The corresponding uncertainties further continuously decrease over time and are significantly less than the initial uncertainties at the end of the assimilation window. The final estimated permeability is almost the same as the true one, except at one cell located on the left boundary (and can be easily identified in the

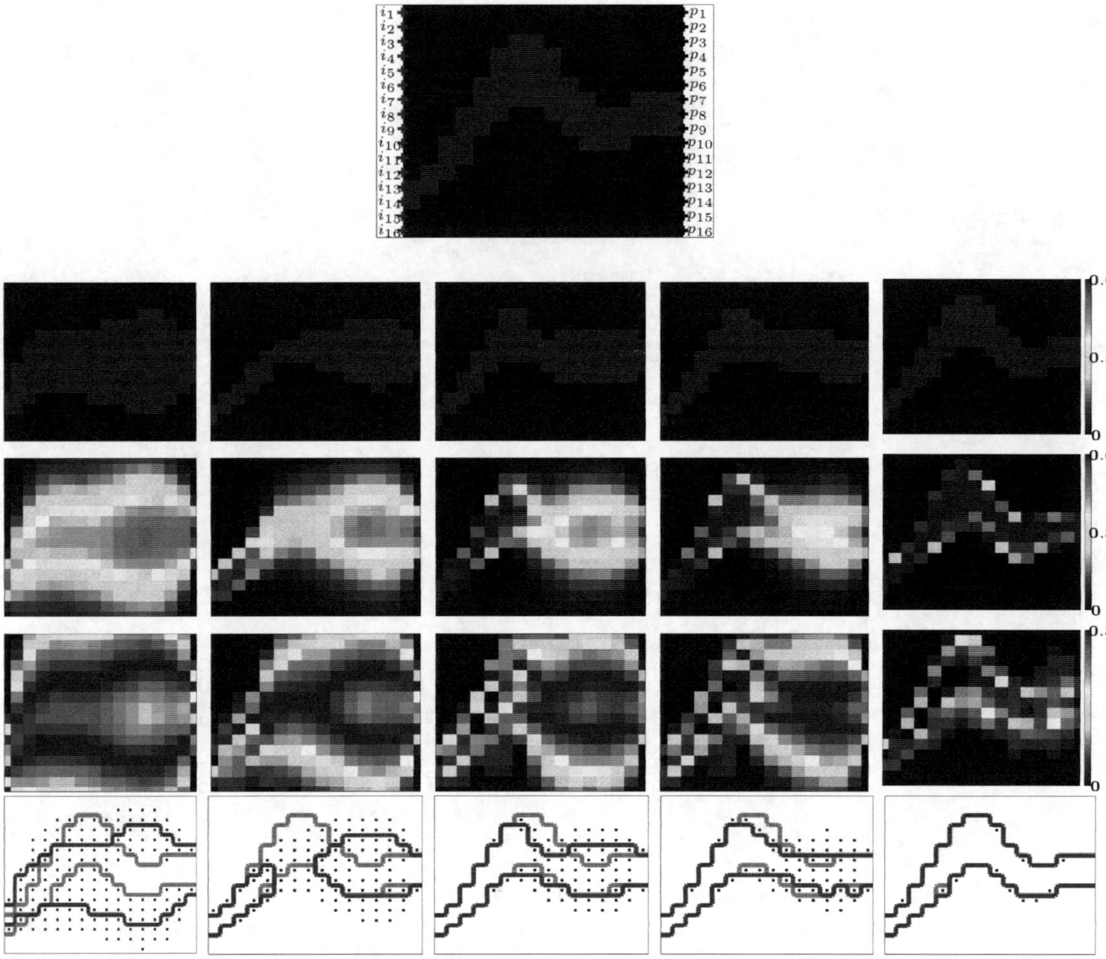

Figure 5. The true log permeability field (first row) and the log permeability field as estimated from the mean of the B-splines (second row), and mean (third row) and standard deviation (fourth row) of the log permeability field and misfit of facies pattern (fifth row) for the curved channel case. Results are shown at the end of 0th, 192th, 384th, 576th and 800th days with 600 realization. For the true log permeability field, the injection well is plotted in blue pluses and the production wells in black pluses. The misfit figures show the true facies pattern (red), the estimated facies pattern (blue), and the cells with facies standard deviation greater than 0.8 (black dots).

misfit figure on day 800 of Fig. 11). At this location, the estimated field is quite different than the truth and lies outside the estimated uncertainty.

Fig. 11 shows the time-evolution of the assimilated processes in the injection well no. 7 and the three production wells no. 3, 14 and 16. Despite the facies mismatch in one of the cells, both history matching and predicting has been achieved very well. The typical behavior of the sequential assimilation system is also observed as before with an uncertainty increase after every prediction step before it gets corrected with the new observations with the Kalman filter analysis step. Similar good results are obtained for the other injection and production wells.

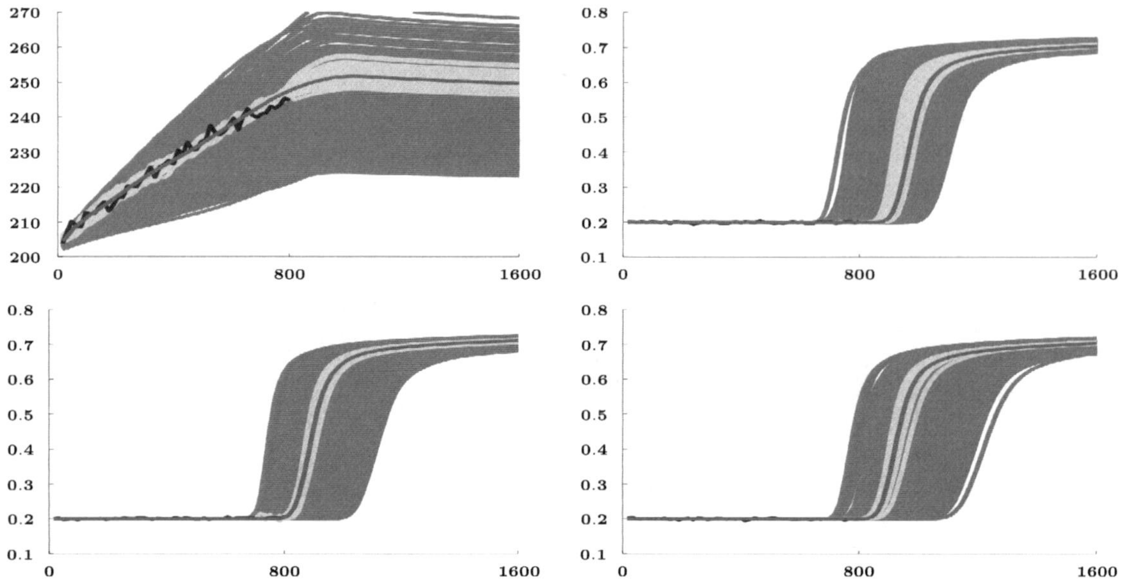

Figure 6. Time-evolution of the estimated pressure field in injection well no. 7 (top left) and saturation field (top right and bottom row) of production well no. 3, 14 and 16 over the production period for the case of a curved channel with 600 realization. The figure shows the true states (red), measurements (blue), simulations of the initial ensembles (magenta) and history matched ensembles (green).

7.3. Sensitivity to Ensemble Size

Application of the EnSRF to computationally demanding models incurs large computational costs for integrating the ensemble forward in time. This sets severe limits on the size of the ensemble that can be used in practice. However, a ensemble which few realizations can impair the performance of the filter analysis in realistic reservoir applications. Balance between the quality of the estimate and acceptable computing resources should be therefore found. Here, we test the assimilation system with a smaller ensemble size, conducting the same experiments as in in sections 7.1 and 7.2 but using only 100 ensemble realizations. One may note that it takes a dual-core Macbook Unibody (P8800, 2.40GHz, 2G RAM) around 4 hours to complete the history matching process when the EnSRF is implemented with 100 realizations and around 23 hours with 600 realizations in Matlab.

Figs. 12-17 are plotted in the same spirit as Figs. 5-8, 10 and 11. One can see from Figs. 12, 14 and 16 that although the structure and the uncertainty of the initial ensemble are almost identical to the corresponding runs with 600 realizations, the evolution of the permeability estimation is very different. In contrast with the assimilation run with 600 realisations, the number of cells with larger facies uncertainty are reduced much faster using 100 realizations. At the end of the history matching period, the estimated permeability fields contain more mismatch cells compared to the estimates obtained with 600 realizations. In Figs. 13, 15 and 17, the uncertainty level (vertical width of the green lines) resulting from the EnSRF with 100 realizations is also much smaller than the one resulting using 600 realizations. The true pressure and saturation, however, lie within the uncertainty level in each well over most of the history matching period.

We consider three measures to quantify and better compare the results of the two runs:

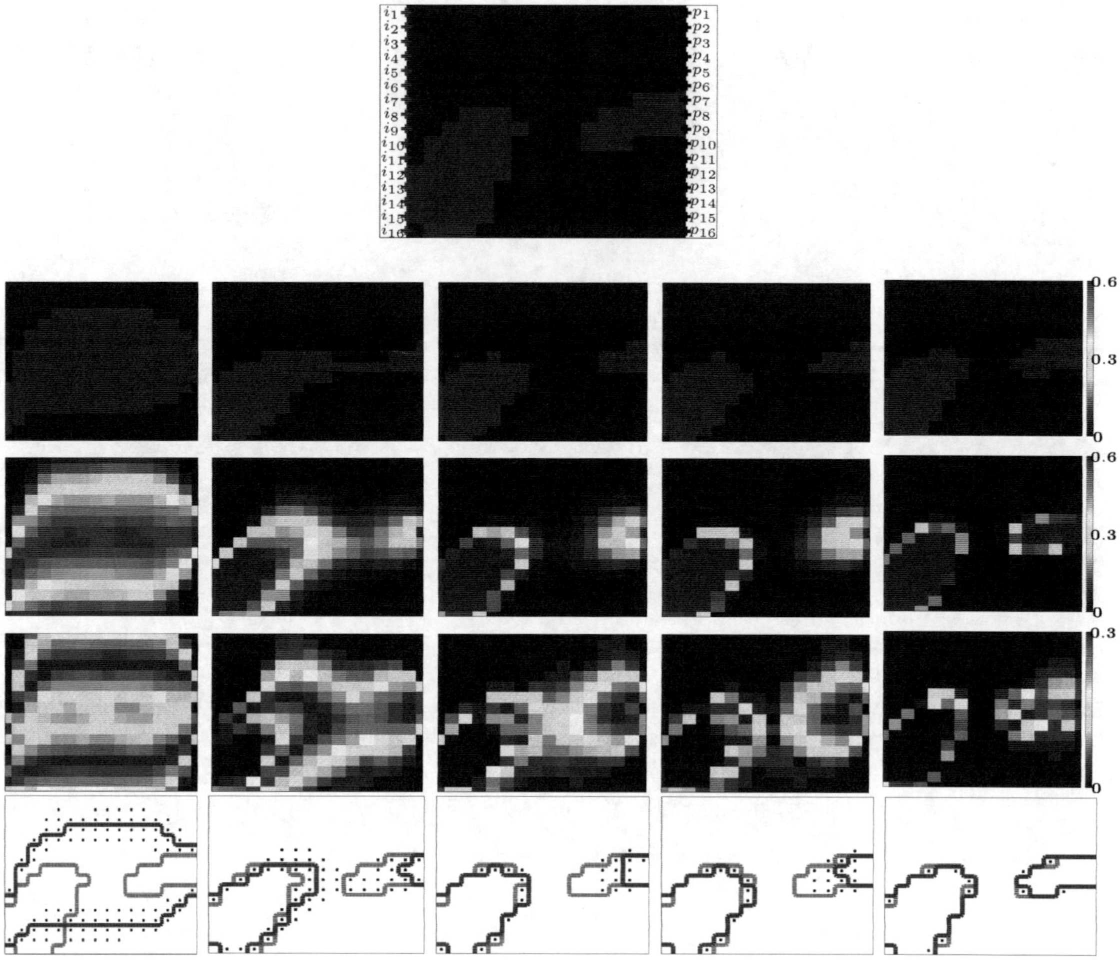

Figure 7. The true log permeability field (first row) and the log permeability field using the mean of the B-splines (second row) and mean (third row) and standard deviation (fourth row) of the log permeability field and misfit of facies pattern (fifth row) for the disconnected channel case. Results are shown at the end of 0th, 192th, 384th, 576th and 800th days with 600 realization. For the true log permeability field, the injection well is plotted in blue pluses and the production wells in black pluses. The misfit figures show the true facies pattern (red), the estimated facies pattern (blue), and the cells with facies standard deviation greater than 0.8 (black dots).

(i) *Permeability Mismatch*:
The number of cells in which the true permeability is different from the EnSRF estimate and lie outside the uncertainty level.

(ii) *Pressure/Saturation Coverage*:
The total number of time steps for which the true pressure in all injection wells (or saturation in all production wells) lie within the estimated uncertainties level are calculated. This number is then divided by the number of injection (or production) wells and the total number of time steps to compute the percentage of pressure (or saturation) coverage.

(iii) *Pressure/Saturation Average Uncertainty*:
The uncertainties (vertical width of the green lines in Figs. 6, 8, 11, 13, 15 and 17) of the

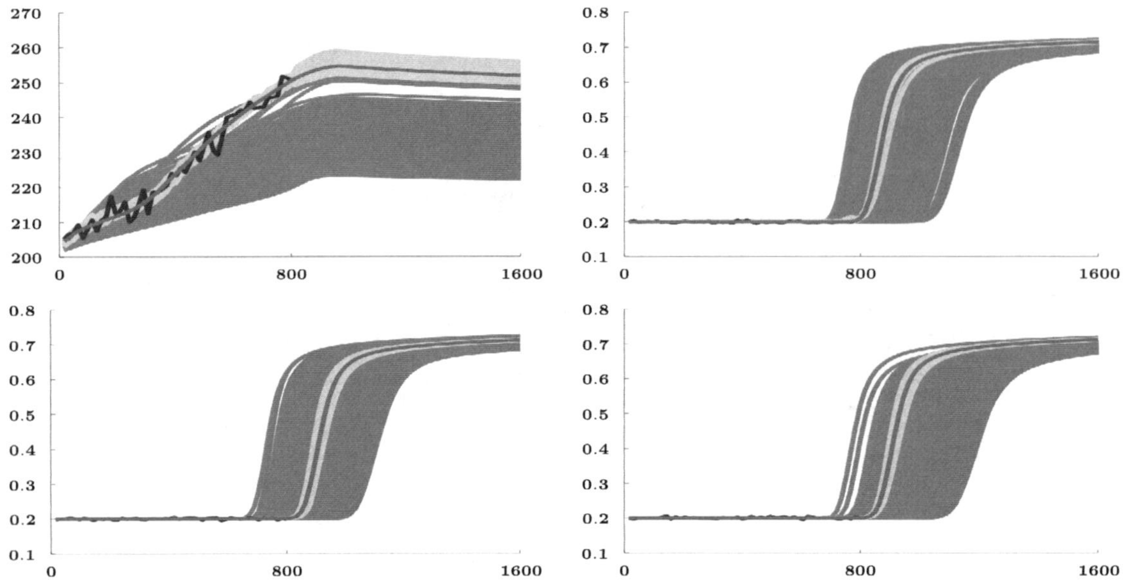

Figure 8. Time-evolution of the estimated pressure in injection well no. 7 (top left) and the saturation (top right and bottom row) of production well no. 3, 14 and 16 over the production period for the case of a disconnected channel with 600 realization. The figure shows the true states (red), measurements (blue), simulations of the initial ensembles (magenta) and history matched ensembles (green).

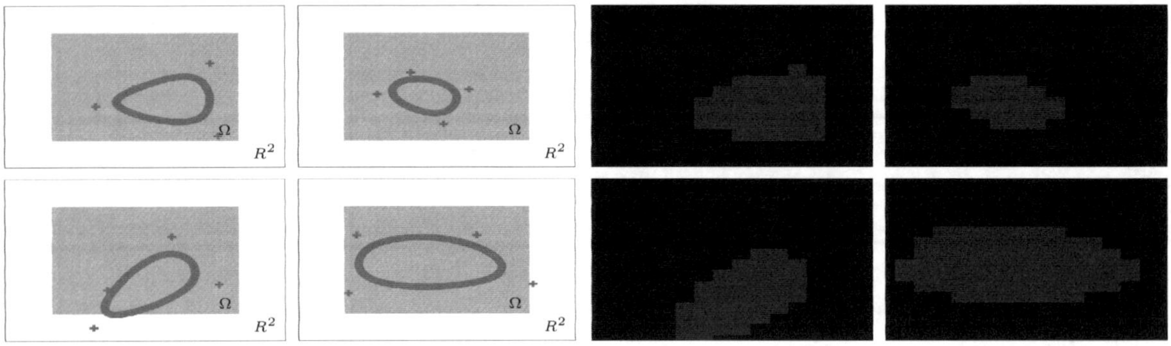

Figure 9. Four typical realizations of the initial ensemble for the case of the two-dimensional closed reservoir. In the left panels, the reservoir domain Ω is the grey rectangle, the B-spline control point domain is the outer white rectangle bounded by black edges, the B-spline control points are indicated in red pluses, the discontinuous curve of the level set function is the red curve). The right panels plot the corresponding permeability fields using equations (19 and 20).

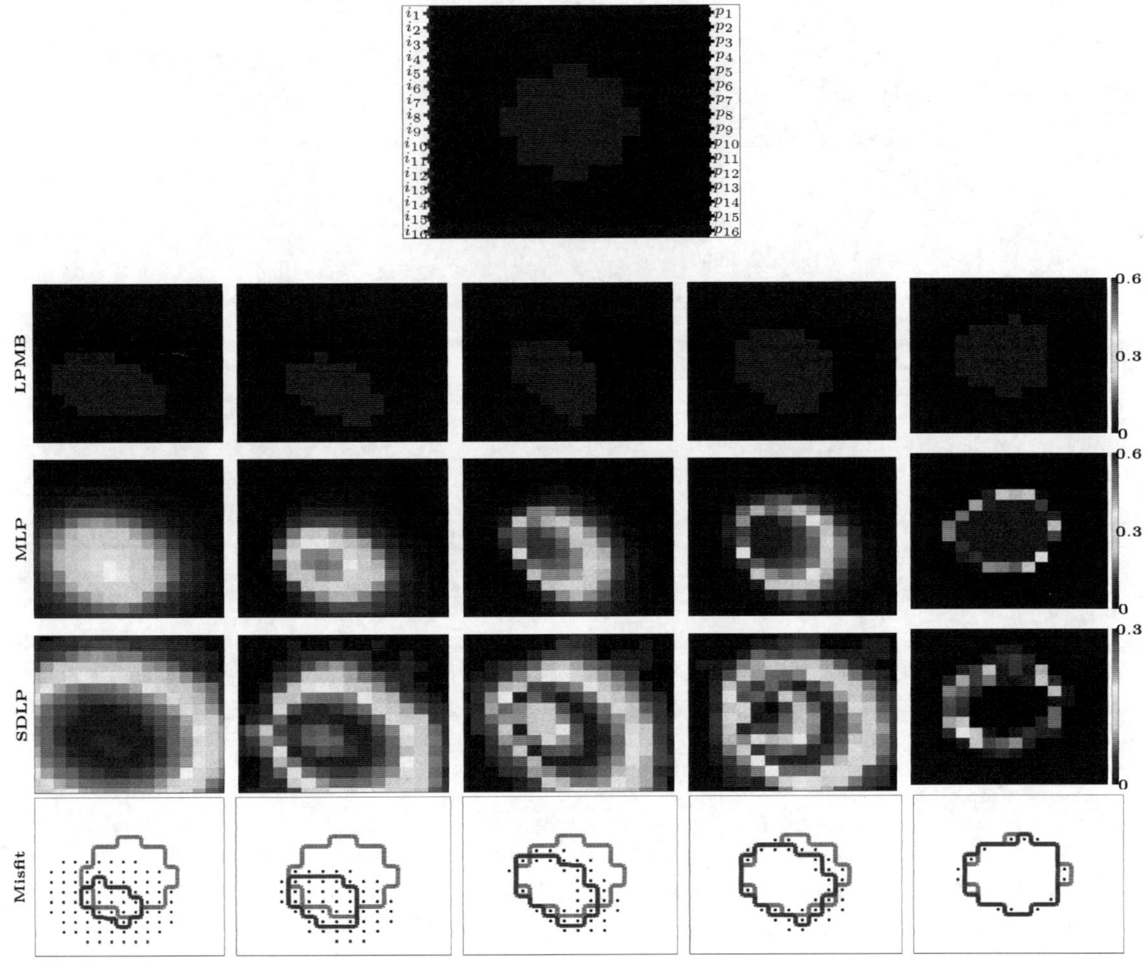

Figure 10. The true log permeability field (first row) and the log permeability field as estimated from the mean of the B-splines (second row), and mean (third row) and standard deviation (fourth row) of the log permeability field and misfit of facies pattern (fifth row) for the two-dimensional closed permeability case. Results are shown at the end of 0th, 192th, 384th, 576th and 800th days with 600 realization. For the true log permeability field, the injection well is plotted in blue pluses and the production wells in black pluses. The misfit figures show the true facies pattern (red), the estimated facies pattern (blue), and the cells with facies standard deviation greater than 0.8 (black dots).

pressure in all injection wells (or saturation in all production wells) are calculated at each time step. The average uncertainty is then computed by averaging these uncertainties over the number of time steps and the number of injection wells (or production wells) .

Comparison results of these measures for the EnSRF runs with 100 and 600 realisations are presented in tables 2 and 3, respectively. In term of predicting the reservoir behaviour, both runs lead to very similar (perfect) results. This is because the data coverages in both cases are above 90%. However, in term of reservoir risk and management, EnSRF with 100 realizations significantly underestimates the pressure and saturation uncertainties in all the wells. For all three different reservoir types, the average pressure and saturation uncertainties are underestimated and are about half and third of the uncertainties estimated using 600 realizations,

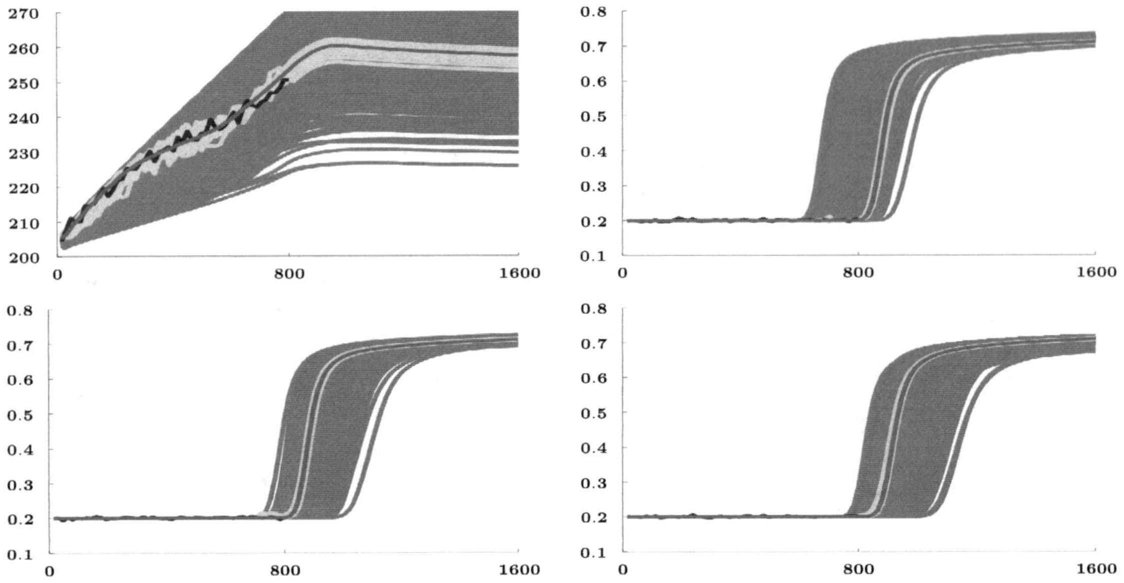

Figure 11. Time-evolution of the estimated pressure field in injection well no. 7 (top left) and saturation field (top right and bottom row) of production well no. 3, 14 and 16 over the production period for the case of two-dimensional closed permeability field with 600 realization. The figure shows the true states (red), measurements (blue), simulations of the initial ensembles (magenta) and history matched ensembles (green).

respectively. Concerning the quality of the permeability estimates, more mismatch cells results from the EnSRF were obtained using 100 realizations.

8. Conclusions

The aim of this paper is to develop an efficient method for using an ensemble Bayesian filter for the problem of history matching of facies pattern in a reservoir model - a problem of which a approach such as an conventional ensemble Kalman filter directly seeking the permeability and other fields might not be an appropriate choice. This is because the variable representing the facies is a discrete variable, and hence non-differentiable and non-Gaussian. By a proper representation of the permeability field using level set functions and a continuous B-spline parameterization, we transformed the history matching problem of the facies pattern to an estimation problem of continuous variables which can be treated by an ensemble Kalman filter. Here we used the ensemble square-root filter (EnSRF).

We conducted numerical experiments with three different types of reservoirs, with two facies types and complicated geological structure and assuming piecewise constant permeability profiles. Starting from vague priors, the new method efficiently recovered the true facies pattern at the end of the history matching period. The uncertainties of the estimated state variables conditional to the data were also substantially reduced compared to the initial uncertainties. The results of the different experiments demonstrate that the proposed method is robust and was able to provide accurate estimates of the reservoir structure. We also studied the the performance of our system using fewer ensemble realizations. Using ensembles of only 100 realizations, the EnSRF provided reasonable estimates of the reservoir state, but failed to accurately estimate the associated uncertainty and the permeability field. A larger ensemble size is then needed to better describe the distribution of the reservoir state vector (here we used 600 realizations).

		Reservoir Type		
		Connected Channel	Disconnected Channel	Closed Feature
Mismatch	Permeability	0	0	1
Coverage	Pressure	100%	100%	100%
	Saturation	100%	100%	100%
Average Uncertainty	Pressure (Bar)	6.03	4.65	5.13
	Saturation	0.0346	0.0267	0.0112

Table 2. Measures of the permeability mismatch, production data coverages, and average uncertainties as they result from the EnSRF using 600 realizations.

		Reservoir Type		
		Connected Channel	Disconnected Channel	Closed Feature
Mismatch	Permeability	9	3	2
Coverage	Pressure	99.8%	99.2%	99.8%
	Saturation	94.5%	90.3%	94.9%
Average Uncertainty	Pressure (Bar)	3.24	2.53	2.62
	Saturation	0.0132	0.00891	0.00498

Table 3. Measure of permeability mismatch, production data coverages, and average uncertainties as they result from EnSRF using 100 realizations.

Future work will further investigate this problem before considering more realistic reservoir applications. The experiments presented in this paper with an idealized reservoir problem were used to test the new method and to demonstrate its ability to tackle the history matching problem of reservoir facies.

9. Acknowledgement

The principal author, Lian Duan, thanks Mr. Yasin Abbas (Institute of Biomedical Engineering, University of Oxford) for technical help on the computational problems. This work was supported in part by Award No KUK-C1-013-04 from King Abdullah University of Science and Technology (KAUST).

References

[1] Jafarpour B and McLaughlin D 2008 *Computational Geosciences* **12** 227–244
[2] Williams M, Keating J and Barghouty M 1998 *SPE Reservoir Evaluation & Engineering* **1** 169–176
[3] Agarwal B, Hermansen H, Sylte J and Thomas L 2000 *SPE Reservoir Evaluation & Engineering* **3** 534–543
[4] Jahns H O 1966 *Soc. Petrol. Eng. J.* **237** 315–327
[5] Farmer C 2005 *Mathematical Methods and Modelling in Hydrocarbon Exploration and Production* 119–212
[6] Evensen G 1994 *Physica D* **77** 108–128
[7] Evensen G 1994 *J Geophys Res* **99** 10,143–10,162
[8] Oliver D 1996 *Mathematical Geology* **28** 811–817
[9] Zhang F and Reynolds A Optimization algorithms for automatic history matching of production data *8th European Conference on the Mathematics of Oil Recovery*
[10] Zhang F, Skjervheim J, Reynolds A and Oliver D 2005 *SPE Reservoir Evaluation & Engineering* **8** 214–223
[11] Oliver D S, Reynolds A C, Bi Z and Abacioglu Y 2001 *Petroleum Geosciences* **7** 65–73
[12] Oliver D and Chen Y 2010 *Computational Geosciences* **15** 185–221
[13] Reading H 1996 *Sedimentary environments: processes, facies, and stratigraphy* (Wiley-Blackwell)
[14] Tai X c and Yao C h 2006 *J. Comput. Math.* **24** 435–443
[15] Sethian J A 1999 *Level set methods and fast marching methods* vol 3 (Cambridge: Cambridge University

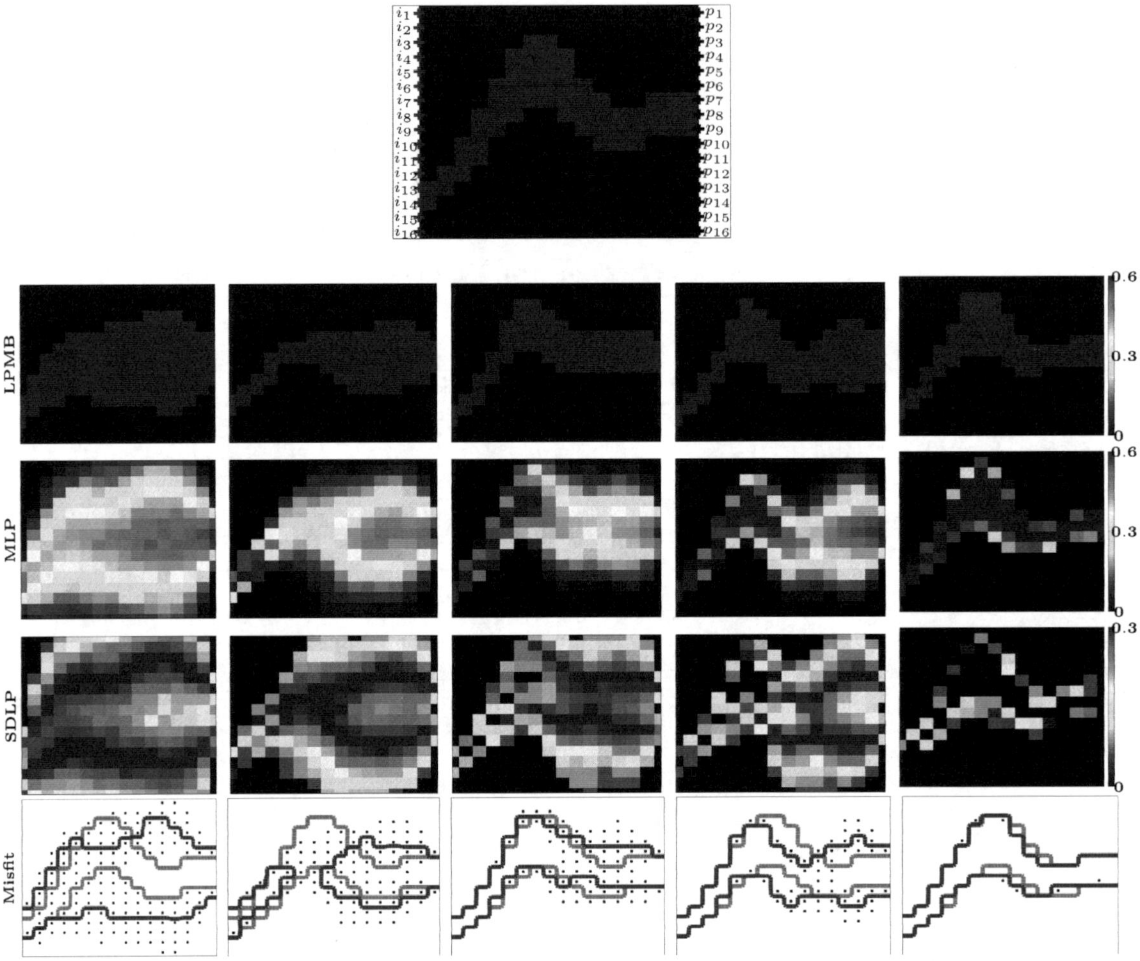

Figure 12. The true log permeability field (first row) and the log permeability field as estimated from the mean of the B-splines (second row), and mean (third row) and standard deviation (fourth row) of the log permeability field and misfit of facies pattern (fifth row) for the mountain-shaped channel case. Results are shown at the end of 0th, 192th, 384th, 576th and 800th days with 100 realization. For the true log permeability field, the injection well is plotted in blue pluses and the production wells in black pluses. The misfit figures show the true facies pattern (red), the estimated facies pattern (blue), and the cells with facies standard deviation greater than 0.8 (black dots).

Press) evolving interfaces in computational geometry, fluid mechanics, computer vision, and materials science
[16] Tsai R and Osher S 2003 *COMM.MATH.SCI* **1 No. 4** 623–656
[17] Chan T F and Tai X C 2003 *J.Comput.Physics* **193** 40–66
[18] Chan T and Vese L A 2001 *IEEE Tran. image Proc.* **10** 266–277
[19] Wang S Y, Lim K M, Khoo B C and Wang M Y 2007 *J. Comput. Phys.* **221** 395–421
[20] Nielsen L, Li H, Tai X, Aanonsen S and Espedal M 2010 *Computing and visualization in science* **13** 41–58
[21] Dorn O and Villegas R 2008 *Inverse Problems* **24** 035015
[22] Han X and Li X 2008 *Remote Sensing of Environment* **112** 1434–1449
[23] Whitaker J S and Hamill T M 2002 *Monthly Weather Review* **130** 1913–1924
[24] Liu N and Oliver D 2005 *Journal of Petroleum Science and Engineering* **47** 147–161

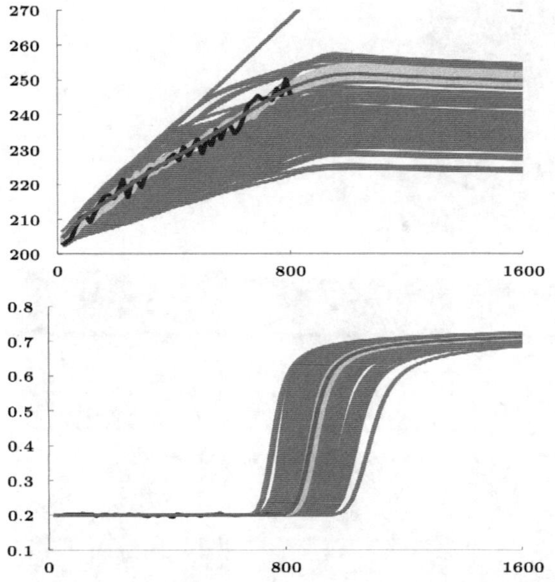

Figure 13. Time-evolution of the estimated pressure field in injection well no. 7 (top left) and saturation field (top right and bottom row) of production well no. 3, 14 and 16 over the production period for the case of a connected channel with 100 realization. The figure shows the true states (red), measurements (blue), simulations of the initial ensembles (magenta) and history matched ensembles (green).

[25] Liu N and Oliver D 2005 *SPE Reservoir Evaluation and Engineering* **8(6)** 470–477
[26] Chang H, Zhang D and Lu Z 2010 *Journal of Computational Physics* **229** 8011–8030
[27] Ren H, Wu Y and Zhu X 2006 *Journal of Forestry Research* **17** 150–152
[28] Tustison N, Avants B and Gee J Improved ffd b-spline image registration *IEEE 11th International Conference on Computer Vision* (IEEE) pp 1–8
[29] Liu X, Huang H and Xu W 2005 *Computational ScienceICCS 2005* 995–1002
[30] Mora M, Andrade J and Tauber C Regularized multilevel b-spline registration: Application to cardiac motion estimation *IEEE international conference on Biomedical imaging: from nano to Macro* (IEEE) pp 512–515
[31] Brouwer D R, Naedal G, Jansen J D, Vefring E H and van Kruijsdijk C P J W 2004 *SPE 90149*
[32] Naedal G, Mannseth T and Vefring E H 2002 *SPE-75235*
[33] Prautzsch H, Boehm W and Paluszny M 2002 *Bzier and B-spline techniques* (Springer Verlag)
[34] Gavalas G, Shah P and Seinfeld J 1976 *Old SPE Journal* **16** 337–350
[35] Tarantola A 2004 *Inverse Problem Theory. Methods for Model Parameter Estimation* Philadelphia,PA:SIAM

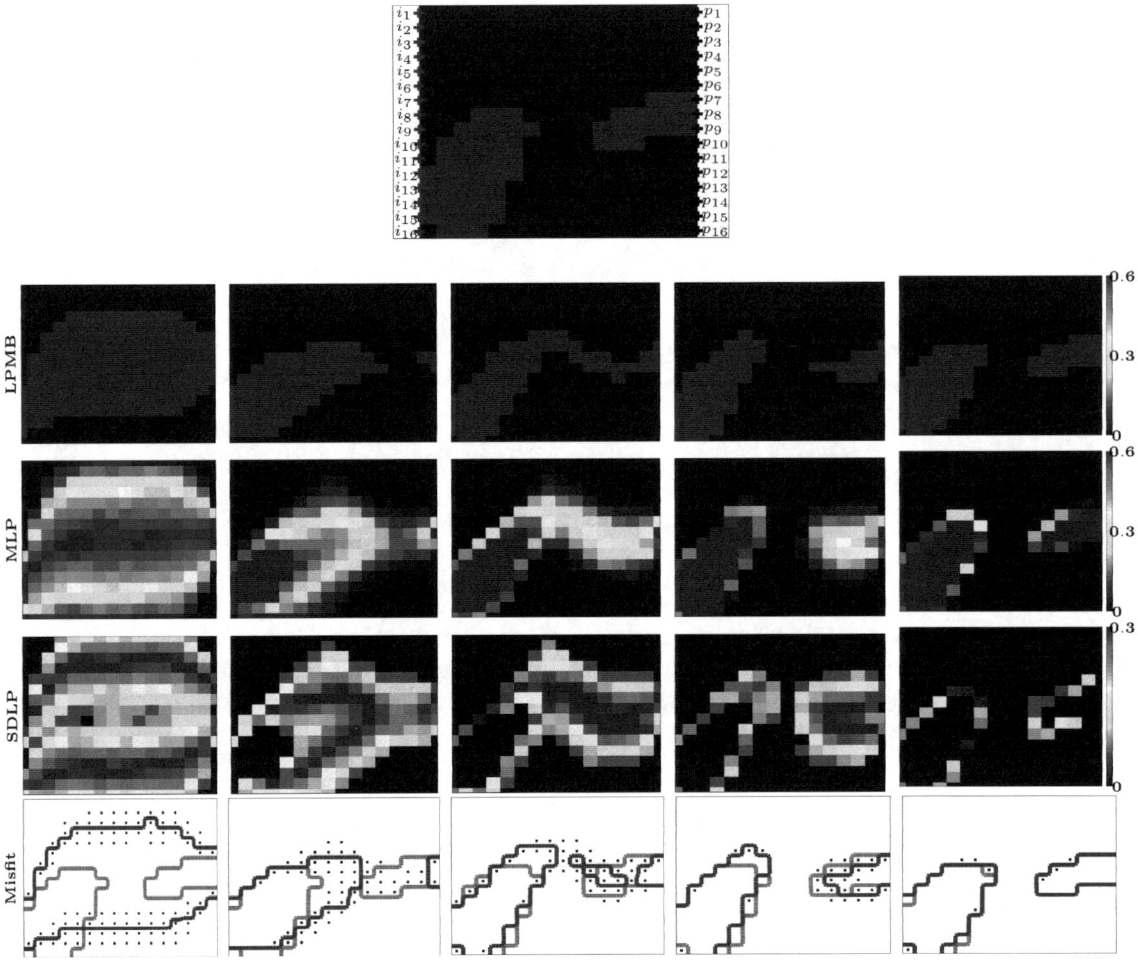

Figure 14. The true log permeability field (first row) and the log permeability field using the mean of the B-splines (second row) and mean (third row) and standard deviation (fourth row) of the log permeability field and misfit of facies pattern (fifth row) for the disconnected channel case. Resutls are shown at the end of 0th, 192th, 384th, 576th and 800th days with 100 realization. For the true log permeability field, the injection well is plotted in blue pluses and the production wells in black pluses. The misfit figures show the true facies pattern (red), the estimated facies pattern (blue), and the cells with facies standard deviation greater than 0.8 (black dots).

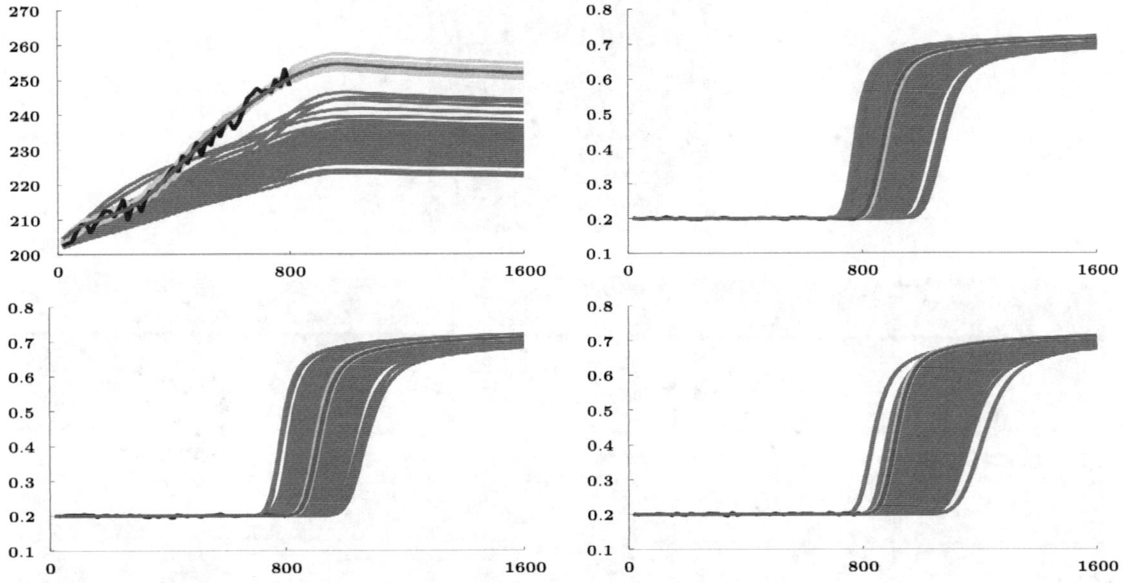

Figure 15. Time-evolution of the estimated pressure in injection well no. 7 (top left) and the saturation (top right and bottom row) of production well no. 3, 14 and 16 over the production period for the case of a disconnected channel with 100 realization. The figure shows the true states (red), measurements (blue), simulations of the initial ensembles (magenta) and history matched ensembles (green).

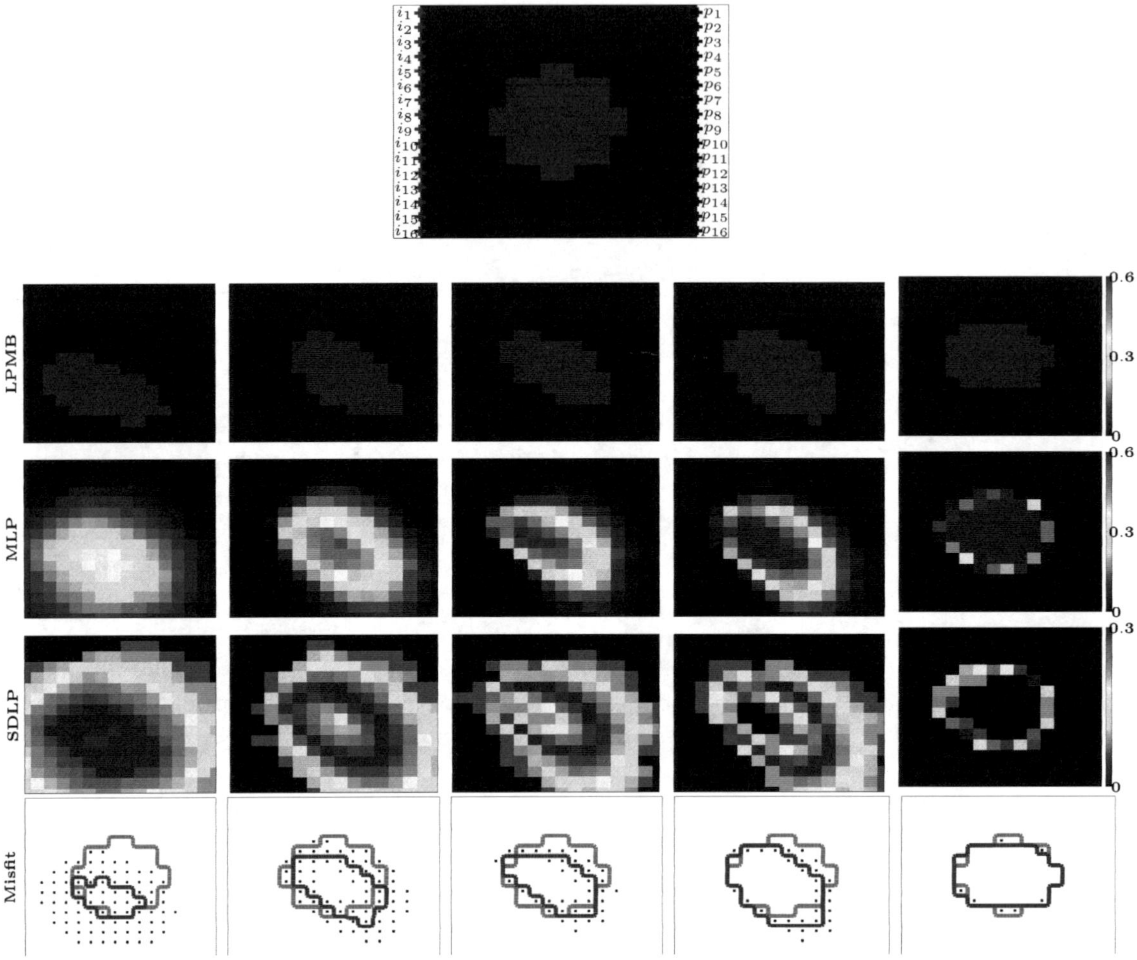

Figure 16. The true log permeability field (first row) and the log permeability field as estimated from the mean of the B-splines (second row), and mean (third row) and standard deviation (fourth row) of the log permeability field and misfit of facies pattern (fifth row) for the two-dimensional closed permeability case. Results are shown at the end of 0th, 192th, 384th, 576th and 800th days with 100 realization. For the true log permeability field, the injection well is plotted in blue pluses and the production wells in black pluses. The misfit figures show the true facies pattern (red), the estimated facies pattern (blue), and the cells with facies standard deviation greater than 0.8 (black dots).

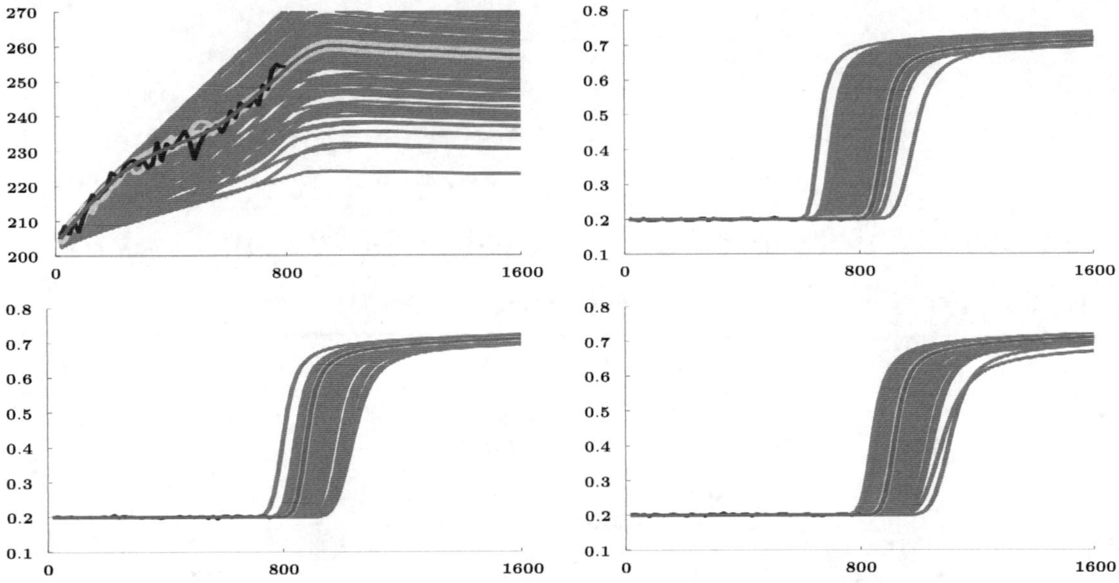

Figure 17. Time-evolution of the estimated pressure field in injection well no. 7 (top left) and saturation field (top right and bottom row) of production well no. 3, 14 and 16 over the production period for the case of two-dimensional closed permeability field with 100 realization. The figure shows the true states (red), measurements (blue), simulations of the initial ensembles (magenta) and history matched ensembles (green).

On reconstruction of an unknown polygonal cavity in a linearized elasticity with one measurement

M Ikehata[1] **and H Itou**[2]

[1,2] Department of Mathematics, Graduate School of Engineering, Gunma University, Kiryu 376-8515, JAPAN

E-mail: [1] ikehata@math.sci.gunma-u.ac.jp

E-mail: [2] h-itou@math.sci.gunma-u.ac.jp

Abstract. In this paper we consider a reconstruction problem of an unknown polygonal cavity in a linearized elastic body [16]. For this problem, an extraction formula of the convex hull of the unknown polygonal cavity is established by means of the enclosure method introduced by Ikehata [2, 4]. The advantages of our method are that it needs only a single set of boundary data and we do not require any a priori assumptions for the unknown polygonal cavity and any constraints on boundary data. The theoretical formula may have possibility of application in nondestructive evaluation.

1. Introduction

Inverse problems have been received a great deal of attention in various fields of science and engineering. Typical examples of inverse problems are to extract information about discontinuity embedded in a medium, such as cracks, cavities, obstacles and inclusions, from observed data. This kind of problems arises in geophysics, medical imaging such as Computed Tomography (CT), Electrical Impedance Tomography (EIT) and Magnetic Resonance Imaging (MRI), Nondestructive testing, etc.

In the mathematical model, the problems are often described as inverse boundary value problems of partial differential equations called the governing equations and the solution is to extract information about location, shape and size of unknown discontinuity from boundary data. However, inverse problems are typically nonlinear and ill posed which means the lack of at least one of three conditions; existence, uniqueness, stability of the solution or solutions. Therefore, one cannot expect to obtain the desired solution in general. Then it is quite important to impose reasonable a priori assumptions on the problems in a practical sense and to extract the useful information of unknown discontinuity from fewer boundary data. From their standpoints we consider the following problem [16].

Let Ω be a bounded domain of $\mathbf{R}^2$ with Lipschitz boundary and represent a homogeneous isotropic linearized elastic plate in both states of plane stress and plane strain. Let D denote cavities such that $\overline{D} \subset \Omega$ and $\Omega \setminus \overline{D}$ is connected. As a priori assumption for unknown D we require that

(A1) $D = D_1 \cup D_2 \cup \cdots \cup D_m$, $\overline{D_j} \cap \overline{D_k} = \emptyset$ for $j \neq k$, where each D_j is a simply connected open set and polygon.

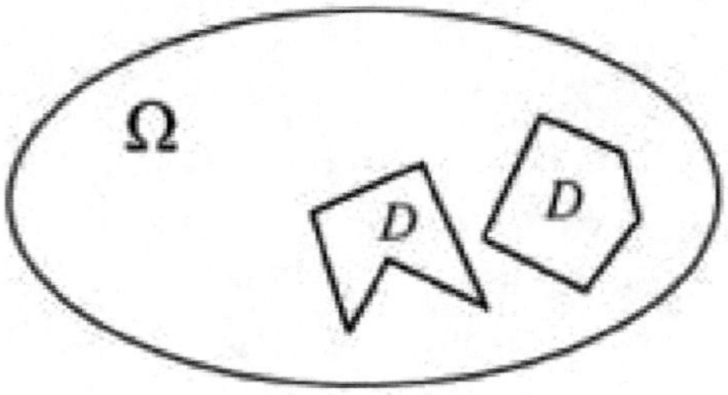

Figure 1. An example of domain

Inverse Problem *Reconstruct unknown D satisfying* (A1) *from a single set of a surface force and the corresponding displacement field on the boundary of* Ω.

In order to solve the Problem we employ the *enclosure method* introduced by Ikehata [4]. By virtue of this method we show that one can reconstruct the convex hull of D uniquely from a single set of boundary data, stated as Theorem in Section 3. As a remark, for the case that D is a linear crack we have already considered the reconstruction problem in an isotropic and anisotropic linearized elasticity, respectively (e.g. [13, 14, 15]).

2. The Enclosure Method

The enclosure method is a methodology in inverse problems for partial differential equations. The method yields a partial information about the location of unknown discontinuity which appears as discontinuity of the coefficients of a partial differential equation or a part of the boundary of the common domain of definition of solutions of the equation. Although the original enclosure method makes use of infinitely many observation data, the single measurement version has been introduced in [2] and can be divided into the following three steps.

(i) Find a special solution of the formal adjoint of the governing equation for the background medium which is parameterized by a large parameter τ and divides the whole space into two parts: in one part the absolute value of the solution decays as $\tau \to \infty$; in another part the solution grows as $\tau \to \infty$.

(ii) Construct an indicator function of independent variable τ by multiplying the governing equation of the medium by the special solution constructed in (i), integrating over the domain of definition and extracting only the integral on the known boundary of the domain.

(iii) Study the asymptotic behaviour of the indicator function as $\tau \to \infty$.

There are several existing applications of the enclosure method to a wide variety of inverse problems. Some typical examples of them are listed below.

- Electrical Impedance Tomography ([2, 3, 4, 5, 9])
- Inverse obstacle scattering problems of acoustic wave ([6, 7, 11]).
- Reconstruction problems of obstacles in an electromagnetic body ([25]).
- Recently the enclosure method has found its applications in inverse problems for heat and wave equations. Therein the data is observed in a limited time. Some results are listed as follows.

- Inverse obstacle scattering problems with dynamical data over a finite time interval ([10]).
- Reconstruction problems of inclusions in a heat conductive body from dynamical data over a finite time interval ([8, 17, 18, 19, 12]).

3. Statement of the main result

Let $\boldsymbol{n}$ be the unit outward normal vector to $\partial(\Omega\backslash\overline{D})$. Let the displacement vector $\boldsymbol{u} = (u_1, u_2)^{\mathrm{T}} \in \{H^1(\Omega \setminus \overline{D})\}^2$ satisfy the Navier equation in the absence of any body forces

$$\frac{\tilde{E}}{2(1+\tilde{\nu})}\Delta \boldsymbol{u} + \frac{\tilde{E}}{2(1-\tilde{\nu})}\nabla(\nabla \cdot \boldsymbol{u}) = \boldsymbol{0} \quad \text{in} \quad \Omega \setminus \overline{D} \tag{3.1}$$

and the traction free boundary condition

$$T\boldsymbol{u} = \boldsymbol{0} \quad \text{on} \quad \partial D \tag{3.2}$$

where $T\boldsymbol{u}$ is the stress vector expressed by

$$T\boldsymbol{u} = \frac{\tilde{\nu}\tilde{E}}{1-\tilde{\nu}^2}(\nabla \cdot \boldsymbol{u})\boldsymbol{n} + \frac{\tilde{E}}{2(1+\tilde{\nu})}\{\nabla \boldsymbol{u} + (\nabla \boldsymbol{u})^{\mathrm{T}}\}\boldsymbol{n}$$

and

$$\tilde{E} = \begin{cases} E & \text{(plane stress)}, \\ \dfrac{E}{1-\nu^2} & \text{(plane strain)}, \end{cases} \qquad \tilde{\nu} = \begin{cases} \nu & \text{(plane stress)}, \\ \dfrac{\nu}{1-\nu} & \text{(plain strain)}. \end{cases}$$

Here E and ν is Young's modulus and Poisson's ratio of the elastic medium, respectively. Since both the shear modulus and the bulk modulus are required to be positive, we suppose $E > 0$ and $-1 < \nu < \frac{1}{2}$.

We denote by h_D the support function of D:

$$h_D(\boldsymbol{\omega}) = \sup_{\boldsymbol{x} \in D} \boldsymbol{x} \cdot \boldsymbol{\omega} \quad \text{for} \quad \boldsymbol{\omega} \in S^1.$$

The values of $h_D(\,\cdot\,)$ give the signed distances from the origin of coordinates to the support lines of D.

To state the main result, we introduce the following assumptions for $\boldsymbol{\omega}$.

(W1) $\boldsymbol{\omega}$ satisfies that the intersection of the line $\boldsymbol{x} \cdot \boldsymbol{\omega} = h_D(\boldsymbol{\omega})$ with ∂D consists of only one point,

(W2) $\boldsymbol{\omega}$ satisfies (W1) and that the interior angle bisector of D at the point $\{\boldsymbol{x} \in \mathbf{R}^2 \mid \boldsymbol{x} \cdot \boldsymbol{\omega} = h_D(\boldsymbol{\omega})\} \cap \partial D$ is not perpendicular to the line $\boldsymbol{x} \cdot \boldsymbol{\omega} = h_D(\boldsymbol{\omega})$.

As shown in Figure 2 in Section 4, (W2) is equivalent to a condition $p + q + \pi \neq 0$. Then our main result is the following formula:

THEOREM ([16]) *Let $\boldsymbol{u}$ be not a rigid displacement. Under the assumptions of* (A1) *and* (W2) *the formula*

$$h_D(\boldsymbol{\omega}) = \lim_{\tau \longrightarrow \infty} \frac{1}{\tau} \log \left| \int_{\partial\Omega} (T\boldsymbol{u} \cdot \boldsymbol{v} - T\boldsymbol{v} \cdot \boldsymbol{u}) \, \mathrm{d}S \right|,$$

is valid, where $\boldsymbol{v}(\boldsymbol{x}) = (\boldsymbol{\omega} + i\boldsymbol{\omega}^{\perp})e^{\tau \boldsymbol{x} \cdot (\boldsymbol{\omega} + i\boldsymbol{\omega}^{\perp})}$, $\tau > 0$; $\boldsymbol{\omega}^{\perp} \in S^1$ *is perpendicular to* $\boldsymbol{\omega}$ *and satisfies* $\det(\boldsymbol{\omega}^{\perp}\ \boldsymbol{\omega}) > 0$.

Some remarks on Theorem are in order.

- For given D satisfying (A1) the set of all $\boldsymbol{\omega}$ that does not satisfy (W2) (or (W1)) is a finite set. Since the support function is continuous, it follows from Theorem that a single set of the data $\boldsymbol{u}$, $T\boldsymbol{u}$ on $\partial\Omega$ uniquely determine $h_D(\boldsymbol{\omega})$ for all $\boldsymbol{\omega} \in S^1$ and thus the convex hull of D via the formula
$$\bigcap_{\boldsymbol{\omega}\in S^1} \{\boldsymbol{x} \in \mathbf{R}^2 \mid \boldsymbol{x}\cdot\boldsymbol{\omega} < h_D(\boldsymbol{\omega})\}.$$
This is the origin of the name "the enclosure method" and simultaneously the solution of Problem.
- In contrast to the related results, we do not require any other a priori assumptions for D excepting (A1) (e.g. [2]) and any constraints on boundary data (e.g. [13, 14, 15]).
- If we assume (W1) instead of (W2), then the formula is valid by replacing $\lim_{\tau\longrightarrow\infty}$ with $\limsup_{\tau\longrightarrow\infty}$, for the detail see the end of Section 4.
- In two-dimensional case rigid displacements can be described in the form
$$F(\boldsymbol{x})\boldsymbol{k} = (k_0 + k_2x_2, k_1 - k_2x_1)^{\mathrm{T}}$$
with an arbitrary constant vector $\boldsymbol{k} = (k_0, k_1, k_2)^{\mathrm{T}}$.
- It follows from the compatibility condition on the data that
$$\int_{\partial\Omega} T\boldsymbol{u}\cdot F(\boldsymbol{x})\boldsymbol{k}\,\mathrm{d}S = 0$$
for arbitrary $\boldsymbol{k} \in \mathbf{R}^3$.

4. Outline of the proof of Theorem

The proof of Theorem proceeds along the same lines as results [2, 13, 14, 15].

The first part of the enclosure method mentioned in Section 2 is to find the special solution for the Navier equation (3.1), which corresponds to $\boldsymbol{v}(\boldsymbol{x})$ in Theorem .

The indicator function in the second part is just
$$\tau \longmapsto \int_{\partial\Omega} (T\boldsymbol{u}\cdot\boldsymbol{v} - T\boldsymbol{v}\cdot\boldsymbol{u})\,\mathrm{d}S.$$

The third part of the enclosure method is to study the asymptotic behaviour of the indicator function as $\tau \longrightarrow \infty$. This part is the most important in this analysis, because if one can see that
$$I_{\boldsymbol{\omega}}(\tau) \equiv e^{-\tau h_D(\boldsymbol{\omega})} \int_{\partial\Omega} (T\boldsymbol{u}\cdot\boldsymbol{v} - T\boldsymbol{v}\cdot\boldsymbol{u})\,\mathrm{d}S$$
is truly algebraic decaying as $\tau \longrightarrow \infty$, then it is easy to see that
$$\lim_{\tau\longrightarrow\infty} \frac{1}{\tau}\log|I_{\boldsymbol{\omega}}(\tau)| = 0 \tag{4.1}$$
and then the formula in Theorem automatically follows.

The assumptions (A1) and (W1) yield that there exists a unique point $\boldsymbol{Q}$ on ∂D such that $\boldsymbol{Q}\cdot\omega = h_D(\omega)$, that is, $\boldsymbol{Q}$ has to be a vertex of a connected component of D. Hereafter we always assume that $\boldsymbol{\omega}$ satisfies (W1). Using the special form of $\boldsymbol{v}$ and integration by parts, we see that $I_{\boldsymbol{\omega}}(\tau)$ coincides with the function
$$e^{-\tau h_D(\boldsymbol{\omega})} \int_{\partial D\cap B_R(\boldsymbol{Q})} T\boldsymbol{v}\cdot\boldsymbol{u}\,\mathrm{d}S, \quad B_R(\boldsymbol{Q}) = \{\boldsymbol{x} \in \mathbf{R}^2 \mid |\boldsymbol{x} - \boldsymbol{Q}| < R\} \quad R > 0,$$
modulo exponentially decaying as $\tau \longrightarrow \infty$. Thus it is quite important to know the behaviour of $\boldsymbol{u}$ in a neighbourhood of $\boldsymbol{Q}$.

4.1. A convergent series expansion near a corner

In order to derive the convergent series expansion of $\boldsymbol{u}$ around $\boldsymbol{Q}$, first let us fix some notation. For $R > 0$ we set $D_R = B_R(\boldsymbol{Q}) \cap (\Omega \setminus \overline{D})$. We see that if $R < 1$ is sufficiently small, then D_R becomes a sector with center angle $2\alpha > \pi$, see Figure 2. We now fix such R.

Next, we identify each point $\boldsymbol{x}$ in D_R with $\boldsymbol{X}$ in a local coordinate system $\boldsymbol{X} = (X_1, X_2)^{\mathrm{T}} = (r\cos\theta, r\sin\theta)^{\mathrm{T}}$ with respect to the origin $\boldsymbol{Q}$ as follows, see Figure 2.

$$\boldsymbol{x} = \boldsymbol{Q} + r(\cos\theta\, \boldsymbol{a} + \sin\theta\, \boldsymbol{a}^{\perp}) \quad 0 < r < R \quad -\alpha < \theta < \alpha,$$

where $\frac{\pi}{2} < \alpha < \pi$,

$$\begin{aligned} \boldsymbol{a} &= \sin(\alpha + p)\,\boldsymbol{\omega} + \cos(\alpha + p)\,\boldsymbol{\omega}^{\perp} = \sin(q - \alpha)\,\boldsymbol{\omega} + \cos(q - \alpha)\,\boldsymbol{\omega}^{\perp}, \\ \boldsymbol{a}^{\perp} &= \cos(\alpha + p)\,\boldsymbol{\omega} - \sin(\alpha + p)\,\boldsymbol{\omega}^{\perp} = \cos(q - \alpha)\,\boldsymbol{\omega} - \sin(q - \alpha)\,\boldsymbol{\omega}^{\perp}, \end{aligned}$$

$-\pi < q < p < 0$ and $p - q = 2\pi - 2\alpha$.

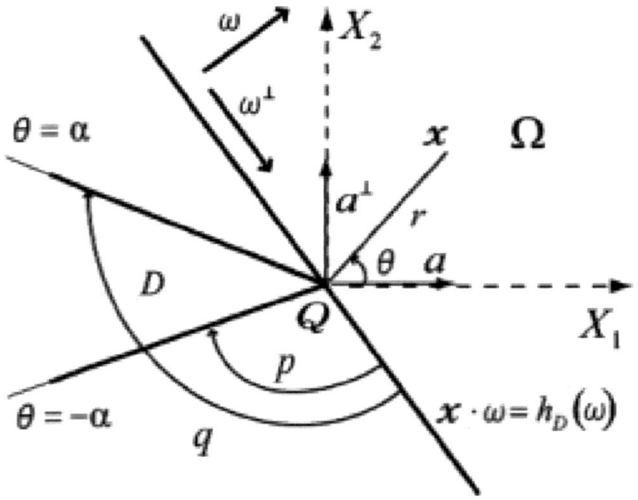

Figure 2. Polar coordinates

Then, we can derive the convergent series expansion of $\boldsymbol{u}$ near $\boldsymbol{Q}$ in D_ρ for $0 < \rho < R$.

PROPOSITION 1 ([16]) *If* $2\alpha \neq \tan 2\alpha$, *then there exist complex numbers* $A_{j,m}$, $B_{j,m}(s)$ *for* $j = 1, 2,\ m = 1, 2, 3, \cdots$ *and a constant vector* $\tilde{\boldsymbol{k}}$ *such that*

$$\boldsymbol{u}(r,\theta) = \sum_{j,m} A_{j,m} r^{-s_{j,m}-1} \boldsymbol{\Phi_j}(s_{j,m},\theta) + \sum_{j,m} \frac{\partial}{\partial s}\left(r^{-s-1} B_{j,m}(s) \boldsymbol{\Phi_j}(s,\theta)\right)\Big|_{s=\tilde{s}_{j,m}} + F(\boldsymbol{X})\tilde{\boldsymbol{k}}, \quad (4.2)$$

where $\boldsymbol{\Phi_j}$ *is the following vector field :*

$$\boldsymbol{\Phi_1}(s,\theta) = \begin{pmatrix} (1+\tilde{\nu})s\sin(s+2)\alpha\sin(s+1)\theta + (\tilde{\nu}-3)\sin s\alpha\sin(s+1)\theta \\ -(1+\tilde{\nu})(s+1)\sin s\alpha\sin(s+3)\theta \\ \\ -(1+\tilde{\nu})s\sin(s+2)\alpha\cos(s+1)\theta + (\tilde{\nu}-3)\sin s\alpha\cos(s+1)\theta \\ +(1+\tilde{\nu})(s+1)\sin s\alpha\cos(s+3)\theta \end{pmatrix},$$

$$\mathbf{\Phi}_2(s,\theta) = \begin{pmatrix} (1+\tilde{\nu})s\cos(s+2)\alpha\cos(s+1)\theta + (\tilde{\nu}-3)\cos s\alpha\cos(s+1)\theta \\ -(1+\tilde{\nu})(s+1)\cos s\alpha\cos(s+3)\theta \\ \\ (1+\tilde{\nu})s\cos(s+2)\alpha\sin(s+1)\theta - (\tilde{\nu}-3)\cos s\alpha\sin(s+1)\theta \\ -(1+\tilde{\nu})(s+1)\cos s\alpha\sin(s+3)\theta \end{pmatrix}$$

and $s_{j,m}$, $\tilde{s}_{j,m}$ which are not -2, are the simple and double roots of $h_j(\alpha,s)=0$;

$$\begin{aligned} h_1(\alpha,s) &= (s+1)\sin 2\alpha - \sin 2(s+1)\alpha, \\ h_2(\alpha,s) &= (s+1)\sin 2\alpha + \sin 2(s+1)\alpha, \end{aligned}$$

respectively, following decreasing of $\mathbf{Re}s$ order for $m=1,2,3,\cdots$ in $\mathbf{Re}s<-1$. Moreover, the series (4.2) is convergent, absolutely in $H^1(D_\rho)$ and uniformly in D_ρ.

REMARK 1 *The coefficients $A_{j,m}$, $B_{j,m}(s)$ in (4.2) have the following properties:*

(i) If $s_{j,m}\in\mathbf{R}$, then the corresponding $A_{j,m}\in\mathbf{R}$.

(ii) If $s_{j,m}\in\mathbf{C}\setminus\mathbf{R}$ such that $s_{j,m+1}=\overline{s}_{j,m}$ which denotes the complex conjugate of $s_{j,m}$, then it holds $A_{j,m+1}=\overline{A}_{j,m}$.

(iii) $B_{j,m}(\tilde{s}_{j,m})\in\mathbf{R}$ and $\dfrac{\partial}{\partial s}B_{j,m}(\tilde{s}_{j,m})\in\mathbf{R}$.

REMARK 2 *If $2\alpha=\tan 2\alpha$, then the following additional term with a real constant A^* should be taken into account to the formula (4.2);*

$$rA^*\begin{pmatrix} (1+\tilde{\nu})(1+\cos 2\alpha)\sin\theta + 2(1-\tilde{\nu})(\cos 2\alpha)\theta\cos\theta + 4\log r\cos 2\alpha\sin\theta \\ (1+\tilde{\nu})(1-\cos 2\alpha)\cos\theta + 2(1-\tilde{\nu})(\cos 2\alpha)\theta\sin\theta - 4\log r\cos 2\alpha\cos\theta \end{pmatrix}.$$

Sketch of the proof of Proposition 1

Here we give the essence of the proof of Proposition 1, for the detail see [16]. First we construct Airy's stress function $U\in H^2(D_R)$ by using Poincaré's lemma. Then, the problem (3.1)–(3.2) in D_R can be reduced to the Dirichlet boundary value problem for the biharmonic equation of U. Next, applying the Mellin transform to the problem for U, we have a boundary value problem for a fourth order ordinary differential equation in the transformed complex domain. For the problem we represent the solution by Green's function (e.g. [24]) and investigate the property. Since according to [1] and [22] it is well known that U multiplying by a cut-off function belongs to a weighted Sobolev space, we can apply the inverse Mellin transform to the solution for $r<\rho$. Furthermore, the residue theorem enable us to change the integration path. Here the residue of the integrand is calculated by the roots of the transcendental equations $h_j(\alpha,s)=0$, which are not explicitly solvable and are complex. This is completely different from that of the case that the governing equation is the Laplace equation [2] and the case that D is a linear crack [13]. In the matter of the transcendental equations the following results are known, e.g. [1], [16] and [23].

LEMMA

*(i) When $\frac{\pi}{2}<\alpha<\alpha_0$, it holds $-2<\mathbf{Re}s_{2,1}<-\frac{3}{2}$.
When $\alpha_0<\alpha<\pi$, it holds $-2<\mathbf{Re}s_{1,1}<-\frac{\pi}{2\alpha}-1<\mathbf{Re}s_{2,1}<-\frac{3}{2}$.*

(ii) In $\mathbf{Re}s<-1$ the lines $\mathbf{Re}s=\ell_{k+1}\equiv -k\frac{\pi}{2\alpha}-1$ $(k=1,2,3,\cdots)$ contain no roots of each $h_j(\alpha,s)=0$.

(iii) If k is an odd number, then the strip $\ell_{k+2}<\mathbf{Re}s<\ell_{k+1}$ contains only two roots of $h_1(\alpha,s)=0$ (including complex conjugate and multiplicity) for $\frac{\pi}{2}<\alpha<\pi$.

(iv) If k is an even number, then the strip contains only two roots of $h_2(\alpha, s) = 0$ for $\frac{\pi}{2} < \alpha < \pi$.

(v) The multiplicity of the roots of each $h_j(\alpha, s) = 0$ is not greater than two. Moreover, the double roots are all real and at most two.

(vi) For $\frac{\pi}{2} < \alpha < \pi$ the roots of each $h_j(\alpha, s) = 0$ in $\mathbf{Re}s < -2$ are not integers.

(vii) For $\frac{\pi}{2} < \alpha < \pi$ the roots of each $h_j(\alpha, s) = 0$ in $s(s+2) \neq 0$ satisfy $\sin s\alpha \cos s\alpha \neq 0$.

Thus we obtain an asymptotic expansion of U with the remainder term expressed by the integral form.

Finally, by using the estimate of the Green's function we have the estimate of the remainder term and we can derive convergent series of U. Since $\boldsymbol{u}$ is uniquely determined by U up to any rigid displacements $F(\boldsymbol{X})\boldsymbol{k}$, we obtain Proposition 1.

4.2. The asymptotic behaviour of $I_{\boldsymbol{\omega}}(\tau)$ as $\tau \longrightarrow \infty$

Next, to show (4.1) we consider the asymptotic behaviour of $I_{\boldsymbol{\omega}}(\tau)$ by using Proposition 1. The arbitrary truncation of the expansion (4.2) together with a remainder estimate yields the complete asymptotic expansion of $I_{\boldsymbol{\omega}}(\tau)$.

PROPOSITION 2 *Assume D contains no corner of angle satisfying $2\alpha = \tan 2\alpha$. For each $k \geq 1$, as $\tau \to \infty$,*

$$\begin{aligned} I_{\boldsymbol{\omega}}(\tau) \quad = \quad & \sum_{\mathbf{Re}s_{j,m} > \ell_{k+1}} \frac{C_j(s_{j,m})A_{j,m}}{\tau^{-s_{j,m}-1}} + \sum_{\mathbf{Re}\tilde{s}_{j,m} > \ell_{k+1}} \frac{\partial}{\partial s}\left(\frac{C_j(s)B_{j,m}(s)}{\tau^{-s-1}}\right)\Bigg|_{s=\tilde{s}_{j,m}} \\ & + O\left(\frac{1}{\tau^{\frac{k\pi}{2\alpha}}}\right), \end{aligned} \tag{4.3}$$

where

$$C_1(s) = 4\frac{\tilde{E}}{1+\tilde{\nu}} \sin s\alpha \, e^{i\tau\boldsymbol{Q}\cdot\boldsymbol{\omega}^\perp} e^{-i\frac{\pi}{2}s} e^{-i(s+2)(p+\alpha)} \Gamma(-s)\left\{1 - e^{is2\pi}\right\},$$

$$C_2(s) = 4i\frac{\tilde{E}}{1+\tilde{\nu}} \cos s\alpha \, e^{i\tau\boldsymbol{Q}\cdot\boldsymbol{\omega}^\perp} e^{-i\frac{\pi}{2}s} e^{-i(s+2)(p+\alpha)} \Gamma(-s)\left\{1 - e^{is2\pi}\right\}.$$

REMARK 3 *If D contains some corners of angle satisfying $2\alpha = \tan 2\alpha$, then* Proposition 2 *should be modified by adding the following term to* (4.3);

$$-A^*\left(8\pi i\frac{\tilde{E}}{1+\tilde{\nu}} \cos 2\alpha \, e^{i\tau\boldsymbol{Q}\cdot\boldsymbol{\omega}^\perp}\right)\frac{1}{\tau}. \tag{4.4}$$

However, since Proposition 2 only implies $I_{\boldsymbol{\omega}}(\tau)$ decays at most algebraically as $\tau \longrightarrow \infty$, this is not enough to prove (4.1) without any restrictions on boundary data in itself.

For this we need to show that there exist a positive constants τ_0, c independent of τ and a real number $\mu_0 < -1$ such that for all $\tau \geq \tau_0$

$$|I_{\boldsymbol{\omega}}(\tau)| \geq c\tau^{\mu_0}. \tag{4.5}$$

Now we set

$$P_1 \equiv \{s_{j,m} \in \mathbf{C} \mid A_{j,m} \neq 0\}, \quad P_2 \equiv \left\{\tilde{s}_{j,m} \in \mathbf{R} \;\middle|\; B_{j,m}(\tilde{s}_{j,m}) \neq 0 \text{ or } \frac{\partial}{\partial s}B_{j,m}(\tilde{s}_{j,m}) \neq 0\right\}.$$

Since $\boldsymbol{u}$ is not a rigid displacement from the hypothesis in Theorem , the unique continuation theorem ensures $P_1 \cup P_2 \neq \emptyset$. Note here that the convergence of (4.2) is indispensable in this part because $P_1 \cup P_2 = \emptyset$ follows $\boldsymbol{u}(r,\theta) = F(\boldsymbol{X})\tilde{\boldsymbol{k}}$ in D_ρ directly.

Journal of Physics: Conference Series **290** (2011) 012005 doi:10.1088/1742-6596/290/1/012005

Accordingly, we can choose $\lambda \in P_1 \cup P_2$ such that $\mathbf{Re}\lambda \geq \mathbf{Re}s$ for any $s \in P_1 \cup P_2$. Then the following three cases are considered.

Case 1. $\lambda = s_{j,m} \in \mathbf{R}$.
In this case, from *6* and *7* in Lemma one sees that $C_j(s_{j,m})$ in Proposition 2 never vanishes. Namely, it holds that

$$I_{\boldsymbol{\omega}}(\tau) = C_j(s_{j,m})A_{j,m}\tau^{s_{j,m}+1} + o(\tau^{s_{j,m}+1}),$$

which simultaneously means (4.5).

Case 2. $\lambda = \tilde{s}_{j,m} \in \mathbf{R}$.
Similarly to *Case 1*, one sees $C_j(\tilde{s}_{j,m}) \neq 0$. From *3* in Remark 1, it holds that

$$I_{\boldsymbol{\omega}}(\tau) = \frac{\partial}{\partial s}\left(C_j(s)B_{j,m}(s)\tau^{s+1}\right)\bigg|_{s=\tilde{s}_{j,m}} + o(\tau^{\tilde{s}_{j,m}+1}),$$

which leads to (4.5).

Case 3. $\lambda = s_{j,m} \in \mathbf{C} \setminus \mathbf{R}$ such that $s_{j,m+1} = \overline{s}_{j,m}$. Namely,

$$I_{\boldsymbol{\omega}}(\tau) = C_j(s_{j,m})A_{j,m}\tau^{s_{j,m}+1} + C_j(s_{j,m+1})A_{j,m+1}\tau^{s_{j,m+1}+1} + o(\tau^{s_{j,m}+1}). \tag{4.6}$$

In this case, it follows from *6*, *7* in Lemma and *2* in Remark 1 that $C_j(s_{j,m})A_{j,m} \neq 0$ and $C_j(s_{j,m+1})A_{j,m+1} \neq 0$. And one sees $|C_j(s_{j,m})A_{j,m}| = |C_j(s_{j,m+1})A_{j,m+1}|$ if and only if $p+q+\pi = 0$. Hence, when $\boldsymbol{\omega}$ satisfies (W2) which means $p+q+\pi \neq 0$, from (4.6) we obtain (4.5).

Summing up the above, (4.1) is verified in the case that D contains no corner of angle satisfying $\tan 2\alpha = 2\alpha$. If $\tan 2\alpha = 2\alpha$, then note that (4.4) in Remark 3 vanishes if and only if $A^* = 0$. In consequence, by the same argument as above, (4.1) is still valid even in the case that some angles of D satisfy $\tan 2\alpha = 2\alpha$ and the proof of Theorem is completed.

Lastly, if we weaken the assumption (W2) to (W1), then the formula in Theorem does not always hold. Now let consider the case $p+q+\pi = 0$ in *Case 3*. Then, it can be easily seen that there exists a sequence $\{\tau_N\}_{N=1}^{\infty}$ such that $\tau_N \to \infty$ as $N \to \infty$ and it holds

$$\lim_{N\to\infty} \tau_N^{-(\mathbf{Re}s_{j,m}+1)} \left|C_j(s_{j,m})A_{j,m}\tau_N^{s_{j,m}+1} + C_j(\overline{s}_{j,m})\overline{A}_{j,m}\tau_N^{\overline{s}_{j,m}+1}\right| > 0.$$

Namely, it follows from (4.6) that

$$\lim_{\tau_N \longrightarrow\infty} \frac{1}{\tau_N} \log |I_{\boldsymbol{\omega}}(\tau_N)| = 0.$$

Hence, one concludes that the formula

$$h_D(\boldsymbol{\omega}) = \limsup_{\tau\longrightarrow\infty} \frac{1}{\tau} \log \left|\int_{\partial\Omega} (T\boldsymbol{u}\cdot\boldsymbol{v} - T\boldsymbol{v}\cdot\boldsymbol{u})\ \mathrm{d}S\right|$$

is valid even when $p+q+\pi = 0$, that is, under the assumption (W1) not (W2).

5. Future works

The enclosure method still has many possibilities for the further applications to inverse problems other than the problems listed in section 2.

Some open problems are in order.

- In the present result consider an inclusion instead of cavity and establish an extraction formula of its support function.
- Consider the scattering problem of an elastic plane wave.
- Extend results in [2], [6] and [16] to the three dimensional case.
- Consider inverse scattering problems for elastic or electromagnetic waves with use of dynamical data.
- Construct and test numerical methods based on our theoretical formulae had established up to now. At present they have succeeded in only a few cases (cf. [20], [21]), therefore further development of studies in this sides is expected.

Acknowledgments

MI was partially supported by Grant-in-Aid for Scientific Research (C)(No. 21540162) of Japan Society for the Promotion of Science. HI was partially supported by Grant-in-Aid for Young Scientists (B)(No. 21740091) of the Ministry of Education, Culture, Sports, Science and Technology(MEXT).

References

[1] Grisvard P 1985 *Elliptic problems in nonsmooth domains* (Boston: Pitman)
[2] Ikehata M 1999 Enclosing a polygonal cavity in a two-dimensional bounded domain from Cauchy data *Inverse Problems* **15** 1231–41
[3] Ikehata M 2000 On reconstruction in the inverse conductivity problem with one measurement *Inverse Problems* **16** 785–93
[4] Ikehata M 2000 Reconstruction of the support function for inclusion from boundary measurements *J. Inv. Ill-Posed Problems* **8** 367–78
[5] Ikehata M 2001 On reconstruction from a partial knowledge of the Neumann-to-Dirichlet operator *Inverse Problems* **17** 45–51
[6] Ikehata M 2004 Inverse scattering problems and the enclosure method *Inverse Problems* **20** 533–51
[7] Ikehata M 2005 The Herglotz wave function, the Vekua transform and the enclosure method *Hiroshima Math. J.* **35** 485–506
[8] Ikehata M 2007 Extracting Discontinuity in A Heat Conductive Body. One-space dimensional case *Applicable Analysis* **86** 963–1005
[9] Ikehata M 2008 A remark on the enclosure method for a body with an unknown homogeneous background conductivity *Cubo* **10** 31–45
[10] Ikehata M 2010 The enclosure method for inverse obstacle scattering problems with dynamical data over a finite time interval *Inverse Problems* **26** 055010
[11] Ikehata M 2010 A note on the enclosure method for an inverse obstacle scattering problem with a single point source *Inverse Problems* **26** 105006
[12] Ikehata M 2011 The framework of the enclosure method with dynamical data and its applications arXiv:1101.4427
[13] Ikehata M and Itou H 2007 Reconstruction of a linear crack in an isotropic elastic body from a single set of measured data *Inverse Problems* **23** 589–607
[14] Ikehata M and Itou H 2008 An inverse problem for a linear crack in an anisotropic elastic body and the enclosure method *Inverse Problems* **24** 025005
[15] Ikehata M and Itou H 2008 Enclosure method and reconstruction of a linear crack in an elastic body *J. Phys.: Conf. Series* **135** 012052
[16] Ikehata M and Itou H 2009 Extracting the support function of a cavity in an isotropic elastic body from a single set of boundary data *Inverse Problems* **25** 105005
[17] Ikehata M and Kawashita M 2009 The enclosure method for the heat equation *Inverse Problems* **25** 075005
[18] Ikehata M and Kawashita M 2010 On the reconstruction of inclusions in a heat conductive body from dynamical boundary data over a finite time interval *Inverse Problems* **26** 095004

[19] Ikehata M and Kawashita M 2010 An inverse problem for a three-dimensional heat equation in thermal imaging and the enclosure method *Preprint*

[20] Ikehata M and Ohe T 2002 Numerical method for finding the convex hull of polygonal cavities using enclosure method *Inverse Problems* **18** 111–24

[21] Ikehata M and Siltanen S 2000 Numerical method for finding the convex hull of an inclusion in conductivity from boundary measurements *Inverse Problems* **16** 1043–52

[22] Kondrat'ev V A and Oleinik O A 1983 Boundary value problems for partial differential equations in nonsmooth domains *Russian Math. Surveys* **38** 3–76

[23] Kozlov V A, Maz'ya V G and Rossmann J 2001 *Spectral Problems Associated with Corner Singularities of Solutions to Elliptic Equations (Mathematical Surveys and Monographs* Vol 85*)* (Providence, RI: American Mathematical Society)

[24] Tsamasphyros G and Theocaris P S 1979 On the solution of the sector problem *J. Elasticity* **9** 271–81

[25] Zhou T 2010 Reconstructing electromagnetic obstacles by the enclosure method *Inverse Probl. Imaging* **4** 547–69

Journal of Physics: Conference Series **290** (2011) 012006 doi:10.1088/1742-6596/290/1/012006

Viscoelastic properties of soft tissues in a living body measured by MR elastography

Yu Jiang and Gen Nakamura

Department of Mathematics, Hokkaido University, Japan

E-mail: jiang@math.sci.hokudai.ac.jp and gnaka@math.sci.hokudai.ac.jp

Abstract. MRE (Magnetic resonance elastography) is a new diagnostic modality to measure the stiffness of soft tissues in a living body. It measures the displacements of waves propagating in the tissues and then this measured data is linked to the stiffness via a proper model for wave propagation in the tissues and by solving some inverse problem under this model. We will show why we can only see transverse waves inside soft tissues even if we inject longitudinal vibrations into the tissues from their surfaces by modeling the soft tissues of a living body as a nearly incompressible isotropic viscoelastic medium. We interpret the nearly incompressibility by an asymptotic analysis. As a consequence, we show that the so-called modified Stokes system is a proper model. Further, by a modified numerical integral method for solving the inverse problem under this PDE model, we can recover the viscoelasticity of soft tissues.

1. Introduction

MRE (Magnetic Resonance Elastography) is a non-destructive testing (NDT) technology which enables us to measure the displacements of attenuating shear waves propagating in soft tissues of a living body. Principally, a MRE system contains an MRI (Magnetic Resonance Imgaging) machine, an external vibration system and data analysis software. By setting up a proper model which describes wave propagation in living tissues and solve some related inverse problem, we can identify the stiffness of tissues. This made MRE very useful for medical diagnostic purposes, e.g., for detection of early stage cancers which have stiffness that are five times more than those of normal tissues and cannot be seen very well by MRI because of the underdevelopment of capillary blood vessels. Another possible application of MRE is in the field of rheology. Existing devices such as rheometers can measure the stiffness and viscosity of high polymer materials only at frequencies up to 100 Hz, whereas MRE can measure those quantities at frequencies higher than 100 Hz. Also, MRE can be used for measurement of material bodies which are of much larger size than that allowed by rheometers. To develop such a kind of "high frequency rheometer" is one of the interests of our MRE research group in Hokkaido University.

The principle of MRE for medical diagnostic purposes was initiated in 1995 by R. Ehman's team in Mayo Clinic, USA (see [17]). After that, several experimental studies and methods to solve the MRE-related inverse problem have been given (cf. [1,4–6,12–17,20–23,25,27–31]). We also gave a method to identify both the shear modulus and viscosity of a viscoelastic medium from finitely many MRE measurements (see [19]).

As mentioned above we need to have a proper model to set up the inverse problem. Since the soft tissues of a living body are usually considered to be isotropic viscoelastic and nearly incompressible, in this paper we model the tissues as a isotropic viscoelastic media (cf. [10, 11, 26]), and we interpret the meaning of nearly incompressible by an asymptotic analysis. As a consequence of this analysis, we show that the so-called *modified Stokes system* approximates very well the nearly incompressible viscoelasticity model in Section 2. Further, to recover the viscoelasticity of soft tissues by MRE-measured data is an *inverse problem* . To solve this inverse problem, we propose a *modified numerical integral method*, which is considered as a stable inversion scheme, with some numerical example in Section 3.

2. Asymptotic Analysis for MRE

2.1. Linear Viscoelasticity

The MRE-measured data are two dimensional data over some cross section of a three dimensional medium and the stack of these data in one direction can give three dimensional data. If the data are uniform in some direction, we only need a two dimensional model for MRE data analysis. Otherwise, we need a three dimensional model. So, to handle both cases at the same time, let n $(n = 2, 3)$ be the space dimension.

Linear viscoelasticty is the usual mechanical model which describes small deformations in the soft tissues of a living body. More precisely, let

$$\mathbf{U}(t, x) = (U_1(t, x), \cdots , U_n(t, x))$$

be the displacement vector of the deformation in the soft tissues of a living body at time $t \geq 0$ and point x of a reference domain $\Omega \subset \mathbb{R}^n$, which is a bounded domain with Lipschitz continuous boundary $\partial\Omega$. If there is no external force acting on the tissues, then $\mathbf{U}(t, x)$ satisfies the equations

$$\rho(x)\partial_t^2 U_i(t, x) = \sum_{l=1}^{n} \frac{\partial}{\partial x_l}\sigma_{il}(\mathbf{U}) \quad (t > 0, x \in \Omega, 1 \leq i \leq n), \tag{1}$$

where (σ_{il}) is the stress tensor.

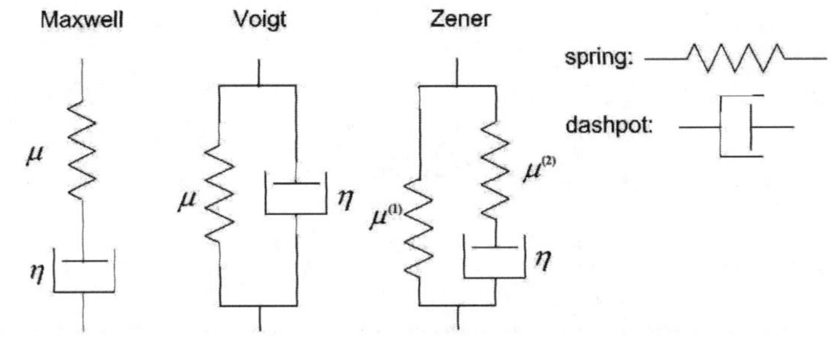

Figure. 1. Viscoelasticity models.

Usually, there are three kinds of viscoelasticity models (see figure 4). The *constitutive equations* of those models are

- Voigt model:

$$\sigma_{il}(\mathbf{U}) = \sum_{k,m=1}^{n} \boldsymbol{\mu}_{ilkm}\varepsilon_{km}(\mathbf{U}) + \sum_{k,m=1}^{n} \boldsymbol{\eta}_{ilkm}\partial_t\varepsilon_{km}(\mathbf{U}).$$

- Maxwell model:

$$\partial_t\varepsilon_{il}(\mathbf{U}) = \sum_{k,m=1}^{n} \boldsymbol{\mu}_{ilkm}^{-1}\partial_t\sigma_{km}(\mathbf{U}) + \sum_{k,m=1}^{n} \boldsymbol{\eta}_{ilkm}^{-1}\sigma_{km}(\mathbf{U}).$$

- Zener model:

$$\begin{aligned}&\sum_{k,m=1}^{n} \mu^{(2)}_{ilkm}\sigma_{km}(\mathbf{U}) + \sum_{k,m=1}^{n} \eta_{ilkm}\partial_t\sigma_{km}(\mathbf{U})\\ =\ &\sum_{k,m=1}^{n} \mu^{(1)}_{ilkm}\mu^{(2)}_{ilkm}\varepsilon_{km}(\mathbf{U}) + \sum_{k,m=1}^{n} (\mu^{(1)}_{ilkm} + \mu^{(2)}_{ilkm})\eta_{ilkm}\partial_t\varepsilon_{km}(\mathbf{U}).\end{aligned}$$

where

$$\varepsilon_{ij}(\mathbf{U}) = \frac{1}{2}\left(\frac{\partial U_i}{\partial x_j} + \frac{\partial U_j}{\partial x_i}\right) \quad (1 \le i,j \le n), \tag{2}$$

$0 < \delta < \rho(x) \in L^\infty(\Omega)$ (usually known as about 1.0×10^3 kg/m^3 and δ is a constant), $\mu_{ilkm}(x)$ $(\mu^{(1)}_{ilkm}, \mu^{(2)}_{ilkm}) \in L^\infty(\Omega)$ and $\eta_{ilkm}(x) \in L^\infty(\Omega)$ are the *strain tensor*, *density*, *elasticity tensor* and *viscosity tensor*, respectively. It is physically natural to assume that those viscoelasticity tensors $0 < \delta < A_{ilkm}$ $(\mu_{ilkm}$ $(\mu^{(1)}_{ilkm}, \mu^{(2)}_{ilkm}), \eta_{ilkm})$ satisfy the *full symmetries*:

$$A_{ilkm}(x) = A_{likm}(x) = A_{ilmk}(x) = A_{kmil}(x)$$

for any $i, l, k, m\,(1 \le i, l, k, m \le n)$, a.e. $x \in \Omega$, and the *strong convexity*: there exists a constant $C > 0$ such that

$$\sum_{i,l,k,m=1}^{n} A_{ilkm}(x)\zeta_{il}\zeta_{km} \ge C\sum_{i,l=1}^{3}\zeta_{il}^2 \tag{3}$$

for any symmetric matrix (ζ_{il}), a.e. $x \in \Omega$, $i, l, k, m\,(1 \le i, l, k, m \le n)$.

2.2. Generation of Time Harmonic Motions

As compared with ultrasound CT, MRI is a slow imaging modality and this is why time harmonic waves can only be used in MRE measurements. A vibration system equipped with MRI generates a time harmonic excitation on a part Γ_D of the surface $\partial\Omega$ of the living body, while the remaining part Γ_N of $\partial\Omega$ is traction free (cf. [31,32]). Mathematically, let Γ_D and Γ_N be open subsets of $\partial\Omega$ with Lipschitz continuous boundaries such that $\partial\Omega = \overline{\Gamma_D} \cup \overline{\Gamma_N}$, $\Gamma_D \ne \emptyset$ and $\Gamma_D \cap \Gamma_N = \emptyset$. A general formulation of the boundary condition and initial condition for MRE measurements is as follows:

$$\begin{cases} \mathbf{U}(t,x) = \chi(t)e^{i\omega t}\mathbf{f}(x) & \text{on } (0,+\infty)\times\Gamma_D,\\ \partial_\nu\mathbf{U}(t,x) = 0 & \text{on } (0,+\infty)\times\Gamma_N \end{cases} \tag{4}$$

and

$$\mathbf{U} = \partial_t\mathbf{U} = 0 \quad \text{on } \{0\}\times\Omega, \tag{5}$$

respectively. Here, ω is a given angular frequency (about 50–500 Hz) of the vibration generated by the vibration system, $\partial_\nu\mathbf{U}(t,x)$ is the conormal derivative along the outward unit normal vector $\nu = (\nu_1, \cdots, \nu_n)$ to $\partial\Omega$ defined by

$$\partial_\nu\mathbf{U}(t,x) := \sum_{l=1}^{n}\sigma_{il}(\mathbf{U})\nu_l \quad (1 \le i \le n),$$

and $\chi(t) \in C^\infty([0,\infty))$ is a cut off function such that

$$\chi(t) = \begin{cases} 0 & (0 \le t \le \frac{1}{2}),\\ 1 & (t \ge 1). \end{cases}$$

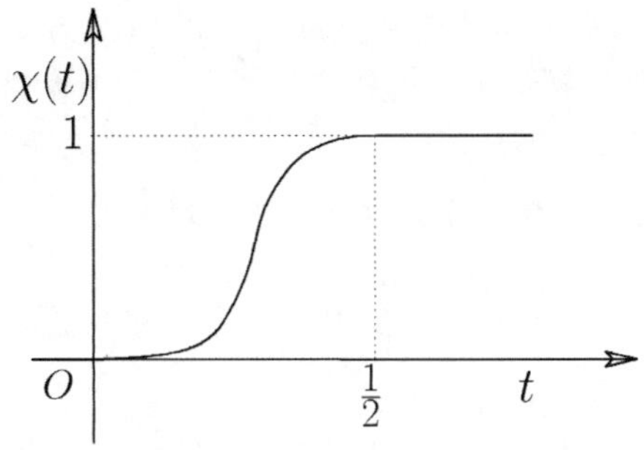

Figure. 2. Cutoff function $\chi(t)$.

(See figure 2)

To prove the well-posedness of the initial-boundary-value problem (1) (4), (5), we need to assume that all the coefficients are of class C^1 near $\partial\Omega$. More precisely, we make the following assumption: there exists a compact set $F \subset \Omega$ such that $F \cap \partial\Omega = \emptyset$ and those viscoelasticity tensors A's are all of class C^1 in $\Omega \setminus F$.

Since the time harmonic vibrations generated by the vibration system take some time to reach the stationary state, we should ascertain how fast the vibrations become stationary. In this section we will show that the solution of the initial-boundary-value problem (1), (4), (5) converges exponentially to a stationary solution. This result is comforting, because it shows that we can quickly have time harmonic vibrations when we vary the angular frequencies of the vibrations so that it is possible to successively conduct MRE measurements with different frequencies.

To begin with we introduce the Sobolev spaces $\dot{H}^s(\overline{\Gamma'_D})$ of fractional order $s = 1/2$ or $3/2$ to which the data of the displacement boundary condition $\mathbf{f}$ belongs. For an open subset Γ'_D of Γ_D with a boundary away from $\partial\Gamma_D$ and of class C^2, we denote by $\dot{H}^s(\overline{\Gamma'_D})$ the set of distributions in the usual fractional Sobolev space $H^s(\Gamma'_D)$ compactly supported in $\overline{\Gamma'_D}$ (cf. [7]). This can be naturally imbedded into $H^s(\Gamma_D)$. Further, we introduce the spaces

$$V := \{\mathbf{u} \in H^1(\Omega) : \ \mathbf{u} = 0 \text{ on } \Gamma_D\}, \quad V_0 := \{\mathbf{u} \in V : \ \nabla \cdot \mathbf{u} = 0 \text{ in } \Omega\}.$$

Associated with these, we define V^* and V_0^* as the dual spaces of V and V_0, respectively, via the continuous extension of the $L^2(\Omega)$ inner product. The norms of V, V_0, V^*, V_0^* are denoted by $\|\cdot\|_V$, $\|\cdot\|_{V_0}$, $\|\cdot\|_{V^*}$, $\|\cdot\|_{V_0^*}$, and for some Sobolev space X, we define $\|f\|_{L^2((0,t);X)} := \int_0^t \|f\|_X dt$ $(t > 0)$.

By applying the argument in [24], we know that the displacement $\mathbf{U}(t,x)$ of the wave generated by the vibration system converges exponentially to a stationary state.

Theorem 2.1. *1) For any given Dirichlet input* $\mathbf{f}(x) \in \dot{H}^{3/2}(\overline{\Gamma'_D})$, *there exists a unique solution* $\mathbf{U}(t,x) \in C^1([0,+\infty); H^1(\Omega)) \cap C^2([0,+\infty); L^2(\Omega))$ *to the mixed problem* (1), (4) *and* (5))

Moreover, this solution $\mathbf{U}(t,x)$ *satisfies*

$$\|\mathbf{U}(t)\|_{H^1(\Omega)} + \|\partial_t \mathbf{U}(t)\|_{L^2(\Omega)} \leq C\|\mathbf{f}\|_{\dot{H}^{3/2}(\overline{\Gamma'_D})},$$

where C *is a positive constant independent of* $\mathbf{f}$.

2) There exist constants $M, \beta > 0$, *such that*

$$\|\mathbf{U}(t,x) - e^{i\omega t}\mathbf{u}(x)\|_{H^1(\Omega)} \leq Me^{-\beta t},$$

where $\mathbf{u}(x) := (u_1(x), \cdots, u_n(x)) \in H^1(\Omega)$ *is the unique solution to the mixed boundary value problem for the stationary viscoelasticity equation (we call it* **stationary viscoelasticity model** *for simplicity):*

$$\begin{cases} \sum_{l=1}^{n} \frac{\partial}{\partial x_l}\sigma'_{il}(\mathbf{u}) + \rho\omega^2\mathbf{u} = 0 & on\ \Omega, \\ \mathbf{u} = \mathbf{f} & on\ \Gamma_D, \quad (1 \le i \le n). \\ \partial_{\mathbf{v}}\mathbf{u} := \sum_{l=1}^{n} \sigma'_{il}(\mathbf{u})\nu_l = 0 & on\ \Gamma_N \end{cases} \tag{6}$$

Here, the *constitutive equations* for this stationary case are as follows:

- Voigt model:

$$\sigma'_{il}(\mathbf{u}) = \sum_{k,m=1}^{n} \boldsymbol{\mu}_{ilkm}\varepsilon_{km}(\mathbf{u}) + i\omega \sum_{k,m=1}^{n} \boldsymbol{\eta}_{ilkm}\varepsilon_{km}(\mathbf{u}).$$

- Maxwell model:

$$\partial_t\varepsilon_{il}(\mathbf{u}) = i\omega \sum_{k,m=1}^{n} \boldsymbol{\mu}^{-1}_{ilkm}\sigma'_{km}(\mathbf{u}) + \sum_{k,m=1}^{n} \boldsymbol{\eta}^{-1}_{ilkm}\sigma'_{km}(\mathbf{u}).$$

- Zener model:

$$\sum_{k,m=1}^{n} \boldsymbol{\mu}^{(2)}_{ilkm}\sigma'_{km}(\mathbf{u}) + i\omega \sum_{k,m=1}^{n} \boldsymbol{\eta}_{ilkm}\sigma'_{km}(\mathbf{u})$$
$$= \sum_{k,m=1}^{n} \boldsymbol{\mu}^{(1)}_{ilkm}\boldsymbol{\mu}^{(2)}_{ilkm}\varepsilon_{km}(\mathbf{u}) + i\omega \sum_{k,m=1}^{n} (\boldsymbol{\mu}^{(1)}_{ilkm} + \boldsymbol{\mu}^{(2)}_{ilkm})\boldsymbol{\eta}_{ilkm}\varepsilon_{km}(\mathbf{u}).$$

Remark.
(i) By Theorem 2.1, $\mathbf{U}(t,x)$ *converges to* $e^{i\omega t}\mathbf{u}(x)$ *exponentially in time. Hence, it is reasonable to say that we know* $\mathbf{u}(x)$ *in* $\overline{\Omega}$ *if we know* $\mathbf{U}(t,x)$ *on* $[0,T] \times \overline{\Omega}$ *for some sufficiently large* $T > 0$.
(ii) The mixed boundary value problem of (6) is well-posed, i.e., for any given $\mathbf{f} \in \dot{H}^{1/2}(\overline{\Gamma'_D})$, *there exists a unique solution* $\mathbf{u}(x) \in H^1(\Omega)$ *to (6) (cf. [8])*

2.3. Asymptotic Interpretation of Isotropic "Nearly Incompressible"

For simplicity, we assume that the soft tissues of a living body are isotropic, i.e. the viscoelasticity tensors $\boldsymbol{\mu}_{ilkm}$ ($\boldsymbol{\mu}^{(1)}_{ilkm}$, $\boldsymbol{\mu}^{(2)}_{ilkm}$ and $\boldsymbol{\eta}_{ilkm}$) are given as

$$\begin{cases} \boldsymbol{\mu}_{ilkm} := \lambda(x)\delta_{il}\delta_{km} + \mu(x)(\delta_{ik}\delta_{lm} + \delta_{im}\delta_{lk}), \\ \boldsymbol{\eta}_{ilkm} := \zeta(x)\delta_{il}\delta_{km} + \eta(x)(\delta_{ik}\delta_{lm} + \delta_{im}\delta_{lk}) \end{cases}$$

with the Kronecker delta δ_{ij} for any $x \in \overline{\Omega}$, i, l, k, m $(1 \le i, l, k, m \le 3)$. Here $\lambda(x)$ and $\mu(x)$ are the Lamé modulus, while $\zeta(x)$ and $\eta(x)$ are the viscosity coefficients. Especially, $\mu(x)$ and $\eta(x)$ are called shear modulus and shear viscosity respectively. Physically well known Possion's ratio ν is given by $\nu = \lambda/2(\lambda + \mu)$. For the isotropic medium, the strong convexity (3) becomes

$$\mu > \delta, \quad \eta > \delta, \quad 3\lambda + 2\mu > \delta, \quad 3\zeta + 2\eta > \delta \quad \text{(a.e. in } \Omega)$$

for some constant $\delta > 0$.

Based on the discussions above, the displacement $\mathbf{u}(x)$ of the wave measured by MRE is governed by the stationary viscoelasticity model (6). The Poisson ratio ν of soft tissues of living bodies has values close to 0.5. In other words, for soft tissues λ is much larger than μ. In this case, we call the stationary viscoelastic model **nearly incompressible**. The typical order of the value for λ is GPa while the order of μ is kPa, which implies $\nu = 0.4999999 < 0.5$. What does this mean in terms of the wave speed and wave length of the shear wave and compressional wave? Take the Voigt type viscoelasticity medium for example, the speed of shear wave v_s and and that of compressional wave v_c are given by

$$v_s = \sqrt{\frac{2(\mu^2+\omega^2\eta^2)}{\rho(\mu+\sqrt{\mu^2+\omega^2\eta^2})}}, \quad v_c = \sqrt{\frac{2\Big[(\lambda+2\mu)^2+\omega^2(\zeta+2\eta)^2\Big]}{\rho\Big[(\lambda+2\mu)+\sqrt{(\lambda+2\mu)^2+\omega^2(\zeta+2\eta)^2}\Big]}}$$

(cf. [3, 10]). Hence, a large λ implies that the wavelength of the compressional wave is much longer than the one of the shear wave. This explains why we cannot observe any compressional wave in MRE measurements visually, but there still remains a question how can we explain it mathematically. The answer is that we can find an approximate model which no longer contains a large λ as a parameter but its solution approximate very well the solution of the nearly incompressible model.

By applying an asymptotic analysis, after taking $\lambda \gg \mu$ into account, we have the following so-called *modified Stokes model*:

$$\begin{cases} \nabla\cdot\Big[2(G'+iG'')\varepsilon(\mathbf{u})\Big] - \nabla p + \rho\omega^2\mathbf{u} = 0 & \text{in } \Omega, \\ \nabla\cdot\mathbf{u} = 0 & \text{in } \Omega, \\ \mathbf{u} = \mathbf{f} & \text{on } \Gamma_D, \\ \partial_\nu\mathbf{u} := \Big[2(G'+iG'')\varepsilon(\mathbf{u}) - p\Big]\nu = 0 & \text{on } \Gamma_N. \end{cases} \tag{7}$$

Here, p is usually called *pressure*. The *storage modulus* and *loss modulus* are defined as follows:

- Voigt model:

$$G' = \mu, \quad G'' = \omega\eta.$$

- Maxwell model:

$$G' = \frac{\mu(\omega\eta)^2}{\mu^2+(\omega\eta)^2} \quad G'' = \frac{\mu^2(\omega\eta)}{\mu^2+(\omega\eta)^2}.$$

- Zener model:

$$G' = \mu^{(1)} + \frac{(\mu^{(2)})^2(\omega\eta)^2}{(\mu^{(2)})^2+(\omega\eta)^2} \quad G'' = \frac{(\mu^{(2)})^2(\omega\eta)}{(\mu^{(2)})^2+(\omega\eta)^2}$$

.

Here, we must remark that the storage modulus and loss modulus are usually the viscoelastic quantities measured by using rheometer.

Further, we can prove the well-posedness of the mixed boundary value problems for modified Stokes systems (7), i.e. for any given $\mathbf{f} \in \dot{H}^{1/2}(\overline{\Gamma'_D})$, there exists a unique $(\mathbf{u}, p) \in H^1(\Omega)\times L^2(\Omega)$ which is the solution to (7) (cf. [8]).

2.4. Numerical Examinations

In what follows we will examine via numerical simulations how well this *modified Stokes model* could recover our experimental data obtained by using the Micro-MRE system we have now in Hokkaido University, Japan. We used the software FreeFem++ (`http://www.freefem.org/ff++/`) for the finite element scheme (cf. [2]) to solve the mixed boundary value problem. Since

MRE can measure the displacement of stationary-state wave motion in any specified plane, say in the plane plane $x_3 = 0$, and our phantom is uniform in the x_3 direction, we used the *plain strain assumption* to simulate the wave in the plane. Of course there may be a modeling error in assuming this, because the injected vibration may not be uniform in the x_3 direction. More details on the plane strain assumption is as follows.

Plane Strain Assumption. Consider a relatively long cylindrical body with the x_3 axis as its central axis. If the body forces and traction on the lateral boundaries are independent of x_3 and have no x_3 component, then the wave displacement field in the body can be put in the reduced form $\mathbf{u} := (u_1(x_1, x_2), u_2(x_1, x_2), 0)$. This deformation is referred to as a state of plane strain in the x_1, x_2-plane, and thus the three dimensional problem is reduced to a two dimensional formulation in the region Ω in the x_1, x_2-plane. Using the strain-displacement relations, the strains corresponding to this plane problem become $\varepsilon_{13} = \varepsilon_{23} = \varepsilon_{33} = 0$.

By the plane strain assumption, we can have the reduced modified Stokes model for $\mathbf{u}(x) = (u_1(x), u_2(x))$ at $x \in \Omega \subset \mathbb{R}^2$. We will notice from our following simulation that this assumption seems to be appropriate in our case. Moreover, by looking at the simulation shown in figure 3, we conclude that our modified Stokes model is a proper PDE model for our MRE problem.

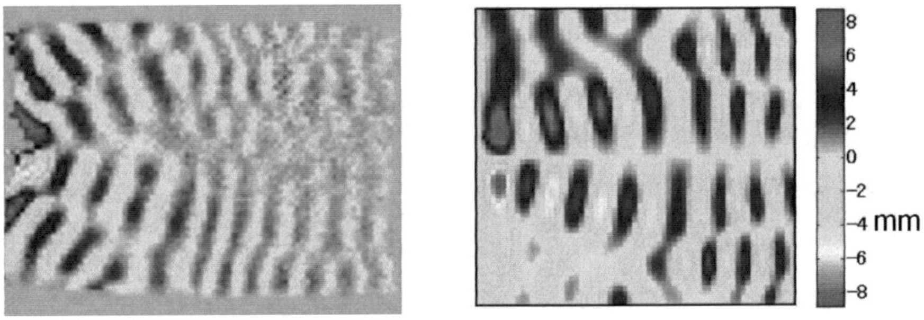

Figure. 3. Numerical simulation (left: experimental data, right: simulated data).

3. Modified Numerical Integral Method

In this section, we will present a modified numerical integral method to recover the viscoelasticity G' and G'' by knowing the interior stationary wave displacement $\mathbf{u}(x)$ in Ω which generated by the boundary input $\mathbf{f}$. For creating standard for MRE, a phantom is used as a sample instead of a living body. In this case we can assume that G', G'' are piecewise constants. Our inversion scheme presented here for is only for such a case. For non piecewise constant inhomogeneous G', G'' we developed a least square method in [9].

3.1. Curl Operator

To filter the pressure p in (7) which cannot be measured, we apply the *curl operator* $\mathbf{w} = \nabla \times \mathbf{u}$ and our PDE model becomes as follows:

$$(G' + iG'')\Delta\mathbf{w} + \rho(2\pi f)^2\mathbf{w} = 0 \quad \text{in } \Omega' \subset\subset \Omega$$

.

3.2. Mollification

To apply the above curl operator and further inversion procedure to the measured data, we need to denoise the data. In order to have the enough smoothness, we employ the mollifier as follows:

$$(S_\epsilon f)(x) = \int_{\mathbb{R}^2} s_\epsilon(y-x)\widehat{f}(y)dy$$

for bounded measurable function f defined on a bounded domain Ω, where the bounded measurable function $\widehat{f}$ is an zero extension of f to a bounded domain Ω_ϵ such that $\Omega \subset \Omega_\epsilon$, $0 < \mathrm{dist}(\Omega, \partial\Omega_\epsilon) \leq \epsilon$. Here, $s_\epsilon(x) = \epsilon^{-2}s(\epsilon^{-1}x)$ $(\epsilon > 0)$. The function s is a nonnegative C^1 function over $\mathbb{R}^2$ such that $\mathrm{supp}\, s \subset \Omega$ and

$$\int_{\mathbb{R}^2} s_\epsilon(x)dx = 1.$$

Then we have $S_\epsilon f \in C^3(\overline{\Omega})$ (cf. [18]).

3.3. Modified Numerical Integral Method

Now, we shall go to the real inversion procedure. We only focus on recovering G', because the major interest here is to diagnose early stage cancer. Hence, we only give the formula for recovering G' as following:

$$G' = -\rho\omega^2 \mathrm{Re}\left(\frac{\int_D |\mathbf{w}|^2 dx}{\int_D \mathbf{w}\overline{\Delta \mathbf{w}} dx}\right),$$

where the "testing function" is $\mathbf{w} = \nabla \times S_\varepsilon \mathbf{u}$ after applying the curl operator to the mollified data $S_\varepsilon \mathbf{u}$, $D \subset\subset \Omega' \subset\subset \Omega$ is the test region whose size is approximately equal to one local wavelength. Because we can prove that $\mathrm{Re}\int_D \mathbf{w}\overline{\Delta \mathbf{w}}dx$ is negative by applying the unique continuation principal, we can say our computation is stable.

3.4. Numerical Result

By applying the above recovery scheme, we can recover the storage modulus G' of the PAAm gel phantom from the 0.3 T MRE-measured data obtained by the Micro-MRE in Hokkaido University (see figure 3.4).

We must remark that because of the size of test region, we cannot have a good recovery near the original boundary. In the interior region, the mean values of recovered storage modulus G' in the dashed circles are shown in table 3.4. The real values below are measured by a rheometer (ARES-2KFRT, TA Instruments, frequency: 0.1 $\sim$ 10 Hz, strain mode: 5%).

Table 1. Mean values of recovered storage modulus G'

	hard (left)	soft (right)
mean value (kPa)	31.100	10.762
standard deviation	0.535	0.201
real value (kPa)	32.5456	9.2472
relative error	0.0444	0.1638

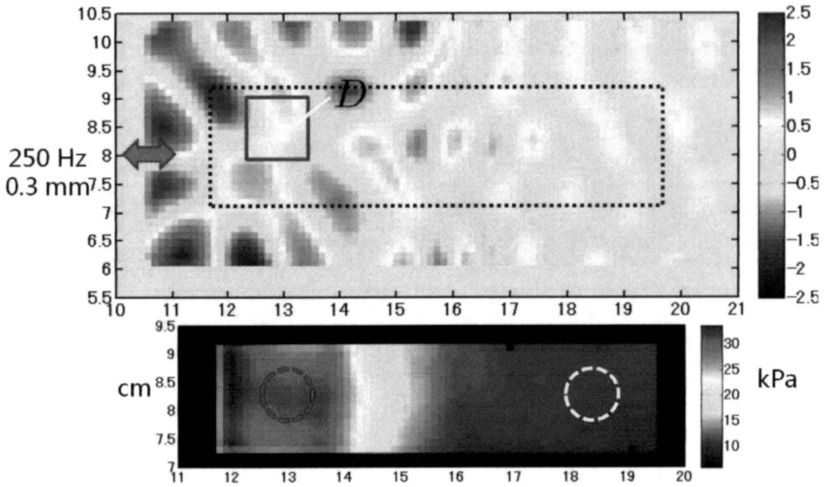

Figure. 4. Recovery of Storage modulus (above: MRE-measured data, below: storage modulus of PAAm phantom).

Acknowledgments

The MRE research in Hokkaido University, Japan is supported by the Japan Science and Technology Agency (JST). The authors thank many people for their contributions to the development of MRE study.

References

[1] Ammari H, Garapon P, Kang H and Lee H 2008 A method of biological tissues elasticity reconstruction using magnetic resonance elastography measurements *Quar. Appl. Math.* **66** 139-175.
[2] Brezzi F and Fortin M 1991 *Mixed and hybrid finite element methods* (New York: Springer).
[3] Caviglia G and Morro A 1992 *Inhomogeneous waves in solids and fluids (Series in theoretical and applied mechanics, v. 7)* (Singapore: World Scientific).
[4] Fowlkes J, Emelianov S, Pipe J, Skovoroda A, Carson P, Adler R and Sarvazyan A 1995 Magnetic resonance imaging techniques for detection of elasticity variation *Medical Physics* **22** 1771–1778.
[5] Gao L, Parker K, Lerner R and Levinson S 1996 Imaging of the elastic properties of tissue – a review *Ultrasound in Medicine and Biology* **22** 959–977.
[6] Higashimori N 2007 Identification of viscoelastic properties by magnetic resonance elastography *J. Phys.: Conf. Ser.* **73** 012009 (6pp) online.
[7] Hörmander L 1985 *The Analysis of Linear Partial Differential Operators III* (Berlin: Springer).
[8] Jiang Y, Fujiwara H and Nakamura G 2010 Asymptotic Analysis for Magnetic Resonance Elastography *submitted to SIAM J. Appl. Math.*.
[9] Jiang Y, Shirota K and Nakamura G 2010 Variational Approach for Recovering Viscoelasticity from MRE Data *submitted to Euro. J. Appl. Math.*.
[10] Kolsky H 2003 *Stress waves in solids* (New York: Dover Publications).
[11] Klatt D, Hamhaber U, Asbach P, Braun J and Sack I 2007 Noninvasive assessment of the rheological behavior of human internal organs using multifrequency MR elastography: a study of brain and liver viscoelasticity *Phys. Med. Biol.* **52** 7281–7294.
[12] Lewa C and de Certaines J 1995 MR imaging of viscoelastic properties *Journal of Magnetic Resonance Imaging* **5** 242–244.
[13] Lin K and McLaughlin J 2009 An error estimate on the direct inversion model in shear stiffness imaging *Inverse Problems* **25** 075003 (19pp).
[14] Lin K, McLaughlin J and Zhang N 2009 Log-elastographic and non-marching full inversion schemes for shear modulus recovery from single frequency elastographic data *Inverse Problems* **25** 075004 (24pp).
[15] Manduca A, Lake D, Kruse S and Ehman R 2003 Spatio-temporal directional filtering for improved inversion of MR elastography images *Med. Image Anal.* **7** 465–473.
[16] Manduca A, Oliphant T, Dresner M, Mahowald J, Kruse S, Amromin E, Felmlee J, Greenleaf J and Ehman

R 2001 Magnetic resonance elastography: noninvasive mapping of tissue elasticity *Med. Image Anal.* **5** 237–254.

[17] Muthupillai R, Lomas D, Rossman P, Greenleaf J, Manduca A. and Ehman R 1995 Magnetic resonance elastography by direct visualization of propagating acoustic strain waves *Science* **269** 1854–1857.

[18] Murio D 2002 Mollification and Space Marching *Inverse Engineering Handbook* (K. Woodbury Editor) (Florida: CRC Press) pp 219–326 .

[19] Nakamura G, Jiang Y, Nagayasu S and Cheng J 2008 Inversion analysis for magnetic resonance elastography *Applicable Analysis* **87** 165–179.

[20] Oliphant T (2001) *Direct methods for dynamic elastography reconstruction: optimal inversion of the interior Helmholtz problem* Thesis (Ph.D.) – Biomedical Sciences – Biomedical Imaging – Mayo Graduate School.

[21] Papazoglou S, Hamhaber U, Braun J and Sack I 2008 Algebraic Helmholtz inversion in planar magnetic resonance elastography *Phys. Med. Biol.* **53** 3147–3158.

[22] Park E and Maniatty A 2006 Shear modulus reconstruction in dynamic elastography: time harmonic case *Phys. Med. Biol.* **51** 3697–3721.

[23] Plewes D, Betty I, Urchuk S and Soutar I 1995 Visualizing tissue compliance with MR imaging. *Journal of Magnetic Resonance Imaging* **5** 733–738.

[24] Rivera J and Lapa E 1995 Decay rates of solutions of an anisotropic inhomogeneous n-dimensional viscoelastic equation with polynomial decaying kernels *Comm. Math. Physics* **177** 583–602.

[25] Romano A, Shirron J and Bucaro J 1998 On the noninvasive determination of material parameters from a knowledge of elastic displacements. *IEEE Trans. Ultra. Ferro. Freq. Cntrl.* **45** 751–7759.

[26] Sack I, Beierbach B, Wuerfel J, Klatt D, Hamhaber U, Papazoglou S, Martus P and Braun J 2009 The impact of aging and gender on brain viscoelasticity *NeuroImage* **46** 652–657.

[27] Sarvazyan A, Skovoroda A, Emelianov S, Fowlkes J, Pipe J, Adler R, Buxton R and Carson P 1995 *Acoustical imaging* (vol. 21, ch. Biophysical bases of elasticity imaging) (New York. Plenum Press) pp 223–240.

[28] Sinkus R, Lorenzen J, Schrader D, Lorenzen M, Dargatz M and Holz D 2000 High resolution tensor MR elastography for breast tumour detection. *Phys. Med. Biol.* **45** 1649–1664.

[29] Sinkus R, Tanter M, Catheline S, Lorenzen J, Kuhl C, Sondermann E and Fink M 2005 Imaging anisotropic and viscous properties of breast tissue by magnetic resonance elastography *Mag. Res. Med.* **53** 372–387.

[30] Sinkus R, Tanter M, Xydeas T, Catheline S, Bercoff J and Fink M 2005 Viscoelastic shear properties of in vivo breast lesions measured by MR elastography *Mag. Res. Imag.* **23** 159–165.

[31] Suga M, Matsuda T, Minato K, Oshiro O, Chihara K, Okamoto J, Takizawa O, Komori M and Takahashi T 2003 Measurement of in vivo local shear modulus using MR elastography multiple phase patchwork offsets *IEEE Transactions on Biomedical Engineering* **50** 908–915.

[32] Suga M, Obata T, Tanaka T, Miura H, Hirano M and Ikehira H 2008 Improvement of magnetic resonance elastography system to measure quantitative shear moduli map *The 57th Japan National Congress for Theoretical and Applied Mechanics* (`http://www.jstage.jst.go.jp/article/japannctam/57/0/192/_pdf/-char/ja/`).

Inverse Boundary Value Problem for Anisotropic Heat Operators

Kyoungsun Kim [1] **and Gen Nakamura**[2]
[1]Department of Mathematics, Ehwa University, Seoul 120-750, Korea.
[2]Department of Mathematics, Hokkaido University, Sapporo 060-0810, Japan.
E-mail: [2]`gnaka@math.sci.hokudai.ac.jp`

Abstract. As a mathematical model of thermography, a reconstruction scheme called the dynamical probe method is given for identifying unknown separated inclusions inside a known anisotropic heat conductor. The heat conductivities of inclusions can be also anisotropic. The measured data is the so called Neumann to Dirichlet map which is a mathematical idealization of many measurements consisting of injecting heat flux and measuring the corresponding heat distribution on the part of the boundary of the known heat conductor by infrared camera for any fixed time interval. This idealization becomes relevant if we have for instance the cooling boundary condition on the other part of the boundary. This is due to the exponential decay of temperature in time which enables to conduct many measurements in a short time.

1. Introduction

We consider a bounded domain Ω in $\mathbb{R}^n$ $(n=2,3)$ with a boundary $\partial\Omega$ of C^2 class and an open subset D of Ω such that $\overline{D}\subset\Omega$, $\Omega\setminus\overline{D}$ is connected and ∂D is of $C^{1,\alpha}$ class $(0<\alpha<1)$. Let the heat conductivity $\gamma(x)$ in $\Omega\times[0,T]$ is given by

$$\gamma(x)=A(x)+(\tilde{A}(x)-A(x))\chi_D \tag{1}$$

where χ_D is the characteristic function of D. We assume that $A\in C^1(\overline{\Omega})$ and $\tilde{A}\in C^1(\overline{D})$ are positive definite symmetric matrices and $\tilde{A}(x)-A(x)$ is always positive definite or negative in an neighborhood of ∂D.

We use the notations $B_{(T_1,T_2)}$ and B_T for $B\times(T_1,T_2)$ and $B\times(0,T)$, respectively. $H^p(\Omega), H^p(\partial\Omega)$ and $H^{q,r}(\Omega_T)$ with $q.r\in\mathbb{Z}_+:=\mathbb{N}\cup\{0\}$ or $p=\frac{1}{2}$ denote the usual Sobolev spaces, where q and r in $H^{q,r}(\Omega_T)$ denote the regularity with respect to x and t, respectively. Also, for an open set $U\subset\mathbb{R}^{n+1}$, $q,r\in\mathbb{Z}_+$, $H^{q,r}(U)$ is defined likewise $H^{q,r}(\Omega_T)$. That is, $g\in H^{q,r}(U)$ if the norm

$$||g||_{H^{q,r}(U)}:=\left(\sum_{\substack{|\alpha|+2k\le q\\ k\le r}}\int_U\left|\partial_x^\alpha\partial_t^k g\right|^2 dtdx\right)^{1/2}$$

is finite. Further, a function $f(x,t)$ is said to be in $L^2((0,T);X)$ if $f(\cdot,t)$ takes a value in a

Hilbert space X for almost all $t \in (0,T)$ and

$$||f||^2_{L^2((0,T);X)} = \int_0^T ||f(\cdot,t)||^2_X dt < \infty.$$

Let $W(\Omega_T) := \{v \in L^2((0,T);H^1(\Omega)); \partial_t v \in L^2((0,T);(H^1(\Omega))^*)\}$, where $(H^1(\Omega))^*$ is the dual space of $H^1(\Omega)$.

In this extended abstract, we also use the Hörmander's notations for function spaces to handle the mixed type boundary condition. Namely, when X is an open subset of $\partial\Omega$, let Y be a subspace of X where $\partial Y \neq \emptyset$ and ∂Y is Lipschitz space. Then a space $\overline{H}^p(Y)$ consists of distributions which have extensions to X and we shall write $\dot{H}^p(\overline{Y})$ for a space of distributions in $H^p(X)$ supported by $\overline{Y}$. It is well known that there are dualities. For dual spaces $(\overline{H}^p(Y))^*$, $(\dot{H}^p(\overline{Y}))^*$ of $\overline{H}^p(Y)$, $\dot{H}^p(\overline{Y})$, we use

$$\dot{H}^{-p}(\overline{Y}) := (\overline{H}^p(Y))^*, \quad \overline{H}^{-p}(Y) := (\dot{H}^p(\overline{Y}))^*.$$

(See Hörmander's book [8].)

Let $\partial\Omega$ consist of two parts. Namely,

$$\partial\Omega = \overline{\Gamma^D} \cup \overline{\Gamma^N},$$

where Γ^D and Γ^N are open subsets of $\partial\Omega$ such that $\Gamma^D \cap \Gamma^N = \emptyset$ and the boundaries $\partial\Gamma^D$ of Γ^D and $\partial\Gamma^N$ of Γ^N are of C^2 class if they are nonempty.

Now consider a boundary value problem which is to find a unique weak solution $u = u(f,g) \in L^2((0,T);H^1(\Omega))$ which satisfies

$$\begin{cases} \mathcal{P}_D u(x,t) := \partial_t u(x,t) - \mathrm{div}_x(\gamma(x)\nabla_x u(x,t)) = 0 \quad \text{in } \Omega_T \\ u(x,t) = f(x,t) \quad \text{on } \Gamma^D_T, \quad (\partial_A u)(x,t) := A\nabla u(x,t)\cdot\nu = g(x,t) \quad \text{on } \Gamma^N_T, \\ u(x,0) = 0 \quad \text{for } x \in \Omega \end{cases} \tag{2}$$

for a given $f \in L^2((0,T);\overline{H}^{\frac{1}{2}}(\Gamma^D))$ and $g \in L^2((0,T);\dot{H}^{-\frac{1}{2}}(\overline{\Gamma^N}))$, where ν is the outward unit normal to $\partial\Omega$. In another words, by assuming the initial temperature of a heat conductive medium Ω is 0 and the temperature on $\partial\Omega_T$ is f, it is to determine the temperature $u = u(f,g)$ induced in Ω_T after applying the heat flux g on Γ^N_T.

It is well known that the boundary value problem (2) is well posed([15]). That is, there exists a unique solution $u = u(f,g) \in L^2((0,T);H^1(\Omega))$ to (2). Further, if $\Gamma^D \neq \emptyset$, and $f(\cdot,t) = 0 = g(\cdot,t)$ $(t > T')$ with $0 < T' < T$, then $u(f,g)$ has the decaying property. That is $u(f,g)$ decays exponentially after $t = T'$. Based on the fact that (2) is well posed, we define the *Neumann-to-Dirichlet map* Λ_D as follows. For fixed $f \in L^2((0,T);\overline{H}^{\frac{1}{2}}(\Gamma^D))$, define

$$\Lambda_D : L^2((0,T);\dot{H}^{-\frac{1}{2}}(\overline{\Gamma^N})) \to L^2((0,T);\overline{H}^{\frac{1}{2}}(\Gamma^N))$$
$$g \mapsto u(f,g)|_{\Gamma^N_T}.$$

Now, we take the Neumann-to-Dirichlet map Λ_D as *measured data.* Then, our *inverse problem* is to reconstruct the unknown inclusion D from Λ_D. This means that by fixing the temperature on Γ^D to f, we repeat conducting a measurement many times which puts a heat flux g on Γ^N and measure the corresponding temperature on Γ^N over the time interval $(0,T)$. However, when $\Gamma^D \neq \emptyset$ and $f = 0$, we can actually repeat the aforementioned measurement many times due to

Journal of Physics: Conference Series **290** (2011) 012007 doi:10.1088/1742-6596/290/1/012007

the decaying property of $u(f,g)$. Hence, we can even superpose the measured data to generate Λ_D approximately.

In [3], they gave a reconstruction scheme named *dynamical probe method* for one space dimensional case by adapting the probe method in a stationary heat conductive medium. Their result is generalized by Isakov, Kim, and Nakamura ([10]). But they are for the domains with isotropic heat conductivities. So, the purpose of this extended abstract is to develop the dynamical method for the heat conductors with anisotropic conductivities.

The rest of this extended abstract is organized as follows. In Section 2 and 3, we will state our main result and a brief idea of its proof.

2. Pre-indicator function

For $(y,s),(y,s') \in \mathbb{R}^n \times \mathbb{R}$ such that $s \neq s'$, let $\Gamma(x,t;y,s)$ and $\Gamma^*(x,t;y,s')$ for $(x,t) \in \Omega_T$ be fundamental solutions for operators $\mathcal{P}_\emptyset := \partial_t - \nabla \cdot (A(x)\nabla)$ and $\mathcal{P}_\emptyset^* := -\partial_t - \nabla \cdot (A(x)\nabla)$ such that $\Gamma(x,t;y,s) = 0$ for $t \leq s$ and $\Gamma^*(x,t;y,s') = 0$ for $t \geq s'$. That is, $\Gamma(x,t;y,s)$ and $\Gamma^*(x,t;y,s')$ satisfy $\mathcal{P}_\emptyset \Gamma(x,t;y,s) = 0$ if $(x,t) \neq (y,s)$ and $\mathcal{P}_\emptyset^* \Gamma^*(x,t;y,s') = 0$ if $(x,t) \neq (y,s')$. Also, let $G(x,t;y,s)$ be a solution of

$$\begin{cases} \mathcal{P}_\emptyset G(x,t;y,s) = \delta(x-y)\delta(t-s) \quad \text{in} \quad \Omega_T, \\ G(\cdot,\cdot;y,s) = 0 \quad \text{on} \quad \Gamma_T^D, \\ G(x,t;y,s) = 0 \quad \text{for} \quad x \in \Omega, \quad t \leq s \end{cases}$$

and $G^*(x,t;y,s')$ a solution of

$$\begin{cases} \mathcal{P}_\emptyset^* G^*(x,t;y,s') = \delta(x-y)\delta(t-s') \quad \text{in} \quad \Omega_T, \\ G^*(\cdot,\cdot;y,s') = 0 \quad \text{on} \quad \Gamma_T^D, \\ G^*(x,t;y,s') = 0 \quad \text{for} \quad x \in \Omega, \quad t \geq s' \end{cases}$$

such that

$$G(x,t;y,s) - \Gamma(x,t;y,s),\ G^*(x,t;y,s') - \Gamma^*(x,t;y,s') \in C^\infty(\overline{\Omega_T}).$$

In [10], authors obatined the following Runge's approximation theorem which is of important for us.

Theorem 2.1. [10] *There exist* $\{v^{0j}_{(y,s)}\}$, $\{\psi^{0j}_{(y',s')}\} \in H^{2,1}(\Omega_{(-\varepsilon,T+\varepsilon)})$ *such that*

$$\begin{cases} \mathcal{P}_\emptyset v^{0j}_{(y,s)} = 0 & \text{in } \Omega_{(-\varepsilon,T+\varepsilon)}, \\ v^{0j}_{(y,s)} = 0 & \text{on } \Gamma^D \times (-\varepsilon, T+\varepsilon), \\ v^{0j}_{(y,s)}(x,t) = 0 & \text{if } -\varepsilon < t \leq 0, \\ v^{0j}_{(y,s)} \to G(\cdot,\cdot;y,s) & \text{in } H^{2,1}(U \times (-\varepsilon', T+\varepsilon')) \text{ as } j \to \infty, \end{cases}$$

$$\begin{cases} \mathcal{P}_\emptyset^* \psi^{0j}_{(y',s')} = 0 & \text{in } \Omega_{(-\varepsilon,T+\varepsilon)}, \\ \psi^{0j}_{(y,s')} = 0 & \text{on } \Gamma^D \times (-\varepsilon, T+\varepsilon), \\ \psi^{0j}_{(y',s')}(x,t) = 0 & \text{if } T \leq t < T+\varepsilon, \\ \psi^{0j}_{(y',s')} \to G^*(\cdot,\cdot;y',s') & \text{in } H^{2,1}(U \times (-\varepsilon', T+\varepsilon')) \text{ as } j \to \infty \end{cases}$$

for all $0 < \varepsilon' < \varepsilon$ *and open* $U \subset \Omega$ *which satisfy* $\overline{U} \subset \Omega$, $\Omega \setminus \overline{U}$ *is connected,* ∂U *is of Lipschitz class,* $\overline{U} \not\ni y,\ y'$ *and* $-\varepsilon < s,\ s' < T+\varepsilon$.

Let v, ψ satisfy

$$\begin{cases} \mathcal{P}_\emptyset v = 0 \text{ in } \Omega_T, \\ v = f \text{ on } \Gamma_T^D, \\ \partial_A v = 0 \text{ on } \Gamma_T^N, \\ v(x,0) = 0 \text{ for } x \in \Omega, \end{cases} \qquad \begin{cases} \mathcal{P}_\emptyset^* \psi = 0 \text{ in } \Omega_T, \\ \psi = 0 \text{ on } \Gamma_T^D, \\ \partial_A \psi = g' \text{ on } \Gamma_T^N, \\ \psi(x,T) = 0 \text{ for } x \in \Omega, \end{cases}$$

where we have taken an arbitrarily fixed function $g' \in L^2((0,T); \dot{H}^{-\frac{1}{2}}(\overline{\Gamma^N}))$. For $j = 1, 2, \cdots$, we define

$$\begin{cases} v^j_{(y,s)} := v + v^{0j}_{(y,s)} \to V_{(y,s)} := v + G(\cdot,\cdot;y,s) \\ \psi^j_{(y,s')} := \psi + \psi^{0j}_{(y,s')} \to \Psi_{(y,s')} := \psi + G^*(\cdot,\cdot;y,s'). \end{cases}$$

in $H^{2,1}(U_T)$ as $j \to \infty$. We call $\{v^j_{(y,s)}\}$, $\{\psi^j_{(y',s')}\}$ the *Runge's approximation functions.*

Now we define the pre-indicator function by using the Neumann to Dirichlet map and Runge's approximation functions.

Definition 2.2. *Let $(y,s),(y,s') \in \Omega_T$ be such that $s \neq s'$, and $\{v^j_{(y,s)}\}, \{\psi^j_{(y,s')}\} \subset H^{2,1}(\Omega_{(-\varepsilon,T+\varepsilon)})$ be sequences of approximate functions given in (2). Then, we define the pre-indicator function $I(y,s';y,s)$ as follows.*

$$I(y,s';y,s) = \lim_{j\to\infty} \int_{\Gamma_T^N} (\Lambda_\emptyset - \Lambda_D)(\partial_A v^j_{(y,s)}|_{\Gamma_T^N})\, \partial_A \psi^j_{(y,s')}|_{\Gamma_T^N} d\sigma_x dt$$

whenever the limit exists, where $\Lambda_\emptyset = \Lambda_D$ with $D = \emptyset$.

This pre-indicator function will be used later to define an indicator function which is a mathematical testing machine to identify the unknown inclusion.

For approximate functions $\{v^j_{(y,s)}\} \subset H^{2,1}(\Omega_{(-\varepsilon,T+\varepsilon)})$ given in the definition above, let $u^j_{(y,s)} := u(f, \partial_A v^j_{(y,s)}|_{\Gamma_T^N})$ and $w^j_{(y,s)} := u^j_{(y,s)} - v^j_{(y,s)}$. Then, $w^j_{(y,s)}$ satisfies the following mixed boundary value problem

$$\begin{cases} \mathcal{P}_D w^j_{(y,s)} = \operatorname{div}((\tilde{A} - A)\chi_D \nabla_x v^j_{(y,s)}) \quad \text{in } \Omega_T, \\ w^j_{(y,s)} = 0 \quad \text{on } \Gamma_T^D, \quad \partial_A w^j_{(y,s)} = 0 \quad \text{on } \Gamma_T^N, \\ w^j_{(y,s)}(x,0) = 0 \quad \text{for } x \in \Omega. \end{cases}$$

The derivation of Lemma 3.2 in [3] gives the following result.

Lemma 2.3. ([3]) *Let $(y,s) \in (\Omega \setminus \overline{D})_T$. Then the sequence $\{w^j_{(y,s)}\}$ has a limit $w_{(y,s)}$ in $W(\Omega_T)$ satisfying*

$$\begin{cases} \mathcal{P}_D w_{(y,s)} = \operatorname{div}((\tilde{A} - A)\chi_D \nabla V_{(y,s)}) \quad \text{in } \Omega_T, \\ w_{(y,s)} = 0 \quad \text{on } \Gamma_T^D, \quad \partial_A w_{(y,s)} = 0 \quad \text{on } \Gamma_T^N, \\ w_{(y,s)}(x,0) = 0 \quad \text{for } x \in \Omega. \end{cases} \tag{3}$$

We call $w_{(y,s)}$ the *reflected solution.* By following the proof in [10], we can represent the pre-indicator function in term of the reflected solution.

Theorem 2.4. *For $(y,s),(y,s') \in (\Omega \setminus \overline{D})_T$ with $s \neq s'$, we have*

$$I(y,s';y,s) = -w_{(y,s)}(y,s') - \int_{\Gamma_T^N} w_{(y,s)}(x,t)\partial_A \Psi_{(y,s')}(x,t) d\sigma_x dt. \tag{4}$$

3. Indicator function and its behavior

Now we define the indicator function as in [10].

Definition 3.1. *Let* $C := \{c(\lambda)\,;\, 0 \le \lambda \le 1\}$ *be a non-selfintersecting* C^1 *curve in* $\overline{\Omega}$ *which joins* $c(0), c(1) \in \partial\Omega$*. Then, for each* $c(\lambda) \in \Omega$ *and each fixed* $s \in (0,T)$*, we define the indicator function* $J(c(\lambda), s)$ *by*

$$J(c(\lambda), s) := \lim_{\epsilon\downarrow 0} \liminf_{\delta\downarrow 0} |I(c(\lambda-\delta), s+\epsilon^2; c(\lambda-\delta), s)| \tag{5}$$

whenever the limit exists.

Then, we have the following theorem.

Theorem 3.2. (Main result) Let C be given as in Definition 3.1 above. Then, for a fixed $s \in (0,T)$, we have the followings.

(i) If the curve C is in $\Omega \setminus \overline{D}$ except $c(0)$ and $c(1)$, then $J(c(\lambda), s) < \infty$ for all λ, $0 \le \lambda \le 1$.

(ii) If $C \cap \overline{D} \ne \emptyset$, let λ_s $(0 < \lambda_s < 1)$ be such that $c(\lambda_s) \in \partial D$ and $c(\lambda) \in \Omega \setminus \overline{D}$ $(0 < \lambda < \lambda_s)$. Namely, $c(\lambda_s)$ is the point at which c first hits ∂D. We call $c(\lambda_s)$ the first hitting point. Then,

$$\lambda_s = \sup\{\, 0 < \lambda < 1\,;\, J(c(\lambda'), s) < \infty \quad \text{for any } 0 < \lambda' < \lambda\,\}. \tag{6}$$

In the rest of this section, we give a brief idea of the proof of this theorem. Since the case $n = 2$ can be handled in the same way, we will give it only for the case $n = 3$. Let $(y, s) \in \Omega_T$. First of all, by Theorem 2.4, we can analyze the behavior of the indicator function $J(c(\lambda), s)$ by the behavior of the reflected solution. If y is not on the boundary of D, then $w_{(y,s)}$ is bounded due to the interior regularity estimate (see, for example [5]). So we only need to know the behavior of $w_{(y,s)}$ when y approaches the boundary of D along the curve C.

If we put $E(x, t; y, s) := w_{(y,s)}(x, t) + V_{(y,s)}(x, t)$, then E is a fundamental solution for the operator $\mathcal{P}_D$.

Now let C be as in Definition 3.1 and $P = c(\lambda_0) \in \partial D$ for some λ_0. Since ∂D is $C^{1,\alpha}$ $(0 < \alpha < 1)$, there is a $C^{1,\alpha}$ diffeomorphism $\Phi : \mathbb{R}^n \to \mathbb{R}^n$ which transforms P to the origin O in $\mathbb{R}^n$, $\Phi(D) \subset \mathbb{R}^n_- = \{x = (x_1, \cdots, x_n) \in \mathbb{R}^n; x_n < 0\}$. We refer the reader to see [10] for an explicit definition of Φ. Put $\xi = \Phi(x)$ and $\eta = \Phi(y)$. Then, $D\Phi(P) = I$.

Let $\gamma_P = A(x) + (\tilde{A}(P) - A(x))\chi_D$ and use Γ_P to denote the fundamental solution for the operator $\mathcal{P}_P := \partial_t - \nabla \cdot \gamma_P \nabla$.

Since $A \in C^1(\overline{\Omega})$, there is a constant r and C satisfying

$$|A(\Phi^{-1}(\xi)) - A(\Phi(O))| \le C|\xi| \quad \text{for} \quad |\xi| \le r.$$

Note that $A(\Phi^{-1}(O)) = A(P)$. We define Γ^0 by

$$\Gamma^0(x, t; y, s) = \begin{cases} \dfrac{1}{[4\pi|A(P)|(t-s)]^{n/2}} \exp\Big[-\dfrac{|(A(P))^*(x-y)|^2}{4(t-s)}\Big], & t > s, \\ 0, \quad t \le s, \end{cases} \tag{7}$$

$$(\Gamma^0)^*(x, t; y, s') = \begin{cases} 0, \quad t \ge s', \\ \dfrac{1}{[4\pi|A(P)|(s'-t)]^{n/2}} \exp\Big[-\dfrac{|(A(P))^*(x-y)|^2}{4(s'-t)}\Big], & t < s', \end{cases} \tag{8}$$

where $(A(P))^*$ is the positive definite symmetric matrix such that $(A(P))^{*2} = (A(P))^{-1}$ and $|A(P)|$ is the determinant of $A(P)$. Then, Γ^0 is the fundamental solution for the operator

$\partial_t - \nabla_x \cdot A(P)\nabla_x$ and $(\Gamma^0)^*$ for $-\partial_t - \nabla_x \cdot A(P)\nabla_x$. Let Γ^0_- be the fundamental solution for the operator $\partial_t - \nabla \cdot ((A(P) + (\tilde{A}(P) - A(P))\chi_-)\nabla)$.

The following decomposition of $w_{(y,s)}$ helps us to analyze its behavior:

$$\begin{aligned} w_{(y,s)}(x,t) &= E(x,t;y,s) - V_{(y,s)}(x,t) \\ &= \{E(x,t;y,s) - \Gamma_P(x,t;y,s)\} + \{\Gamma_P(x,t;y,s) - \Gamma_-(\Phi(x),t;\Phi(y),s)\} \\ &\quad + \{\Gamma_-(\Phi(x),t;\Phi(y),s) - \Gamma^0_-(\Phi(x),t;\Phi(y),s)\} \\ &\quad + \{\Gamma^0_-(\Phi(x),t;\Phi(y),s) - \Gamma^0(\Phi(x),t;\Phi(y),s)\} + \{\Gamma^0(\Phi(x),t;\Phi(y),s) - \Gamma^0(x,t;y,s)\} \\ &\quad + \{\Gamma^0(x,t;y,s) - \Gamma(x,t;y,s)\} + \{\Gamma(x,t;y,s) - V_{(y,s)}(x,t)\}, \end{aligned}$$

where Γ_- is the fundamental solution for the operator $\partial_t - \mathrm{div}((A(\Phi^{-1}(\xi))) + (\tilde{A}(P) - A(\Phi^{-1}(\xi)))\chi_-)\nabla)$ in $\mathbb{R}^{n+1}$ with the characteristic function χ_- of the space $\mathbb{R}^n_-$ such that $\Gamma_-(x,t;y,s) = 0$ for $t \le s$ and Γ^0_- is the fundamental solution for the operator $\partial_t - \mathrm{div}((A(P) + (\tilde{A}(P) - A(P))\chi_-)\nabla)$ in $\mathbb{R}^{n+1}$ such that $\Gamma^0_-(x,t;y,s) = 0$ for $t \le s$. We introduced many fundamental solutions for various parabolic differential operators. For reader's convenience, we list fundamental solutions and corresponding operators.

$$\begin{aligned} E &: \partial_t - \nabla \cdot ((A(x) + (\tilde{A}(x) - A(x))\chi_D)\nabla) \\ \Gamma_P &: \partial_t - \nabla \cdot ((A(x) + (\tilde{A}(P) - A(x))\chi_D)\nabla) \\ \Gamma_- &: \partial_t - \nabla \cdot ((A(\Phi^{-1}(\xi)) + (\tilde{A}(P) - A(\Phi^{-1}(\xi)))\chi_-)\nabla) \\ \Gamma^0_- &: \partial_t - \nabla \cdot ((A(P) + (\tilde{A}(P) - A(P))\chi_-)\nabla) \\ \Gamma^0 &: \partial_t - \nabla \cdot (A(P)\nabla) \\ \Gamma &: \partial_t - \nabla \cdot (A(x)\nabla). \end{aligned}$$

Let $\varepsilon > 0$ be given. We take $x = y = y(\delta) = c(\lambda_0 - \delta) \in C \setminus \overline{D}$ so that $y(\delta) \to P$ $(\delta \downarrow 0)$. Then, as $\delta \downarrow 0$, $\xi = \eta \to \Phi(P) = O$, we can show that the fourth term in the above decomposition of $w_{(y,s)}$ is the dominant term by showing the following behavior of each term :

(1)
$$\limsup_{\delta \to 0} |E(x, s+\varepsilon^2; y, s) - \Gamma_P(x, s+\varepsilon^2; y, s)| = O(\varepsilon^{\mu-3}),$$
as $\varepsilon \to 0$.

(2)
$$\limsup_{\delta \downarrow 0} |(\tilde{\Gamma}_p - \Gamma_-)(\xi, s+\varepsilon^2; \eta, s)| = O(\varepsilon^{\alpha-3}) \quad \text{as} \quad \varepsilon \to 0.$$

(3)
$$\limsup_{\delta \downarrow 0} |\Gamma_-(\xi, t+\varepsilon^2; \eta, s) - \Gamma^0_-(\xi, t+\varepsilon^2; \eta, s)| = O(\varepsilon^{\mu-3}) \quad \text{as} \quad \varepsilon \to 0.$$

(In 1,2,3, we used a pointwise space gradient estimate for a fundamental solution of parabolic equation with discontinuos coefficients [6].)

(4) Put
$$W(\xi, t; \eta, s) := \Gamma^0_-(\xi, t; \eta, s) - \Gamma^0(\xi, t; \eta, s)$$
and let $W^{\pm}(\xi, t; \eta, s)$ be $W(\xi, t; \eta, s)$ for $\pm \xi_n > 0$. Then, there exist a constant $C > 0$ such that
$$\lim_{\delta \downarrow 0} |W^+(\eta, s+\varepsilon^2; \eta, s)| \ge C\varepsilon^{-3} \quad \text{as} \quad \varepsilon \to 0.$$

(5)

$$\limsup_{\delta \downarrow 0} |\Gamma^0(\Phi(x), t; \Phi(y), s) - \Gamma^0(x, t; y, s)| = 0.$$

(6) Let $G(x, t; y, s) = \Gamma^0(x, t; y, s) - \Gamma(x, t; y, s)$. Then,

$$\limsup_{\delta \downarrow 0} |G(y, s + \varepsilon^2; y, s)| = O(\varepsilon^{-2}) \quad \text{as} \quad \varepsilon \to 0.$$

(7) It follows from the definitions of Γ and v that

$$\Gamma(x, t; y, s) - V_{(y,s)}(x, t)$$

is C^∞ at (y, s) and so bounded in some closed neighborhood of (y, s).

References

[1] Alessandrini G and Di Cristo M 2005 Stable determination of an inclusion by boundary measurements *SIAM J. Math. Anal.* **37** 200–217.
[2] Di Cristo M, Kim K and Nakamura G 2009 Estimate of the fundamental solution for parabolic operators with discontinuous coefficients (*Preprint* http://arxiv.org/pdf/0906.4438).
[3] Daido Y, Kang H and Nakamura G 2007 A probe method for the inverse boundary value problem of non-stationary heat equations *Inverse Problems* **23** 1787–1800.
[4] Elayyan A and Isakov V 1997 On uniqueness of recovery of the discontinuous conductivity coefficient of a parabolic equation *SIAM J. Math. Anal.* **28** 49–59.
[5] Friedman A 1964 *Partial Differential Equations of Parabolic Type* (Englewood Cliffs, N.J.: Prentice-Hall Inc.).
[6] Fan J, Kim K, Nagayasu S and Nakamura G A gradient estimate for solutions to parabolic equations with discontinuous coefficients (*Preprint*).
[7] Fabes E and Stroock D 1986 A New proof of Moser's parabolic Harnack inequality using the old ideas of Nash *Arch. Rational Mech. Anal.* **96** 327–338.
[8] Hörmander L 1985 *The Analysis of Linear Partial Differential Operator III (A series of comprehensive study in Mathematics)* (Berlin: Springer-Verlag).
[9] Ikehata M 2002 Reconstruction of inclusion from boundary measurements *J. Inverse and Ill-Posed Problems* **10** 37–65.
[10] Isakov V, Kim K and Nakamura G 2010 Reconstruction of an unknown inclusion by thermal imaging *Annali della Scu. Norm. Sup. di Pisa, Classe di Scienze, Vol. IX* **4** 34 pp.
[11] Isakov V 1988 *Inverse Problems for Partial Differential Equations (Applied Mathematical Sciences Vol. 127)* (New York: Springer).
[12] Kupradze V 1979 *Three-dimensional Problems of the Mathematical Theory of Elasticity and Thermoelasticity (North-Holland series in Applied Mathematics and Mechanics, vol 25)* (Amsterdam : North-Holland).
[13] Lions J and Magenes E 1972 *Non-homogeneous Boundary Value Problems and Applications II* (Berlin: Springer-Verlag).
[14] Ladyženskaja O, Solonnikov V and Uralceva N 1968 *Linear and Quasi-linear Equations of Parabolic Type (American Mathematical Society, Translations of Mathematical Monographs Vol. 23)* (Providence, R.I. : American Mathematical Society).
[15] Wloka J 1987 *Partial Differential Equations* (New York : Cambridge University Press).

Journal of Physics: Conference Series **290** (2011) 012008 doi:10.1088/1742-6596/290/1/012008

Two-point boundary value problems and exact controllability for several kinds of linear and nonlinear wave equations

De-Xing Kong [a] **and Qing-You Sun** [b]

[a] Department of Mathematics, Zhejiang University, Hangzhou 310027, China
[b] Center of Mathematical Sciences, Zhejiang University, Hangzhou 310027, China

E-mail: qysun@cms.zju.edu.cn

Abstract. All articles *must* In this paper we introduce some new concepts for second-order hyperbolic equations: two-point boundary value problem, global exact controllability and exact controllability. For several kinds of important linear and nonlinear wave equations arising from physics and geometry, we prove the existence of smooth solutions of the two-point boundary value problems and show the global exact controllability of these wave equations. In particular, we investigate the two-point boundary value problem for one-dimensional wave equation defined on a closed curve and prove the existence of smooth solution which implies the exact controllability of this kind of wave equation. Furthermore, based on this, we study the two-point boundary value problems for the wave equation defined on a strip with Dirichlet or Neumann boundary conditions and show that the equation still possesses the exact controllability in these cases. Finally, as an application, we introduce the hyperbolic curvature flow and obtain a result analogous to the well-known theorem of Gage and Hamilton for the curvature flow of plane curves.

1. Introduction

Consider the following seconder-order hyperbolic partial differential equation

$$\mathscr{P}(t, x, u, Du, D^2u) = 0, \tag{1.1}$$

where t is the time variable, $x = (x_1, \cdots x_n)$ stand for the spacial variables, $u = u(t, x)$ is the unknown function, $\mathscr{P}$ is a given smooth function of the independent of variables $t, x_1, \cdots, x_n$, the unknown function u, the first-order partial derivatives $Du = (u_t, u_{x_1}, \cdots, u_{x_n})$ and the second-order partial derivatives $D^2u = (u_{tt}, u_{tx_1}, \cdots)$. Since we only consider the hyperbolic case, the initial data associated with the equation (1.1) read

$$t = 0: \quad u = u_0(x), \quad u_t = u_1(x), \tag{1.2}$$

where $u_0(x)$ and $u_1(x)$ are two given functions which stand for the initial position and the initial velocity, respectively. The equation (1.1) and the initial data (1.2) constitute the famous Cauchy problem. Another important problem related to (1.1) is the following so-called *two-point boundary value problem* for the equation (1.1):

Two-point Boundary Value Problem (TBVP): *Given two suitable smooth functions $u_0(x), u_T(x)$ and a positive constant T, can we find a C^2-smooth function $u = u(t,x)$ defined on the strip $[0,T] \times \mathbb{R}^n$ such that the function $u = u(t,x)$ satisfies the equation (1.1) on the domain $[0,T] \times \mathbb{R}^n$, the initial condition*

$$u(0,x) = u_0(x), \quad \forall\, x \in \mathbb{R}^n \tag{1.3}$$

and the terminal condition

$$u(T,x) = u_T(x), \quad \forall\, x \in \mathbb{R}^n? \tag{1.4}$$

Another Statement of TBVP: *Given two suitable smooth functions $u_0(x), u_T(x)$ and a positive constant T, can we find an initial velocity $u_t(0,x) = u_1(x)$ such that the Cauchy problem (1.1)-(1.2) has a solution $u = u(t,x) \in C^2([0,T] \times \mathbb{R}^n)$ which satisfies the terminal condition (1.4)?*

The system (1.1) together with (1.3)-(1.4) can be viewed as a distributed parameter control system when the initial velocity function $u_1(x)$ is considered as a control input.

We now give the following definition.

Definition 1.1. *For any given $T > 0$, if the TBVP (1.1), (1.3)-(1.4) has a C^2 solution on the strip $[0,T] \times \mathbb{R}^n$, then the equation (1.1) is called to possess the global exact controllability; If the TBVP (1.1), (1.3)-(1.4) admits a C^2 solution on the strip $[0,T] \times \mathbb{R}^n$ for some given $T > 0$, then we say that the equation (1.1) possesses the exact controllability.*

Remark 1.1. *For the system of seconder-order hyperbolic partial differential equations, we have similar concepts and definitions; For higher-order hyperbolic partial differential equations, we have a similar discussion.*

The wave equations play an important role in both theoretical and applied fields, they include two classes: linear wave equations and nonlinear wave equations. The classical wave equation is an important second-order linear partial differential equation of waves, such as sound waves, light waves and water waves. It arises in fields such as acoustics, electromagnetics, fluid dynamics, and general relativity. Historically, the problem of a vibrating string such as that of a musical instrument was studied by Jean le Rond d'Alembert, Leonhard Euler, Daniel Bernoulli, and Joseph-Louis Lagrange.

The wave equation is the prototypical example of a hyperbolic partial differential equation. In its simplest form, the wave equation refers to a unknown scalar function $y = y(t,x)$ which satisfies

$$y_{tt} - c^2 \Delta y = 0, \tag{1.5}$$

where $\Delta = \sum_{i=1}^{n} \frac{\partial^2}{\partial x_i^2}$ denotes the Laplacian and c is a fixed constant which stands for the propagation speed of the wave. One of aims of the present is to investigate the TBVP for (1.5), more precisely we study the following TBVP:

TBVP for (1.5): *Given two functions $f(x), g(x) \in C^{[n/2]+2}(\mathbb{R}^n)$ and a positive constant T, can we find a C^2-smooth function $y = y(t,x)$ defined on the strip $[0,T] \times \mathbb{R}^n$ such that the function $y = y(t,x)$ satisfies the equation (1.5) on the domain $[0,T] \times \mathbb{R}^n$, the initial condition*

$$y(0,x) = f(x), \quad \forall\, x \in \mathbb{R}^n \tag{1.6}$$

and the terminal condition

$$y(T,x) = g(x), \quad \forall\, x \in \mathbb{R}^n? \tag{1.7}$$

Another Statement: *Given two functions $f(x), g(x) \in C^{[n/2]+2}(\mathbb{R}^n)$ and a positive constant T, can we find an initial velocity $v = v(x)$ such that the Cauchy problem for the wave equation* (1.5) *with the initial data*

$$t = 0: \quad y = f(x), \quad y_t = v(x) \tag{1.8}$$

has a solution $y = (t, x) \in C^2([0, T] \times \mathbb{R}^n)$ which satisfies the terminal condition (1.7)*?*

In this paper we shall show the global exact controllability of the equation (1.5) and some nonlinear wave equations arising from geometry and the theory of relativistic strings.

It is well-known that there are many deep and beautiful results on the TBVP for ordinary differential equations, however, according to the authors' knowledge, few of results on the TBVP for hyperbolic equations, even for (linear or nonlinear) wave equations have been known. Therefore, we can say that the result presented in this paper is the first result on this research topic.

On the other hand, the study on *boundary* control problems for hyperbolic systems was initiated by D.L. Russell in the 1960s. In [15], using the characteristic method, he showed that a class of $n \times n$ first order linear hyperbolic systems is exactly boundary controllable. This work led to an intensive investigation of controllability and stabilization of linear hyperbolic systems for more than 30 years. The literature pertaining to this study is now absolutely enormous, we refer to two excellent review papers Russell [16] and Lions [14]. However, while it may be fair to say that the study of boundary control of linear hyperbolic systems is now nearly complete, the study of nonlinear hyperbolic systems is still vastly open. Up to now, some results on the exact boundary controllability of (abstract) semilinear wave equations have been obtained (see [1], [2], [4], [12]-[13], [19]-[21] and references cited therein). As for boundary control of quasilinear hyperbolic systems, there have been few results so far. Motivated by Ruessell's work, Cirinà [3] studied boundary control of general quasilinear hyperbolic systems. Using a different approach from that of Russell, he proved that the system is *locally* exactly boundary controllable in the sense that the C^1 norms of both initial and terminal states are required to be small.

All the above results are obtained under the assumption that the initial and terminal states are smooth, and they are discussed in the framework of classical solutions. For the global exact controllability of system (1.1) in the case that the initial and terminal states maybe contain discontinuity points of the first kind, up to now only a few of results have been known. In Kong [7], the author investigated the global exact boundary controllability of 2×2 quasilinear hyperbolic system of conservation laws with linearly degenerate characteristics and proved that the system with nonlinear boundary conditions is globally exactly boundary controllable in the class of piecewise C^1 functions. Later, by a new constructive method, Kong and Yao reproved the global exact boundary controllability of a class of quasilinear hyperbolic systems of conservation laws with linearly degenerate characteristics, shown that the system with nonlinear boundary conditions is globally exactly boundary controllable in the class of piecewise C^1 functions, in particular, gave the optimal control time of the system (see [9]).

Here we would like to remark that the TBVP and the boundary control problems are essentially different two kinds of problems. Both of them play an important role in both theoretical and applied aspects.

We now state the first result in this paper.

Theorem 1.1. *The TBVP* (1.5)-(1.7) *admits a C^2-smooth solution $y = y(t, x)$ defined on the strip $[0, T] \times \mathbb{R}^n$.*

Theorem 1.1 implies that the wave equation (1.5) possesses the global exact controllability.

Remark 1.2. *The solution of the TBVP of* (1.5)-(1.7) *does not possess the uniqueness (see Remarks 2.2 and 3.2 for the details).*

Remark 1.3. *Theorem 1.1 can be generalized to the case of inhomogeneous wave equations, i.e.,*

$$y_{tt} - c^2 \Delta y = F(t, x), \tag{1.9}$$

where $F(t, x)$ is a given function which stands for the source term of the system.

Remark 1.4. *We have similar results for some nonlinear wave equations including a wave map equation arising from geometry and the equations for the motion of relativistic strings in the Minkowski space-time $\mathbb{R}^{1+n}$ (see Section 7 for the details). These nonlinear wave equations play an important role in both mathematics and physics.*

However, some problems arising from engineering, control theory etc. can be reduced to the TBVP for wave equations defined on a closed curve, say, a circle. The typical example is the vibration of a closed elastic string. In this case, since the wave equation is defined on a closed curve and then the solution must possess the periodicity, the method used in the proof of Theorem 1.1 will no longer work. It needs some new ideas and new technologies to study such a kind of problems.

In this paper we also investigate this kind of problems mentioned above. For simplicity, we consider the following TBVP[1]

$$\begin{cases} y_{tt} - y_{\theta\theta} = 0, & \forall\, (t, \theta) \in [0, T] \times \mathbb{R}, \\ y(0, \theta) = f(\theta), & \forall\, \theta \in \mathbb{R}, \\ y(T, \theta) = g(\theta), & \forall\, \theta \in \mathbb{R}, \end{cases} \tag{1.10}$$

where $y = y(t, \theta)$ is the unknown function of the variables t and θ, T is a given positive constant, $f(\theta)$ and $g(\theta)$ are two given periodic C^3 functions of $\theta \in \mathbb{R}$, say, the period is L, in which L is a positive real number. The second result in the present paper is the following theorem.

Theorem 1.2. *The TBVP* (1.10) *admits a global L-periodic C^2 solution $y = y(t, \theta)$ defined on the domain $[0, T] \times \mathbb{R}$, provided that $\frac{T}{L}$ is a rational number with $\frac{2T}{L} \notin \mathbb{N}$.*

Remark 1.5. *Theorem 1.2 implies that the wave equation in* (1.10) *possesses the exact controllability. On the other hand, in general, the solution of the TBVP* (1.10) *is not unique (see the proof of Theorem 1.2 in Section 4 for the details).*

Remark 1.6. *In Theorem 1.2, if $\frac{2T}{L} \in \mathbb{N}$, then there exists a relationship between $f(\theta)$ and $g(\theta)$. This means that $f(\theta)$ and $g(\theta)$ can not be given arbitrarily. See Remark 4.2 for the details.*

As a consequence, we consider the following TBVP defined on a circle

$$\begin{cases} y_{tt} - y_{\theta\theta} = 0, & \forall\, (t, \theta) \in [0, T] \times \mathbb{S}^1, \\ y(0, \theta) = f(\theta), & \forall\, \theta \in \mathbb{S}^1, \\ y(T, \theta) = g(\theta), & \forall\, \theta \in \mathbb{S}^1, \end{cases} \tag{1.11}$$

where $\mathbb{S}^1$ stands for the unit circle, $f(\theta)$ and $g(\theta)$ are two given C^3 functions of $\theta \in \mathbb{S}^1$. By Theorem 1.2, we have

Theorem 1.3. *The TBVP* (1.11) *admits a global C^2 solution $y = y(t, \theta)$ defined on the cylinder $[0, T] \times \mathbb{S}^1$, provided that $\frac{T}{2\pi}$ is a rational number and $\frac{T}{\pi} \notin \mathbb{N}$.*

[1] In fact, by scaling, any wave equation $y_{tt} - c^2 y_{\theta\theta} = 0$ with the propagation speed c can be reduced to the wave equation in (1.10). Therefore, in this paper, without loss of generality, we only consider the wave equation with the propagation speed $c = 1$.

Remark 1.7. *Theorem 1.3 implies that the wave equation defined on a circle possesses the exact controllability.*

Remark 1.8. *For the* $(1+n)$*-dimensional wave equation* (1.5)*, we have similar results, provided that the initial/terminal data* $f(x)$ *and* $g(x)$ *in* (1.6)-(1.7) *are* $C^{[n/2]+3}$ *smooth functions and are periodic in* $r = \sqrt{x_1^2 + \cdots + x_n^2}$*. In fact, in the present situation, using Theorem 1.2, by a way similar to the proof of Theorem 1.1 we can prove that the TBVP* (1.5)-(1.7) *admits a global* C^2 *solution* $y = y(t,x)$ *defined on the strip* $[0,T] \times \mathbb{R}^n$*, provided that* $\frac{T}{L}$ *is a rational number with* $\frac{2T}{L} \notin \mathbb{N}$*, where* L *is the period of* $f(x)$*,* $g(x)$ *with respect to* r*. In other words, in this case the equation* (1.5) *still possesses the exact controllability.*

Based on the results mentioned above, we further study the two-point boundary value problems for the wave equation defined on a strip with Dirichlet or Neumann boundary conditions and show that the equation still possesses the exact controllability in these cases. Finally, as an application of the results mentioned above, we introduce the hyperbolic curvature flow and obtain a result analogous to the well-known theorem of Gage and Hamilton [5] for the curvature flow of plane curves.

This paper is organized as follows. Section 2 is devoted to the study on the global exact controllability of one-dimensional wave equation. Based on Section 2, in Section 3 we investigate the global exact controllability of linear wave equations in several space variables and we prove Theorem 1.1. In Section 4, we give the proof of Theorem 1.2 and Theorem 1.3, respectively. As some applications of Theorem 1.2, in Sections 5 and 6 we study the two-point boundary value problems for the wave equation defined on a strip with (homogeneous or inhomogeneous) Dirichlet or Neumann boundary conditions and show that the equation still possesses the exact controllability in these cases. The global exact controllability of some nonlinear wave equations arising from geometry and physics has been investigated in Section 7. In Section 8 we introduce the hyperbolic curvature flow and prove a result analogous to the one shown by Gage and Hamilton in [5] for curvature flow of plane curves. In Section 9 we give a summary and some discussions and then present several open problems.

2. One-dimensional wave equation

This section concerns the global exact controllability of one-dimensional wave equation, which is a basis of the present paper.

Consider the following TBVP for one-dimensional wave equation

$$\begin{cases} y_{tt} - y_{xx} = 0, \\ y(0,x) = f(x), \\ y(T,x) = g(x), \end{cases} \tag{2.1}$$

where T is an arbitrary fixed positive constant, $f(x)$ and $g(x)$ are two given functions of $x \in \mathbb{R}$. We have the following theorem which is a special case of Theorem 1.1 but fundamental in this paper.

Theorem 2.1. *Suppose that* $f(x)$ *and* $g(x)$ *are two given* C^2*-smooth functions of* $x \in \mathbb{R}$*. Then the TBVP* (2.1) *admits a* C^2*-smooth solution* $y = y(t,x)$ *defined on the strip* $[0,T] \times \mathbb{R}$*.*

Proof. Introduce

$$\tilde{y}(t,x) = y(t,x) - \frac{f(x-t) + f(x+t)}{2}, \tag{2.2}$$

and

$$\tilde{f}(x) = g(x) - \frac{f(x-T) + f(x+T)}{2}. \tag{2.3}$$

Obviously, $\tilde{f}(x)$ is a C^2-smooth function of $x \in \mathbb{R}$. By means of $\tilde{y}(t,x)$ and $\tilde{f}(x)$, the TBVP (2.1) can be equivalently rewritten as

$$\begin{cases} \tilde{y}_{tt} - \tilde{y}_{xx} = 0, \\ \tilde{y}(0,x) = 0, \\ \tilde{y}(T,x) = \tilde{f}(x), \end{cases} \tag{2.4}$$

Therefore, in order to prove Theorem 2.1, it suffices to show that the TBVP (2.4) has a C^2-smooth solution $\tilde{y} = \tilde{y}(t,x)$ defined on the strip $[0,T] \times \mathbb{R}$.

By d'Alembert formula, the solution of the following Cauchy problem

$$\begin{cases} \tilde{y}_{tt} - \tilde{y}_{xx} = 0, \\ t = 0: \quad \tilde{y} = 0, \quad \tilde{y}_t = v(x) \end{cases} \tag{2.5}$$

reads

$$\tilde{y}(t,x) = \frac{1}{2} \int_{x-t}^{x+t} v(\tau) d\tau, \tag{2.6}$$

where $v(x)$ is a C^1-smooth function to be determined, which stands for the initial velocity.

We next show that there indeed exists an initial velocity $v(x)$ such that the solution $\tilde{y} = \tilde{y}(t,x)$ of the Cauchy problem (2.5), defined by (2.6), satisfies the terminal condition in the TBVP (2.4), i.e.,

$$\tilde{y}(T,x) = \tilde{f}(x). \tag{2.7}$$

To do so, we construct a C^1 function $u(x)$ on $[-T,T]$ which satisfies

$$\int_{-T}^{T} u(\tau) d\tau = 2\tilde{f}(0), \quad u(T) - u(-T) = 2\tilde{f}'(0) \quad \text{and} \quad u'_-(T) - u'_+(-T) = 2\tilde{f}''(0). \tag{2.8}$$

Define

$$v(x) = \begin{cases} u(x - 2NT) + 2\sum_{i=1}^{N} \tilde{f}'(x - (2i-1)T), & \forall\, x \geq 0, \\ u(x + 2NT) - 2\sum_{i=1}^{N} \tilde{f}'(x + (2i-1)T), & \forall\, x < 0, \end{cases} \tag{2.9}$$

where N is given by

$$N = \left[\frac{|x| + T}{2T} \right].$$

In (2.9), the terms including the summation disappear in the case of $N = 0$.

We claim that the function $v(x)$ defined by (2.9) is C^1-smooth.

In what follows, we distinguish two cases to show this fact.

Case A: $x \geq 0$. In this case, for every $x \neq (2N-1)T$ $(N \in \mathbb{N})$, by (2.9) it is easy to check that $v(x)$ is C^1-smooth at such a point x.

When $x = (2N-1)T$, it follows from (2.9) that

$$\lim_{\alpha \to x^+} v(\alpha) = v(x) = u(-T) + 2\sum_{i=0}^{N-1} \tilde{f}'(2iT) \tag{2.10}$$

and

$$\lim_{\alpha\to x^-} v(\alpha) = \lim_{\alpha\to x^-} u\big(\alpha - 2(N-1)T\big) + 2\sum_{i=1}^{N-1} \lim_{\alpha\to x^-} \tilde{f}'\big(\alpha - (2i-1)T\big) = u(T) + 2\sum_{i=1}^{N-1} \tilde{f}'(2iT). \tag{2.11}$$

Combining (2.10) and (2.11) gives

$$v(x) - \lim_{\alpha\to x^-} v(\alpha) = u(-T) - u(T) + 2\tilde{f}'(0). \tag{2.12}$$

Noting the second equation in (2.8) yields the continuity of $v(x)$ for all $x \geq 0$.

On the other hand, it follows from (2.9) that

$$v'_-(x) = u'_-\big(x - 2(N-1)T\big) + 2\sum_{i=1}^{N-1} \tilde{f}''\big(x - (2i-1)T\big) = u'_-(T) + 2\sum_{i=1}^{N-1} \tilde{f}''\big((2N-2i)T\big) \tag{2.13}$$

and

$$v'_+(x) = u'_+(x - 2NT) + 2\sum_{i=1}^{N} \tilde{f}''\big(x - (2i-1)T\big) = u'_+(-T) + 2\sum_{i=1}^{N} \tilde{f}''\big((2N-2i)T\big). \tag{2.14}$$

Combining (2.13) and (2.14) gives

$$v'_+(x) - v'_-(x) = u'_+(-T) - u'_-(T) + 2\tilde{f}''(0). \tag{2.15}$$

Noting the third equation in (2.8), we have

$$v'_+(x) = v'_-(x). \tag{2.16}$$

Summarizing the above argument yields that the function $v(x)$ defined by (2.9) is C^1-smooth for all $x \geq 0$.

Case B: $x < 0$. Similarly, we can prove that the function $v(x)$ is C^1-smooth in the present situation.

Combining Cases A and B, we have shown that the function $v(x)$ defined by (2.9) is a C^1-smooth function of $x \in \mathbb{R}$, and then the solution $\tilde{y} = \tilde{y}(t,x)$, defined by (2.6), of the Cauchy problem (2.5) is C^2-smooth on the whole upper plane $\mathbb{R}^+ \times \mathbb{R}$.

We now claim that, for the initial velocity $v(x)$ defined by (2.9), the solution $\tilde{y} = \tilde{y}(t,x)$ defined by (2.6) satisfies the terminal condition (2.7).

In fact, we verify this statement by distinguishing the following two cases:

Case 1: $x \geq 0$. In the present situation, it holds that

$$x \leq 2NT + T \leq x + 2T.$$

Thus it follows from (2.9) that

$$\begin{aligned}
\frac{1}{2}\int_x^{x+2T} v(\tau)d\tau &= \frac{1}{2}\left[\int_x^{2NT+T} v(\tau)d\tau + \int_{2NT+T}^{x+2T} v(\tau)d\tau\right] \\
&= \int_x^{2NT+T}\left[\frac{1}{2}u(\tau-2NT) + \sum_{i=1}^{N}\tilde{f}'(\tau-(2i-1)T)\right]d\tau + \\
&\quad \int_{2NT+T}^{x+2T}\left[\frac{1}{2}u(\tau-2(N+1)T) + \sum_{i=1}^{N+1}\tilde{f}'(\tau-(2i-1)T)\right]d\tau \\
&= \frac{1}{2}\left[\int_x^{2NT+T} u(\tau)d\tau + \int_{2NT+T}^{x+2T} u(\tau)d\tau\right] + \\
&\quad \sum_{i=1}^{N}\Big[\tilde{f}((2N-2i+2)T) - \tilde{f}(x-(2i-1)T)\Big] + \\
&\quad \sum_{i=1}^{N+1}\Big[\tilde{f}(x-(2i-3)T) - \tilde{f}((2N-2i+2)T)\Big] \\
&= \frac{1}{2}\left[\int_{x-2NT}^{T} u(\tau)d\tau + \int_{-T}^{x-2NT} u(\tau)d\tau\right] + \\
&\quad \sum_{i=1}^{N}\Big[\tilde{f}(x-(2i-3)T) - \tilde{f}(x-(2i-1)T)\Big] + \\
&\quad \tilde{f}(x-(2N-1)T) - \tilde{f}(0) \\
&= \frac{1}{2}\int_{-T}^{T} u(\tau)d\tau + \tilde{f}(x+T) - \tilde{f}(0). \qquad (2.17)
\end{aligned}$$

Noting the first condition of (2.8), we obtain

$$\frac{1}{2}\int_x^{x+2T} v(\tau)d\tau = \tilde{f}(x+T). \qquad (2.18)$$

This leads to

$$\tilde{y}(T,x) = \frac{1}{2}\int_{x-T}^{x+T} v(\tau)d\tau = \tilde{f}(x). \qquad (2.19)$$

This is the desired terminal condition (2.7) for the case of $x \geq 0$.

Case 2: $x < 0$**.** In a similar manner, we can prove the terminal condition (2.7) holds for all $x < 0$.

The above discussion shows that $\tilde{y} = \tilde{y}(t,x)$ defined by (2.6) is a C^2-smooth solution of the TBVP (2.4) defined on the strip $[0,T] \times \mathbb{R}$. This proves Theorem 2.1. □

Remark 2.1. *In order to illustrate that it is easy to construct the function u satisfying* (2.8)*, in this remark we present two examples. The first example reads*

$$u = \frac{\tilde{f}''(0)}{2T}x^2 + \frac{\tilde{f}'(0)}{T}x + \frac{\tilde{f}(0)}{T} - \frac{\tilde{f}''(0)T}{6}. \qquad (2.20)$$

It is easy to check that the function u defined by (2.20) *satisfies all conditions in* (2.8)*. The second example is*

$$u(x) = \begin{cases} \tilde{h} + \dfrac{h}{2} + \dfrac{h}{2}\sin\left(\dfrac{\pi}{T}x + \dfrac{\pi}{2}\right), & \forall\, x \in [-T, 0), \\ \tilde{h} + h + \dfrac{1}{T}\tilde{f}''(0)x^2, & \forall\, x \in [0, T], \end{cases} \qquad (2.21)$$

where

$$h = 2\tilde{f}'(0) - T\tilde{f}''(0) \quad and \quad \tilde{h} = \frac{\tilde{f}(0)}{T} + \left(\frac{7}{12}T\tilde{f}''(0) - \frac{3}{2}\tilde{f}'(0)\right).$$

It is easy to verify that the function u defined by (2.21) also satisfies all conditions in (2.8).

Remark 2.2. *The solution of the TBVP (2.4) (equivalently, (2.2)) is not unique. In fact, by the definition of $u(x)$ we observe that such a function u is not unique. This results the non-uniqueness of the initial velocity $v(x)$ and then the solution $\tilde{y} = \tilde{y}(t,x)$. For example, choose a C^1-smooth function $v_1(x)$ satisfying*

$$\int_{x-T}^{x+T} v_1(\tau)d\tau = 0. \tag{2.22}$$

This implies that $v_1(x)$ is a $2T$-periodic function and satisfies

$$\int_{-T}^{T} v_1(\tau)d\tau = 0. \tag{2.23}$$

Obviously, the function

$$\tilde{y}(t,x) = \frac{1}{2}\int_{x-t}^{x+t} \big(v(\tau) + v_1(\tau)\big)d\tau \tag{2.24}$$

also gives a C^2-smooth solution of the TBVP (2.4).

The idea to construct $v(x)$ (see (2.9)) essentially comes from the *characteristic-quadrilateral identity* given in [10]. In fact, using the characteristic-quadrilateral identity, we have

$$\tilde{y}(A) + \tilde{y}(D) = \tilde{y}(B_1) + \tilde{y}(C_1). \tag{2.25}$$

See Figure 1. By the the initial condition in the TBVP (2.4), we get

$$\tilde{y}(A) = \tilde{y}(B_1) + \tilde{y}(C_1). \tag{2.26}$$

Similarly, we obtain

$$\tilde{y}(B_1) = \tilde{y}(B_2) + \tilde{y}(C_2), \quad \cdots, \quad \tilde{y}(B_{N-1}) = \tilde{y}(B_N) + \tilde{y}(C_N). \tag{2.27}$$

On the other hand, it follows from (2.6) that

$$\tilde{y}(A) = \frac{1}{2}\int_{-T}^{x} v(\tau)d\tau, \quad \tilde{y}(B_N) = \frac{1}{2}\int_{-T}^{x-2NT} v(\tau)d\tau. \tag{2.28}$$

Combining (2.26)-(2.28) gives the definition of $v(x)$ shown by (2.9).

Remark 2.3. *In fact, the existence of the C^2-solution of the TBVP (2.4) is equivalent to the existence of the C^1-solution of the following integral equation*

$$\int_{x-T}^{x+T} v(\tau)d\tau = \tilde{f}(x). \tag{2.29}$$

(2.29) is a typical example of Volterra integral equations of the first kind. By the theory of Volterra integral equations, we can obtain the existence of the C^1-solution of the integral equation (2.29). However, by this theory we do not know how to construct the desired solution. But, in our theory we present a direct method to construct the desired solution of the TBVP (2.1).

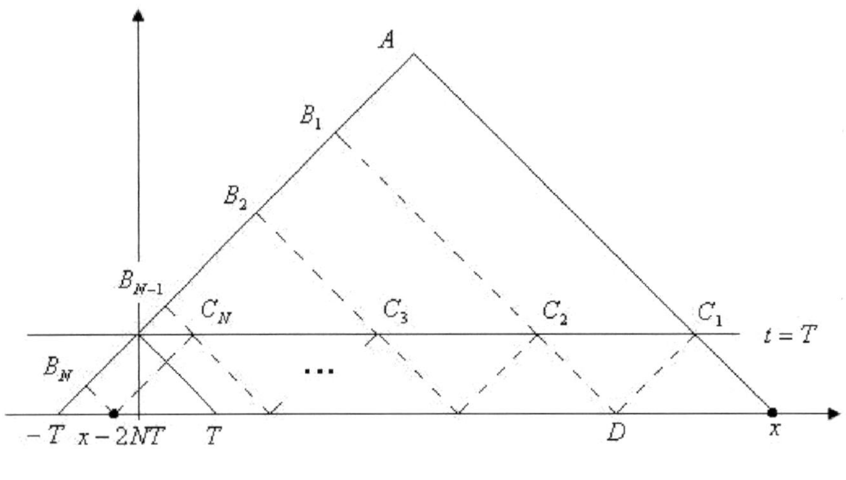

Figure 1. The method to construct the initial velocity $v(x)$ by characteristic-quadrilaterals

3. Wave equations in several space variables

In this section, we study the global exact controllability of wave equations in several space variables and prove Theorem 1.1.

Consider the TBVP (1.5)-(1.7), i.e., the following TBVP

$$\begin{cases} \Box y(t,x) = 0, \\ y(0,x) = f(x), \\ y(T,x) = g(x), \end{cases} \tag{3.1}$$

where $(t,x) \in \mathbb{R}^+ \times \mathbb{R}^n$, the symbol $\Box$ stands for the d'Alembert's operator, i.e.,

$$\Box = \partial_t^2 - \sum_{i=1}^{n} \partial_i^2 = \partial_t^2 - \Delta_x,$$

and $f(x), g(x) \in C^{[n/2]+2}(\mathbb{R}^n)$ are two given functions defined on $\mathbb{R}^n$. In (3.1), without loss of generality, we assume that the propagation speed c of wave is 1. Thus, in order to prove Theorem 1.1, it suffice to show the following theorem.

Theorem 3.1. *The TBVP (3.1) has a C^2-smooth solution $y = y(t,x)$ defined on the strip $[0,T] \times \mathbb{R}^n$.*

Remark 3.1. *Theorem 3.1 implies the global exact controllability of wave equations in several space variables. Similar result is true for the following inhomogeneous wave equations*

$$\Box y(t,x) = F(t,x).$$

We next give the proof of Theorem 3.1 (equivalently, Theorem 1.1).

Proof. We distinguish two cases to prove Theorem 3.1.

Case A: $n = 2k+1$ $(k \in \mathbb{N})$

Define the spherical mean of any given function $h(x)$ defined on $\mathbb{R}^n$:

$$A_r h(x) = \frac{1}{\omega_{n-1}} \int_{S^{n-1}} h(x+ry) d\sigma(y), \tag{3.2}$$

where ω_{n-1} denotes the area of the unit sphere $S^{n-1} \subset \mathbb{R}^n$ and $d\sigma(y)$ stands for the Lebesgue measure on the unit sphere S^{n-1}. Let

$$w(t,r) = \left(\frac{1}{r}\frac{\partial}{\partial r}\right)^{k-1} \left(r^{2k-1}\mathcal{A}_r y(t,x)\right). \tag{3.3}$$

It is easy to check that the function $w(t,r)$ satisfies the following TBVP

$$\begin{cases} w_{tt} - w_{rr} = 0, \\ w(0,r) = \left(\frac{1}{r}\frac{\partial}{\partial r}\right)^{k-1} \left(r^{2k-1}\mathcal{A}_r f(x)\right), \\ w(T,r) = \left(\frac{1}{r}\frac{\partial}{\partial r}\right)^{k-1} \left(r^{2k-1}\mathcal{A}_r g(x)\right). \end{cases} \tag{3.4}$$

Thus, by Theorem 2.1, the TBVP (3.4) has a C^2-smooth solution defined on the domain $[0,T] \times \mathbb{R}$.

Denote

$$c_0 = \prod_{i=1}^{k} (2i-1). \tag{3.5}$$

Then we have

$$y(t,x) = \lim_{r\to 0} \mathcal{A}_r y(t,x) = \lim_{r\to 0} \frac{1}{c_0 r} w(t,r). \tag{3.6}$$

It is easy to verify that the function $y = y(t,x)$ defined by (3.6) is a C^2-smooth solution of the TBVP (3.1) on the strip $[0,T] \times \mathbb{R}^n$.

Case B: $n = 2k$ $(k \in \mathbb{N})$

By Hadamard's method of descent, we can also prove that the TBVP (3.1) has a C^2-smooth solution on the strip $[0,T] \times \mathbb{R}^n$. The main idea here is that if y solves a wave equation with n space variables, then it is also a solution of the corresponding wave equation with $n+1$ space variables, which happens to be independent of the last variable x_{n+1}. Here we omit the details. Thus the proof of Theorem 3.1 is completed. □

Remark 3.2. *Noting Remark 2.2, we know that the solution of the TBVP (3.1) does not possesses the uniqueness.*

4. Wave equation defined on a closed curve

This section concerns the exact controllability of the wave equation defined on a circle. In other words, in this section we prove Theorem 1.2 and then Theorem 1.3.

To do so, we need the following Lemma (see Kong [8]).

Lemma 4.1. *Suppose that $F(x)$ is a L-periodic C^2 function of $x \in \mathbb{R}$, and its derivative of third order, i.e., $F'''(x)$, is piecewise smooth. Suppose furthermore that the Fourier series of $F(x)$ is given by*

$$F(x) = \frac{1}{2}A_0 + \sum_{k=1}^{\infty} A_k \cos\left(\frac{2k\pi}{L}x\right) + \sum_{k=1}^{\infty} B_k \sin\left(\frac{2k\pi}{L}x\right), \tag{4.1}$$

where

$$A_0 = \frac{2}{L}\int_0^L f(x)dx, \quad A_k = \frac{2}{L}\int_0^L f(x)\cos\left(\frac{2k\pi}{L}x\right)dx, \quad B_k = \frac{2}{L}\int_0^L f(x)\sin\left(\frac{2k\pi}{L}x\right)dx.$$

Then the series $\sum_{k=1}^{\infty} |k^2 A_k|$ and $\sum_{k=1}^{\infty} |k^2 B_k|$ are convergent.

Proof. It follows from (4.1) that

$$\begin{cases} \sum_{k=1}^{\infty} B_k \sin\left(\frac{2k\pi}{L}x\right) = \frac{1}{2}(F(x) - F(-x)) \triangleq G(x), \\ \frac{1}{2}A_0 + \sum_{k=1}^{\infty} A_k \cos\left(\frac{2k\pi}{L}x\right) = \frac{1}{2}(F(x) + F(-x)) \triangleq H(x). \end{cases} \tag{4.2}$$

Since $G(x)$ is a L-periodic odd function, its derivative of third order, i.e., $G'''(x)$, is a L-periodic piecewise continuous even function. So the form of the Fourier series of $G'''(x)$ should be

$$G'''(x) = \frac{1}{2}B_0^{(3)} + \sum_{k=1}^{\infty} B_k^{(3)} \cos\left(\frac{2k\pi}{L}x\right), \tag{4.3}$$

where $B_0^{(3)}$ and $B_k^{(3)}$ $(k = 1, 2, \cdots)$ stand for the coefficients of the Fourier series. By Parseval inequality, we obtain

$$\frac{1}{2}\left(B_0^{(3)}\right)^2 + \sum_{k=1}^{\infty}\left(B_k^{(3)}\right)^2 = \frac{2}{L}\int_0^L \left[G'''(x)\right]^2 dx < \infty. \tag{4.4}$$

On the other hand,

$$\begin{aligned} B_k^{(3)} &= \frac{2}{L}\int_0^L G'''(x)\cos\left(\frac{2k\pi}{L}x\right)dx \\ &= \frac{2}{L}\left[G''(x)\cos\left(\frac{2k\pi}{L}x\right)\right]\Big|_0^L + \frac{2k\pi}{L}\frac{2}{L}\int_0^L G''(x)\sin\left(\frac{2k\pi}{L}x\right)dx \\ &= \frac{2k\pi}{L}\frac{2}{L}\left[G'(x)\sin\left(\frac{2k\pi}{L}x\right)\right]\Big|_0^L - \left(\frac{2k\pi}{L}\right)^2\frac{2}{L}\int_0^L G'(x)\cos\left(\frac{2k\pi}{L}x\right)dx \\ &= -\left(\frac{2k\pi}{L}\right)^2\frac{2}{L}\left[G(x)\cos\left(\frac{2k\pi}{L}x\right)\right]\Big|_0^L - \left(\frac{2k\pi}{L}\right)^3\frac{2}{L}\int_0^L G(x)\sin\left(\frac{2k\pi}{L}x\right)dx \\ &= -\left(\frac{2k\pi}{L}\right)^3 B_k. \end{aligned} \tag{4.5}$$

Here we have made use of the fact that $G(x)$ and $G''(x)$ are L-periodic odd functions. So

$$\sum_{k=1}^{\infty}\left(k^3 B_k\right)^2 = \left(\frac{L}{2\pi}\right)^6 \sum_{k=1}^{\infty}\left(B_k^{(3)}\right)^2 < \infty. \tag{4.6}$$

Then by Cauchy inequality, it yields

$$\sum_{k=1}^{\infty}\left|k^2 B_k\right| \le \sqrt{\sum_{k=1}^{\infty}(k^3 B_k)^2 \cdot \sum_{k=1}^{\infty}\frac{1}{k^2}} < \infty. \tag{4.7}$$

Similarly, we can show that $\sum_{k=1}^{\infty}\left|k^2 A_k\right|$ is convergent.

Thus, the proof of Lemma 4.1 is completed. □

Remark 4.1. *Obviously, it follows from Lemma 4.1 that the series*

$$\sum_{k=1}^{\infty} |A_k|, \quad \sum_{k=1}^{\infty} |B_k|, \quad \sum_{k=1}^{\infty} |kA_k|, \quad \sum_{k=1}^{\infty} |kB_k|$$

are convergent.

We now prove Theorem 1.2.

Proof. Suppose that the Fourier series of the solution $y = y(t, \theta)$ to (1.10) reads

$$y(t,\theta) = \frac{1}{2}a_0(t) + \sum_{k=1}^{\infty} a_k(t) \cos\left(\frac{2k\pi}{L}\theta\right) + \sum_{k=1}^{\infty} b_k(t) \sin\left(\frac{2k\pi}{L}\theta\right), \tag{4.8}$$

where $a_0(t)$, $a_k(t)$, $b_k(t)$ stand for the coefficients of the Fourier series. Then by the first equation in (1.10), we obtain

$$\begin{cases} a_0''(t) = 0, \\ a_k''(t) + \left(\frac{2k\pi}{L}\right)^2 a_k(t) = 0, \\ b_k''(t) + \left(\frac{2k\pi}{L}\right)^2 b_k(t) = 0. \end{cases} \tag{4.9}$$

In particular,

$$\begin{cases} f(\theta) = \frac{1}{2}a_0(0) + \sum_{k=1}^{\infty} a_k(0) \cos\left(\frac{2k\pi}{L}\theta\right) + \sum_{k=1}^{\infty} b_k(0) \sin\left(\frac{2k\pi}{L}\theta\right), \\ g(\theta) = \frac{1}{2}a_0(T) + \sum_{k=1}^{\infty} a_k(T) \cos\left(\frac{2k\pi}{L}\theta\right) + \sum_{k=1}^{\infty} b_k(T) \sin\left(\frac{2k\pi}{L}\theta\right), \end{cases} \tag{4.10}$$

in which

$$\begin{cases} a_0(0) = \frac{2}{L}\int_0^L f(x)dx, \\ a_0(T) = \frac{2}{L}\int_0^L g(x)dx, \end{cases} \tag{4.11}$$

$$\begin{cases} a_k(0) = \frac{2}{L}\int_0^L f(x) \cos\left(\frac{2k\pi}{L}x\right)dx, \\ a_k(T) = \frac{2}{L}\int_0^L g(x) \cos\left(\frac{2k\pi}{L}x\right)dx, \end{cases} \tag{4.12}$$

and

$$\begin{cases} b_k(0) = \frac{2}{L}\int_0^L f(x) \sin\left(\frac{2k\pi}{L}x\right)dx, \\ b_k(T) = \frac{2}{L}\int_0^L g(x) \sin\left(\frac{2k\pi}{L}x\right)dx. \end{cases} \tag{4.13}$$

It follows from (4.9) that

$$\begin{cases} a_0(t) = \alpha_0 + \beta_0 t, \\ a_k(t) = \alpha_k \cos\left(\frac{2k\pi}{L}t\right) + \beta_k \sin\left(\frac{2k\pi}{L}t\right), \\ b_k(t) = \bar{\alpha}_k \cos\left(\frac{2k\pi}{L}t\right) + \bar{\beta}_k \sin\left(\frac{2k\pi}{L}t\right), \end{cases} \tag{4.14}$$

where α_0, β_0, α_k, β_k, $\bar{\alpha}_k$ and $\bar{\beta}_k$ are some constants. Comparing (4.11)-(4.13) with (4.14) gives

$$\begin{cases} \alpha_0 = \dfrac{2}{L}\displaystyle\int_0^L f(x)dx, \\ \alpha_0 + \beta_0 T = \dfrac{2}{L}\displaystyle\int_0^L g(x)dx, \end{cases} \tag{4.15}$$

$$\begin{cases} \alpha_k = \dfrac{2}{L}\displaystyle\int_0^L f(x)\cos\left(\frac{2k\pi}{L}x\right)dx, \\ \alpha_k \cos\left(\dfrac{2k\pi}{L}T\right) + \beta_k \sin\left(\dfrac{2k\pi}{L}T\right) = \dfrac{2}{L}\displaystyle\int_0^L g(x)\cos\left(\frac{2k\pi}{L}x\right)dx \end{cases} \tag{4.16}$$

and

$$\begin{cases} \bar{\alpha}_k = \dfrac{2}{L}\displaystyle\int_0^L f(x)\sin\left(\frac{2k\pi}{L}x\right)dx, \\ \bar{\alpha}_k \cos\left(\dfrac{2k\pi}{L}T\right) + \bar{\beta}_k \sin\left(\dfrac{2k\pi}{L}T\right) = \dfrac{2}{L}\displaystyle\int_0^L g(x)\sin\left(\frac{2k\pi}{L}x\right)dx. \end{cases} \tag{4.17}$$

That is,

$$\begin{cases} \alpha_0 = \dfrac{2}{L}\displaystyle\int_0^L f(x)dx, \\ \beta_0 = \dfrac{2}{LT}\displaystyle\int_0^L [g(x) - f(x)]dx, \end{cases} \tag{4.18}$$

$$\begin{cases} \alpha_k = \dfrac{2}{L}\displaystyle\int_0^L f(x)\cos\left(\frac{2k\pi}{L}x\right)dx, \\ \beta_k = \begin{cases} 0, & \text{if } \dfrac{2kT}{L} \in \mathbb{N}, \\ \dfrac{2}{L\sin\left(\frac{2k\pi}{L}T\right)}\displaystyle\int_0^L \left[g(x) - f(x)\cos\left(\frac{2k\pi}{L}T\right)\right]\cos\left(\frac{2k\pi}{L}x\right)dx, & \text{if } \dfrac{2kT}{L} \notin \mathbb{N} \end{cases} \end{cases} \tag{4.19}$$

and

$$\begin{cases} \bar{\alpha}_k = \dfrac{2}{L}\displaystyle\int_0^L f(x)\sin\left(\frac{2k\pi}{L}x\right)dx, \\ \bar{\beta}_k = \begin{cases} 0, & \text{if } \dfrac{2kT}{L} \in \mathbb{N}, \\ \dfrac{2}{L\sin\left(\frac{2k\pi}{L}T\right)}\displaystyle\int_0^L \left[g(x) - f(x)\cos\left(\frac{2k\pi}{L}T\right)\right]\sin\left(\frac{2k\pi}{L}x\right)dx, & \text{if } \dfrac{2kT}{L} \notin \mathbb{N}. \end{cases} \end{cases} \tag{4.20}$$

Here we would like to point out that, when $\dfrac{2kT}{L} \in \mathbb{N}$, i.e., $\sin\left(\dfrac{2k\pi}{L}T\right) = 0$, in (4.19)-(4.20) we take the corresponding coefficients β_k and $\bar{\beta}_k$ as zero. In fact, we may also take β_k and $\bar{\beta}_k$ as some series satisfying

$$\sum_{k=1}^{\infty} |k^2\beta_k| < \infty, \quad \sum_{k=1}^{\infty} |k^2\bar{\beta}_k| < \infty. \tag{4.21}$$

The different choice of β_k and $\bar{\beta}_k$ can give the different solution. This implies that the solution of the TBVP under consideration is not unique.

Since $\dfrac{T}{L}$ is a rational number and $\dfrac{2T}{L} \notin \mathbb{N}$, $\dfrac{2T}{L}$ can be expressed as a fraction $\dfrac{p}{q}$, where p and q are two irreducible integers and q is not less than 2, i.e., $q \geq 2$. By the property of sinusoid, we have

$$\left|\sin\left(\frac{2k\pi}{L}T\right)\right| = \left|\sin\left(\frac{kp\pi}{q}\right)\right| \geq \left|\sin\left(\frac{\pi}{q}\right)\right| \triangleq C_s, \quad \text{if } k \text{ is not the multiple of } q. \tag{4.22}$$

So for the given T and L, C_s is a constant. Then it follows from (4.14), (4.19), (4.21) and Remark 4.1 that

$$\begin{aligned}\sum_{k=1}^{\infty}|a_k(t)| &\leq \sum_{k=1}^{\infty}(|\alpha_k| + |\beta_k|)\\ &\leq \left(1+\frac{1}{C_s}\right)\sum_{k=1}^{\infty}\left|\frac{2}{L}\int_0^L f(x)\cos\left(\frac{2k\pi}{L}x\right)dx\right| + \frac{1}{C_s}\sum_{k=1}^{\infty}\left|\frac{2}{L}\int_0^L g(x)\cos\left(\frac{2k\pi}{L}x\right)dx\right|\\ &< \infty.\end{aligned} \tag{4.23}$$

Similarly, we obtain from (4.14), (4.20), (4.21) and Remark 4.1 that

$$\sum_{k=1}^{\infty}|b_k(t)| < \infty. \tag{4.24}$$

Combining (4.23), (4.24) and (4.15) gives the convergence of the Fourier series (4.8) of the solution $y = y(t, \theta)$ immediately.

Moreover, by Lemma 4.1, Remark 4.1 and (4.21) we obtain

$$\sum_{k=1}^{\infty}|k\alpha_k| < \infty, \quad \sum_{k=1}^{\infty}|k\beta_k| < \infty, \quad \sum_{k=1}^{\infty}|k\bar{\alpha}_k| < \infty, \quad \sum_{k=1}^{\infty}|k\bar{\beta}_k| < \infty \tag{4.25}$$

and

$$\sum_{k=1}^{\infty}|k^2\alpha_k| < \infty, \quad \sum_{k=1}^{\infty}|k^2\beta_k| < \infty, \quad \sum_{k=1}^{\infty}|k^2\bar{\alpha}_k| < \infty, \quad \sum_{k=1}^{\infty}|k^2\bar{\beta}_k| < \infty. \tag{4.26}$$

Using (4.25) and (4.26), by a similar argument as used above, we can prove

$$\sum_{k=1}^{\infty}|ka_k(t)| < \infty, \quad \sum_{k=1}^{\infty}|kb_k(t)| < \infty, \quad \sum_{k=1}^{\infty}|a_k'(t)| < \infty, \quad \sum_{k=1}^{\infty}|b_k'(t)| < \infty \tag{4.27}$$

and

$$\begin{cases}\displaystyle\sum_{k=1}^{\infty}|k^2a_k(t)| < \infty, \quad \sum_{k=1}^{\infty}|k^2b_k(t)| < \infty, \quad \sum_{k=1}^{\infty}|ka_k'(t)| < \infty, \quad \sum_{k=1}^{\infty}|kb_k'(t)| < \infty,\\ \displaystyle\sum_{k=1}^{\infty}|a_k''(t)| < \infty, \quad \sum_{k=1}^{\infty}|b_k''(t)| < \infty.\end{cases} \tag{4.28}$$

Obviously, (4.27) and (4.28) imply that the Fourier series of the first-order derivatives of y (i.e., y_t and y_θ) and second-order derivatives of y (i.e., y_{tt}, $y_{t\theta}$ and $y_{\theta\theta}$) are also convergent, respectively.

Thus, the proof of Theorem 1.2 is completed. □

Journal of Physics: Conference Series **290** (2011) 012008 doi:10.1088/1742-6596/290/1/012008

Remark 4.2. *By (4.10), (4.14), (4.18), the first equation in (4.19) and the first equation in (4.20), if* $\frac{2T}{L} \in \mathbb{N}$*, then* $g(\theta)$ *satisfies*

$$\begin{aligned} g(\theta) &= \frac{1}{2}(\alpha_0 + \beta_0 T) + \sum_{k=1}^{\infty} \alpha_k \cos\left(\frac{2k\pi T}{L}\right) \cos\left(\frac{2k\pi}{L}\theta\right) + \sum_{k=1}^{\infty} \bar{\alpha}_k \cos\left(\frac{2k\pi T}{L}\right) \sin\left(\frac{2k\pi}{L}\theta\right) \\ &= \frac{1}{L}\int_0^L g(x)dx + \sum_{k=1}^{\infty} \frac{2}{L}\int_0^L f(x)\cos\left(\frac{2k\pi}{L}x\right)dx \cos\left(\frac{2k\pi}{L}\theta\right)\cos\left(\frac{2k\pi T}{L}\right) + \\ &\quad \sum_{k=1}^{\infty} \frac{2}{L}\int_0^L f(x)\sin\left(\frac{2k\pi}{L}x\right)dx \sin\left(\frac{2k\pi}{L}\theta\right)\cos\left(\frac{2k\pi T}{L}\right). \end{aligned} \tag{4.29}$$

Then in this case, for any given $f(\theta)$*, the terminal value* $g(\theta)$ *can not be arbitrary. In particular, if* $\frac{T}{L} \in \mathbb{N}$*, then it should hold that*

$$g(\theta) = \frac{1}{L}\int_0^L [g(x) - f(x)]dx + f(\theta). \tag{4.30}$$

Remark 4.3. *If* $\frac{T}{L}$ *is a irrational number, then (4.22) is incorrect, and then we can not get the convergence in (4.23), etc.*

Remark 4.4. *Theorem 1.2 shows that, for some* T *satisfied the assumptions in Theorem 1.2, there exist some initial velocity* $v(\theta)$ *such that the Cauchy problem for the wave equation in (1.10) with the initial data*

$$t = 0: \quad y = f(\theta), \quad y_t = v(\theta) \tag{4.31}$$

has a solution $y = y(t,\theta) \in C^2([0,T] \times \mathbb{R})$ *which satisfies the terminal condition in (1.10), i.e., the third equation in (1.10).*

Theorem 1.3 comes from Theorem 1.2 directly.

5. Wave equation with homogeneous Dirichlet or Neumann boundary conditions

In this section, we investigate the TBVP for the wave equation defined on the domain $[0,T] \times [0,L]$ with homogeneous Dirichlet boundary conditions and homogeneous Neumann boundary conditions, respectively, where $T,\ L > 0$ are two given real numbers.

More precisely, we consider the following problem for the wave equation with the homogeneous Dirichlet boundary conditions

$$\begin{cases} y_{tt} - y_{xx} = 0, \\ y(0,x) = f(x), \\ y(T,x) = g(x), \\ y(t,0) = y(t,L) = 0, \end{cases} \tag{5.1}$$

and the problem for the wave equation with the homogeneous Neumann boundary conditions

$$\begin{cases} y_{tt} - y_{xx} = 0, \\ y(0,x) = f(x), \\ y(T,x) = g(x), \\ y_x(t,0) = y_x(t,L) = 0, \end{cases} \tag{5.2}$$

where $y = y(t,x)$ is the unknown function of $(t,x) \in [0,T] \times [0,L]$, $f(x)$ and $g(x)$ are two given C^3 functions of $x \in [0,L]$. Moreover, f and g satisfy the compatibility conditions

$$f(x)\Big|_{x=L}^{x=0} = g(x)\Big|_{x=L}^{x=0} = 0, \quad f''(x)\Big|_{x=L}^{x=0} = g''(x)\Big|_{x=L}^{x=0} = 0, \quad \text{for the Dirichlet conditions;} \tag{5.3}$$

$$f'(x)\Big|_{x=L}^{x=0} = g'(x)\Big|_{x=L}^{x=0} = 0, \quad \text{for the Neumann conditions.} \tag{5.4}$$

We have the following theorem which can be viewed as consequences of Theorem 1.2.

Theorem 5.1. *(A) Suppose that the compatibility conditions in (5.3) are satisfied, $\frac{T}{L}$ is a rational number and $\frac{T}{L} \notin \mathbb{N}$. Then the problem (5.1) admits a C^2 solution $y = y(t,x)$ on the domain $[0,T] \times [0,L]$.*

(B) Suppose that the compatibility conditions in (5.4) are satisfied, $\frac{T}{L}$ is a rational number and $\frac{T}{L} \notin \mathbb{N}$. Then the problem (5.2) admits a C^2 solution $y = y(t,x)$ on the domain $[0,T] \times [0,L]$.

Proof. For the case of the Dirichlet boundary conditions, we extend any C^2 solution $y = y(t,x)$ to the domain $[0,T] \times [-L,L]$ by

$$y(t,x) = -y(t,-x), \quad \text{for } x \in [-L,0], \tag{5.5}$$

and then extend $y = y(t,x)$ to be $2L$-periodic. See [10] and [11]. One easily checks that if the given initial/terminal data have the form in (5.1), the extended initial/terminal data are $2L$-periodic and given by (see [10]-[11])

$$\tilde{f}(x) \triangleq y(0,x) = \begin{cases} -f(-x), & \text{for } x \in [-L,0], \\ f(x), & \text{for } x \in [0,L], \end{cases} \tag{5.6}$$

$$\tilde{g}(x) \triangleq y(T,x) = \begin{cases} -g(-x), & \text{for } x \in [-L,0], \\ g(x), & \text{for } x \in [0,L], \end{cases} \tag{5.7}$$

respectively. Thus, we obtain an extended TBVP for the wave equation defined on the strip $[0,T] \times \mathbb{R}$. When the compatibility conditions in (5.3) are satisfied, this extended $y = y(t,x)$ is a C^2 solution of the extended TBVP with the extended initial/terminal data $\big(\tilde{f}(x), \tilde{g}(x)\big)$. Therefore, we may make use of Theorem 1.2 and obtain the solution, denoted by $\tilde{y} = \tilde{y}(t,x)$, to the extended TBVP defined on the strip $[0,T] \times \mathbb{R}$. Obviously, $\tilde{y} = \tilde{y}(t,x)$ is a $2L$-periodic odd C^2 function. So the Dirichlet boundary conditions are satisfied naturally. Let $y = y(t,x)$ be the restriction of $\tilde{y} = \tilde{y}(t,x)$ on the region $[0,T] \times [0,L]$. It is easy to see that $y = y(t,x)$ is the C^2 solution to (5.1).

Similarly, for the case of Neumann boundary conditions, we can extend $y = y(t,x)$ by

$$y(t,x) = y(t,-x), \quad \text{for } x \in [-L,0]. \tag{5.8}$$

See [10]-[11]. Then, $y = y(t,x)$ can be extended to be a classical $2L$-periodic C^2 solution of the extended TBVP for the wave equation with $2L$-periodic initial/terminal data given by

$$\tilde{f}(x) \triangleq y(0,x) = \begin{cases} f(-x), & \text{for } x \in [-L,0], \\ f(x), & \text{for } x \in [0,L], \end{cases} \tag{5.9}$$

$$\tilde{g}(x) \triangleq y(T,x) = \begin{cases} g(-x), & \text{for } x \in [-L,0], \\ g(x), & \text{for } x \in [0,L], \end{cases} \tag{5.10}$$

respectively. When the compatibility conditions in (5.4) are satisfied, by a similar argument as used above, we can prove the part (B) in Theorem 5.1.

Thus, the proof of Theorem 5.1 is completed. □

Remark 5.1. *Theorem 5.1 shows that the wave equation defined on the domain* $[0,T] \times [0,L]$ *still possesses the exact controllability in the cases of homogeneous Dirichlet boundary conditions and homogeneous Neumann boundary conditions, provided that the corresponding compatibility conditions are satisfied.*

6. Wave equation with inhomogeneous Dirichlet or Neumann boundary conditions

This section is devoted to the exact controllability of the wave equation defined on the domain $[0,T] \times [0,L]$ with inhomogeneous Dirichlet boundary conditions and inhomogeneous Neumann boundary conditions, respectively, where T, L are two given positive real numbers.

More precisely, we consider the following TBVP for the wave equation with the inhomogeneous Dirichlet boundary conditions

$$\begin{cases} y_{tt} - y_{xx} = 0, \\ y(0,x) = f(x), \\ y(T,x) = g(x), \\ y(t,0) = h(t), \\ y(t,L) = l(t), \end{cases} \tag{6.1}$$

and the TBVP for the wave equation with the inhomogeneous Neumann boundary conditions

$$\begin{cases} y_{tt} - y_{xx} = 0, \\ y(0,x) = f(x), \\ y(T,x) = g(x), \\ y_x(t,0) = H(t), \\ y_x(t,L) = K(t), \end{cases} \tag{6.2}$$

where $y = y(t,x)$ is the unknown function of $(t,x) \in [0,T] \times [0,L]$, $f(x), g(x)$ are two given C^3 functions of $x \in [0,L]$ which stands for the initial data and terminal data, respectively, $h(t)$, $l(t)$ are two given C^3 functions of $t \in [0,T]$, and $H(t)$, $K(t)$ are two given C^2 functions of $t \in [0,T]$. Moreover, we assume that f, g, h, l, H and K satisfy the following compatibility conditions

$$\begin{cases} f(0) = h(0), & f''(0) = h''(0), \\ f(L) = l(0), & f''(L) = l''(0), \\ g(0) = h(T), & g''(0) = h''(T), \\ g(L) = l(T), & g''(L) = l''(T), \end{cases} \quad \text{for the Dirichlet conditions;} \tag{6.3}$$

$$\begin{cases} f'(0) = H(0), \\ f'(L) = K(0), \\ g'(0) = H(T), \\ g'(L) = K(T), \end{cases} \quad \text{for the Neumann conditions.} \tag{6.4}$$

We have

Theorem 6.1. *(A) Suppose that the compatibility conditions in (6.3) are satisfied,* $\frac{T}{L}$ *is a rational number and* $\frac{T}{L} \notin \mathbb{N}$*. Then the problem (6.1) admits a* C^2 *solution* $y = y(t,x)$ *on the domain* $[0,T] \times [0,L]$*.*

(B) Suppose that the compatibility conditions in (6.4) are satisfied, $\frac{T}{L}$ *is a rational number and* $T < L$*. Then the problem (6.2) admits a* C^2 *solution* $y = y(t,x)$ *on the domain* $[0,T] \times [0,L]$*.*

Proof. For the problem (6.1), noting that $\frac{T}{L} \notin \mathbb{N}$, without loss of generality, we may assume that $T < L$ (otherwise, we only need exchange the t-axes and the x-axes and then reduce the problem to the case of $T < L$). In this situation, we can extend $h(t)$ to the interval $[-T-L, T+L]$ by

$$h(t+L) + h(t-L) = 2l(t), \quad \text{for } t \in [0,T]. \tag{6.5}$$

Moreover we require that the extended $h(t)$ is a C^3 function on the interval $[-T-L, T+L]$.

Introduce

$$\tilde{y}(t,x) = y(t,x) - \frac{h(t+x) + h(t-x)}{2}. \tag{6.6}$$

Then by (6.5), the problem (6.1) becomes

$$\begin{cases} \tilde{y}_{tt} - \tilde{y}_{xx} = 0, \\ \tilde{f}(x) \triangleq \tilde{y}(0,x) = f(x) - \dfrac{h(x) + h(-x)}{2}, \\ \tilde{g}(x) \triangleq \tilde{y}(T,x) = g(x) - \dfrac{h(T+x) + h(T-x)}{2}, \\ \tilde{h}(t) \triangleq \tilde{y}(t,0) = 0, \\ \tilde{l}(t) \triangleq \tilde{y}(t,L) = 0. \end{cases} \tag{6.7}$$

And by (6.3), (6.5), the boundary data $\tilde{f}$, $\tilde{g}$, $\tilde{h}$ and $\tilde{l}$ satisfy the compatibility conditions

$$\begin{cases} \tilde{f}(0) = \tilde{f}(L) = 0, & \tilde{f}''(0) = \tilde{f}''(L) = 0, \\ \tilde{g}(0) = \tilde{g}(L) = 0, & \tilde{g}''(0) = \tilde{g}''(L) = 0. \end{cases} \tag{6.8}$$

Therefore, we can make use of Theorem 5.1 (A) and obtain the C^2 solution to the problem (6.7), and then the C^2 solution to (6.1). This proves the part (A) in Theorem 6.1.

Similarly, for the problem (6.2), noting $T < L$, we can extend $H(t)$ to the interval $[-L, T+L]$ by

$$H(t+L) + H(t-L) = 2K(t), \quad \text{for } t \in [0,T]. \tag{6.9}$$

As before, we require that the extended $H(t)$ is a C^2 function on $[-L, T+L]$.

Let

$$\tilde{y}(t,x) = y(t,x) - \frac{1}{2}\int_{t-x}^{t+x} H(\xi)d\xi. \tag{6.10}$$

Then by (6.9), the problem (6.2) becomes

$$\begin{cases} \tilde{y}_{tt} - \tilde{y}_{xx} = 0, \\ \tilde{f}(x) \triangleq \tilde{y}(0,x) = f(x) - \dfrac{1}{2}\displaystyle\int_{-x}^{x} H(\xi)d\xi, \\ \tilde{g}(x) \triangleq \tilde{y}(T,x) = g(x) - \dfrac{1}{2}\displaystyle\int_{T-x}^{T+x} H(\xi)d\xi, \\ \tilde{H}(t) \triangleq \tilde{y}_x(t,0) = 0, \\ \tilde{K}(t) \triangleq \tilde{y}_x(t,L) = 0. \end{cases} \tag{6.11}$$

And by (6.4), (6.9), the boundary data $\tilde{f}$, $\tilde{g}$, $\tilde{H}$ and $\tilde{K}$ satisfy the compatibility conditions

$$\begin{cases} \tilde{f}'(0) = \tilde{f}'(L) = 0, \\ \tilde{g}'(0) = \tilde{g}'(L) = 0. \end{cases} \tag{6.12}$$

Therefore, we can make use of Theorem 5.1 (B) and obtain the C^2 solution to the problem (6.11), and then the C^2 solution to (6.2). This proves the part (B) in Theorem 6.1.

Thus, the proof of Theorem 6.1 is completed. □

Remark 6.1. *Theorem 6.1 shows that the wave equation defined on the domain $[0, T] \times [0, L]$ still possesses the exact controllability even in the cases of inhomogeneous Dirichlet boundary conditions and inhomogeneous Neumann boundary conditions, provided that the corresponding compatibility conditions are satisfied.*

7. Some nonlinear wave equations

This section is devoted to the global exact controllability for some nonlinear wave equations including a wave map equation arising from geometry and the equations for the motion of relativistic strings in the Minkowski space-time $\mathbb{R}^{1+n}$.

7.1. A wave map equation

The theory of wave maps plays an important role in both mathematics and theoretical physics. The wave map equation is highly geometrical, and can be rewritten in many different ways. It is also related to the Einstein equations in general relativity. In this subsection, we consider the following TBVP for a kind of wave map equation[2]

$$\begin{cases} y_{tt} - y_{xx} = y_t^2 - y_x^2, \\ y(0, x) = f(x), \\ y(T, x) = g(x), \end{cases} \tag{7.1}$$

where T is a given positive real number, and $f(x), g(x)$ are two given C^2-smooth functions which stand for the initial and terminal states, respectively.

By making the following transformation on the unknown

$$z(t, x) = e^{-y(t,x)}, \tag{7.2}$$

the TBVP (7.1) can be rewritten as

$$\begin{cases} z_{tt} - z_{xx} = 0, \\ z(0, x) = e^{-f(x)} \triangleq \hat{f}(x), \\ z(T, x) = e^{-g(x)} \triangleq \hat{g}(x), \end{cases} \tag{7.3}$$

Obviously, the TBVP (7.1) has a C^2-smooth solution on the strip $[0, T] \times \mathbb{R}$ if and only if the TBVP (7.3) has a C^2-smooth positive solution on $[0, T] \times \mathbb{R}$. Therefore, in order to prove the existence of a C^2-smooth solution of the TBVP (7.1) on $[0, T] \times \mathbb{R}$, it suffices to show the existence of a C^2-smooth positive solution of the TBVP (7.3) on this domain. In fact, if so, $y(t, x) = -\ln z(t, x)$ is a C^2-smooth solution to the TBVP (7.1).

[2] In fact, the equation in (7.1) is nothing but the wave map from the Minkowski space $\mathbb{R}^{1+1}$ to the Riemmannian manifold $(\mathbb{R}, g)$, where $ds^2 = g dy^2 \triangleq e^{-2y} dy^2$.

To do so, we suppose that

$$\inf f(x) > \sup g(x). \tag{7.4}$$

Similar to (2.2) and (2.3), we introduce

$$\tilde{z}(t,x) = z(t,x) - \frac{\hat{f}(x-t) + \hat{f}(x+t)}{2} \tag{7.5}$$

and

$$\begin{aligned} \tilde{f}(x) &= \hat{g}(x) - \frac{\hat{f}(x-T) + \hat{f}(x+T)}{2} \\ &= e^{-g(x)} - \frac{1}{2}\left[e^{-f(x-T)} + e^{-f(x+T)}\right] > 0. \end{aligned} \tag{7.6}$$

In (7.6), we have made use of the assumption (7.4). Thus, the TBVP (7.3) can be rewritten as

$$\begin{cases} \tilde{z}_{tt} - \tilde{z}_{xx} = 0, \\ \tilde{z}(0,x) = 0, \\ \tilde{z}(T,x) = \tilde{f}(x). \end{cases} \tag{7.7}$$

As shown in Section 2, we may construct a non-negative C^1 function $u(x)$ defined on the interval $[-T,T]$ which satisfies the condition (2.8), and define an initial velocity $v(x)$ as shown in (2.9). By d'Alembert formula, the solution of the Cauchy problem for the wave equation in (7.7) with the initial data

$$\tilde{z}(0,x) = 0, \quad \tilde{z}_t(0,x) = v(x) \tag{7.8}$$

reads

$$\tilde{z}(t,x) = \frac{1}{2}\int_{x-t}^{x+t} v(\tau)d\tau. \tag{7.9}$$

It follows from (7.9), (7.5) and the fact that $\hat{f}(x) > 0$ that the TBVP (7.3) has a C^2-smooth positive solution $z = z(t,x)$, provided that $v(x) \geq 0$.

By the above argument, the key point to show the existence of a C^2-smooth positive solution of the TBVP (7.3) on $[0,T] \times \mathbb{R}$ is to construct a non-negative initial velocity $v(x)$. By (2.9), we have

Lemma 7.1. *Suppose that the function $\tilde{f}(x)$ defined by (7.6) satisfies*

$$\begin{cases} \sum_{i=1}^{N} \tilde{f}'(x-(2i-1)T) \geq 0, & \forall\, x > T, \\ \sum_{i=1}^{N} \tilde{f}'(x+(2i-1)T) \leq 0, & \forall\, x < -T. \end{cases} \tag{7.10}$$

Then the function $v(x)$ defined by (2.9) is non-negative.

Lemma 7.1 is obvious, here we omit its proof.

Summarizing the above argument yields the following theorem.

Theorem 7.1. *Suppose that $f(x)$ and $g(x)$ are two C^2-smooth functions and satisfy (7.4) and (7.10). Then the TBVP (7.1) admits a C^2-smooth solution $y = y(t,x)$ defined on the strip $[0,T] \times \mathbb{R}$.*

In order to understand the condition (7.10) clearly, we present the following two examples.

Example 7.1. *Choose the functions $f(x)$ and $g(x)$ such that the function $\tilde{f}(x)$ defined by (7.6) satisfies*

$$\begin{cases} \tilde{f}'(x) \geq 0, & \forall\, x > 0, \\ \tilde{f}'(x) \leq 0, & \forall\, x < 0. \end{cases} \tag{7.11}$$

In this case, it is easy to see that $\tilde{f}(x)$ satisfies the assumption (7.10), and then the TBVP (7.1) has a C^2-smooth solution on the domain $[0,T]\times\mathbb{R}$. For example, we may choose $f(x)$ and $g(x)$ satisfying

$$\tilde{f}(x) = x^{2n} + c,$$

where $\tilde{f}(x)$ is defined by (7.6), c is a constant and n is a positive integer.

Example 7.2. *Choose the functions $f(x)$ and $g(x)$ such that the function $\tilde{f}(x)$ defined by (7.6) satisfies*

$$\begin{cases} \tilde{f}'(x) \geq 0, & \forall\, x \in [0, 2T], \\ \tilde{f}'(x) = -\tilde{f}'(x+2T). \end{cases} \tag{7.12}$$

In the present situation, we have

$$\sum_{i=1}^{N} \tilde{f}'(x-(2i-1)T) \begin{cases} \geq 0, \ if\ N\ is\ odd, \\ = 0, \ if\ N\ is\ even, \end{cases} \quad if\ \ x \geq 0 \tag{7.13}$$

and

$$\sum_{i=1}^{N} \tilde{f}'(x+(2i-1)T) \begin{cases} \leq 0, \ if\ N\ is\ odd, \\ = 0, \ if\ N\ is\ even, \end{cases} \quad if\ \ x < 0. \tag{7.14}$$

This implies that $\tilde{f}(x)$ satisfies the assumption (7.10), and then the TBVP (7.1) has a C^2-smooth solution on the domain $[0,T]\times\mathbb{R}$. As an example, we may choose $f(x)$ and $g(x)$ satisfying

$$\tilde{f}'(x) = \frac{\pi}{2T}\sin\frac{\pi x}{2T}, \quad i.e., \quad \tilde{f}(x) = -\cos\frac{\pi x}{2T} + c,$$

where $\tilde{f}(x)$ is defined by (7.6) and c is a constant.

7.2. The equations for the motion of relativistic strings in the Minkowski space-time $\mathbb{R}^{1+n}$

Let $X = (t, x_1, \cdots, x_n)$ be a position vector of a point in the $(1+n)$-dimensional Minkowski space $\mathbb{R}^{1+n}$. Consider the motion of a relativistic string and let $x_i = x_i(t,\theta)$ $(i = 1,\cdots,n)$ be the local equation of its world surface, where (t,θ) are the the surface parameters. The equations governing the motion of the string read (see [11])

$$|x_\theta|^2 x_{tt} - 2\langle x_t, x_\theta\rangle x_{t\theta} + (|x_t|^2 - 1)x_{\theta\theta} = 0. \tag{7.15}$$

The system (7.15) contains n nonlinear partial differential equations of second order. These equations also describe the extremal surfaces in the Minkowski space $\mathbb{R}^{1+n}$. Kong et al [11] considered the Cauchy problem for the equations (7.15) with the following initial data

$$t = 0: \ \ x = p(\theta), \quad x_t = q(\theta), \tag{7.16}$$

where p is a given C^2 vector-valued function and q is a given C^1 vector-valued function. The Cauchy problem (7.15)-(7.16) describes the motion of a free relativistic string in the Minkowski space $\mathbb{R}^{1+n}$ with the initial position $p(\theta)$ and initial velocity (in the classical sense) $q(\theta)$. In particular, when $p(\theta)$ and $q(\theta)$ are periodic, the string under consideration is a closed one. It has

Journal of Physics: Conference Series **290** (2011) 012008 doi:10.1088/1742-6596/290/1/012008

shown that the global smooth solution of the Cauchy problem (7.15)-(7.16) exists and is unique (see [11]). On the other hand, it is well known that closed form representations of solutions for partial differential equations are very important and fundamental in both mathematical analysis and physical understanding. Unfortunately, nonlinear partial differential equations in general do not possess representations of solutions in closed form. Surprisingly, in the paper [10], we discovered a general solution formula in closed form for the nonlinear wave equations (7.15). By introducing a new concept of *generalized time-periodic function*, we proved that, if the initial data is periodic, then the smooth solution of the Cauchy problem (7.15)-(7.16) is *generalized time-periodic*, namely, the space-periodicity also implies the time-periodicity. This fact yields an interesting physical phenomenon: *the motion of closed strings is always generalized time-periodic.*

However, up to now there is not any result on the TBVP for the equations (7.15). In this subsection we investigate this problem and prove the global exact controllability of the equations (7.15).

By [6], under a very natural assumption which is a necessary and sufficient condition guaranteeing the motion is physical, there exists a globally diffeomorphic transformation of variables

$$\tau = t, \quad \vartheta = \vartheta(t, \theta) \tag{7.17}$$

such that the system (7.15) become

$$\tilde{x}_{\tau\tau} - \tilde{x}_{\vartheta\vartheta} = 0, \tag{7.18}$$

where

$$\tilde{x}(\tau, \vartheta) = x\big(t(\tau, \vartheta), \theta(\tau, \vartheta)\big), \tag{7.19}$$

in which $t = t(\tau, \vartheta)$, $\theta = \theta(\tau, \vartheta)$ is the inverse of the transformation (7.17).

By Theorem 2.1, the system (7.18) possesses the global exact controllability. Noting the transformation (7.17) is globally diffeomorphic, we obtain the global exact controllability of the system (7.15).

In physics, the global exact controllability of the system (7.15) implies some interesting physical phenomena. For example, if we take the periodic initial data $p(\theta)$ which stands for a closed string, and a non-periodic terminal condition (e.g., $x(T, \theta) = \theta$) which denotes an open string, then the global exact controllability of the system (7.15) means that a closed string may become an open string. This fact implies that the topological structure of the string may change in its motion process.

8. Hyperbolic curvature flow and Gage-Hamilton's theorem

Let $\gamma(t)$ be a one parameter family of closed convex smooth curves in the plane. The position vector $X(t, s)$ parameterizes the curve and the curvature is $k(t, s)$. The hyperbolic curvature flow is described by the following wave equation

$$k_{tt} - k_{ss} = 0, \tag{8.1}$$

where the subscript ν denotes partial differentiation with respect ν.

Consider the TBVP for the wave equation (8.1) with the following initial condition and terminal condition

$$k(0, s) = f(s), \quad k(T, s) = k_0 > 0, \tag{8.2}$$

where $f(s)$ is a given non-negative periodic smooth function with the period L, T is a positive real number, and k_0 is a positive constant. We have

Theorem 8.1. *There exists a positive constant k_0 such that the TBVP (8.1)-(8.2) admits a non-negative smooth solution $k = k(t, s)$, provided that $\frac{T}{L}$ is a rational number with $\frac{2T}{L} \notin \mathbb{N}$.*

Remark 8.1. *By the basic theorem on plane curves, Theorem 8.1 implies that, under the hyperbolic curvature flow, any closed convex smooth curve in the plane can evolve into a circle in a finite time, and remains convex for sequent times. This is a result analogous to the well-know theorem of Gage and Hamilton for the mean curvature flow of plane curves.*

We now prove Theorem 8.1.

Proof. Consider the TBVP for the wave equation (8.1) with the following initial condition and terminal condition

$$k(0,s) = f(s), \quad k(T,s) = k_*, \tag{8.3}$$

where k_* is an arbitrary given positive constant. By Theorem 1.2, the TBVP (8.1), (8.3) has a global L-periodic solution $k = k(t,s)$. Define

$$v(s) = \partial_t k(0,s). \tag{8.4}$$

Clearly, $v(s)$ is the initial velocity corresponding to the solution $k = k(t,s)$. That is to say, the solution of the Cauchy problem for the wave equation (8.1) with the initial data $k(0,s) = f(s)$, $k_t(0,s) = v(s)$ is nothing but $k = k(t,s)$, moreover this solution satisfies the terminal data $k(T,s) = k_*$.

Notice that $v(s)$ is L-periodic. Let

$$M = \min_{s\in[0,L]} v(s). \tag{8.5}$$

If $M \geq 0$, then by D'Alembert formula

$$k(t,s) = \frac{f(s+t)+f(s-t)}{2} + \frac{1}{2}\int_{s-t}^{s+t} v(\tau)d\tau \geq 0, \quad \forall\ (t,s) \in [0,T]\times \mathbb{R}. \tag{8.6}$$

Taking $k_0 = k_*$, we finish the proof of Theorem 8.1.

Otherwise, it holds that $v(s) + |M| \geq 0$. Thus, it is easy to see that $\bar{k} \triangleq k(t,s) + |M|t$ is a global L-periodic smooth solution of the TBVP

$$\begin{cases} \bar{k}_{tt}(t,s) - \bar{k}_{ss}(t,x) = 0, \\ \bar{k}(0,s) = f(s), \\ \bar{k}(T,s) = k_* + |M|T. \end{cases} \tag{8.7}$$

Obviously, the initial velocity corresponding to the solution $\bar{k}(t,s)$ reads

$$\bar{v}(s) \triangleq \partial_t \bar{k}(0,s) = v(s) + |M| \geq 0. \tag{8.8}$$

This implies that the solution is non-negative, i.e., $\bar{k}(t,s) \geq 0$ for all $(t,s) \in [0,T]\times\mathbb{R}$. Taking $k_0 = k_* + |M|T$ gives the conclusion of Theorem 8.1. Thus, the proof of Theorem 8.1 is completed. □

9. Summary and discussions

In the present paper we introduce three new concepts for second-order hyperbolic equations, they read two-point boundary value problem, global exact controllability and exact controllability, respectively. These second-order hyperbolic equations considered here include many important partial differential equations arising from both theoretical aspects and applied fields, e.g., mechanics (fluid mechanics and elasticity), physics, engineering, control theory and geometry, etc., the typical examples are wave equation, hyperbolic Monge-Ampère equation, wave map.

For several kinds of important linear and nonlinear wave equations, in this paper we prove the existence of smooth solutions of the two-point boundary value problems and show the global exact controllability of these wave equations. In particular, we investigate the two-point boundary value problem for one-dimensional wave equation defined on a closed curve and prove the existence of smooth solution, this implies the exact controllability of this kind of wave equation. Furthermore, based on this, we study the two-point boundary value problems for the wave equation defined on a strip with Dirichlet or Neumann boundary conditions and show that the equation still possesses the exact controllability in these cases. Finally, as an application of Theorem 1.2, we introduce the hyperbolic curvature flow and prove a result analogous to the well-known theorem of Gage and Hamilton [5] for the curvature flow of plane curves. This result can be viewed as a simple application of hyperbolic partial differential equation to both geometry and topology.

Usually, "two-point boundary value problem" has another meaning: it applies to a boundary problem for a second order ODE in a bounded interval. The present paper generalizes this concept to the case of second-order hyperbolic partial differential equations. The two-point boundary value problem for partial differential equations is a new research topic. According to the authors' knowledge, up to now, few of results on the two-point boundary value problems for partial differential equations, in particular, hyperbolic partial differential equations (even for linear or nonlinear wave equations) have been known. Therefore, the present paper can be viewed as the first work in this new research topic.

It is well-known that, there are a lot of deep and beautiful results on the two-point boundary value problems for ordinary differential equations and for second-order differential inclusions. On the other hand, there are many important results on the *boundary* control problems for wave equations and hyperbolic systems. However, the two-point boundary value problems and the boundary control problems are essentially different two kinds of problems. Both of them play an important role in both theoretical and applied aspects.

The main aim of this paper is to introduce the concept "two-point boundary value problem" for second-order hyperbolic equations and to show the global exact controllability or exact controllability for several kinds of important linear and nonlinear wave equations. Although the results obtained in this paper are restrictive in some sense (they only apply to cases in which the solution to the Cauchy problem can be found explicitly: linear (classical) wave equations and their reformulations), these results shed further light on the study of this new research topic. In the future we will investigate the following open problems which seem to us more interesting and important: (i) what happens if we consider a general equation of the form

$$y_{tt} - y_{xx} = f(y)$$

with a regular and bounded function $f(y)$? (ii) what happens if we include regular coefficients and/or lower order terms? (iii) what happens if we consider a quasilinear wave equation of the form

$$y_{tt} - (\mathscr{C}(y_x))_x = 0$$

with a smooth and increasing function $\mathscr{C}(\nu)$, e.g., $\mathscr{C}(\nu) = \frac{\nu}{\sqrt{1+\nu^2}}$? (iv) what happens if we consider other models, particularly some nonlinear models arising from applied fields such as control theory, fluid dynamics, elasticity as well as engineering, etc.

Acknowledgements. This work was supported in part by the NNSF of China (Grant No. 10971190) and the Qiu-Shi Chair Professor Fellowship from Zhejiang University.

References

[1] G. Chen, "Energy decay estimates and exact boundary controllability of the wave equations in a bounded domain", *J. Math. Pures Appl.*, **58** (1979), 249-274.

[2] W. C. Chewning, "Controllability of the nonlinear wave equation in several space variables", *SIAM J. Control Optim.*, **14** (1976), 19-25.
[3] M. Cirinà, "Boundary Controllability of nonlinear hyperbolic systems", *SIAM J. Control*, **7** (1969), 198-212.
[4] H. O. Fattorini, "Local controllability of nonlinear wave equation", *Math. Systems Theory*, **L9** (1975), 363-366.
[5] M. Gage and R. S. Hamilton, "The heat equation shrinking convex plane curves", *J. Differential Geom.*, **23** (1986), 69-96.
[6] C.-L. He and D.-X. Kong, "The global existence of smooth solutions of relativistic string equations in the Schwarzschild space-time", to appear, arXiv:1002.1357v2.
[7] D.-X. Kong, "Global exact boundary controllability of a class of quasilinear hyperbolic systems of conservation laws", *Systems & Control Letters*, **47** (2002), 287-298.
[8] D.-X. Kong, "Partial Differential Equations", Higher Education Press, Beijing, 2010.
[9] D.-X. Kong and H. Yao, "Global exact boundary controllability of a class of quasilinear hyperbolic systems of conservation laws II", *SIAM J. Control Optim.*, **44** (2005), 140-158.
[10] D.-X. Kong and Q. Zhang, "Solution formula and time-periodicity for the motion of relativistic strings in the Minkowski space $\mathbb{R}^{1+n}$", *Physica D: Nonlinear Phenomena*, **238** (2009), 902-922.
[11] D.-X. Kong, Q. Zhang and Q. Zhou, "The dynamics of relativistic strings moving in the Minkowski space $\mathbb{R}^{1+n}$", *Communications in Mathematical Physics*, **269** (2007), 153-174.
[12] I. Lasiecka and R. Triggiani, "Exact controllability for the wave equation with Neumann boundary control", *Appl. Math. Optim.*, **19** (1989), 243-290.
[13] I. Lasiecka and R. Triggiani, "Exact controllability of semilinear abstract systems with applications to wave and plates boundary control problems", *Appl. Math. Optim.*, **23** (1991), 109-154.
[14] J. L. Lions, "Exact controllability, stabilization and perturbations for distributed systems", *SIAM Rev.*, **30** (1988), 1-68.
[15] D. L. Russell, "On boundary-value controllability of linear symmetric hyperbolic systems", *Proc. Conference on Mathematical Theory of Control at the University of Southern California*, Academic Press, New York, 1967, pp. 312-321.
[16] D. L. Russell, "Controllability and stabilizability theory for linear partial differential equations: Recent progress and open questions", *SIAM Rev.*, **20** (1978), 639-739.
[17] P. F. Yao, "On the observability inequalities for the exact controllability of the wave equation with variable coefficients", *SIAM J. Control Optim.*, **37** (1999), 1568-1599.
[18] P. F. Yao, "Boundary controllability for the quasilinear wave equation", *Appl. Math. Optim.*, to appear.
[19] E. Zuazua, "Exact controllability for the semilinear wave equation", *J. Math. Pures Appl.*, **69** (1990), 1-32.
[20] E. Zuazua, "Exact boundary controllability for the semilinear wave equation", in "Nonlinear Partial Differential Equaitons and their Applications", Collège de France Seminar, vol. **X** (Pairs, 1987-1988), 357-391, H. Brézis and J. L. Lions Eds., Pitman Research Notes in Mathematics, Ser. **220**, Longman Sci. Tech., Harlow, 1991.
[21] E. Zuazua, "Exact controllability for semilinear wave equations in one space dimension", *Ann. Inst. H. Poincaré Anal. Non Linéaire*, **10** (1993), 109-129.

Journal of Physics: Conference Series **290** (2011) 012009 doi:10.1088/1742-6596/290/1/012009

Conformal mapping and impedance tomography

Rainer Kress

University of Göttingen, 37083 Göttingen, Germany

E-mail: kress@math.uni-goettingen.de

Abstract. Over the last decade Akduman, Haddar and Kress [1, 3, 5] have employed a conformal mapping technique for the inverse problem to recover a perfectly conducting or a non-conducting inclusion in a homogeneous background medium from Cauchy data on the accessible exterior boundary. More recently, Haddar and Kress [4] proposed an extension of this approach to two-dimensional inverse electrical impedance tomography with piecewise constant conductivities. A main ingredient of this extension is the incorporation of the transmission condition on the unknown interior boundary via a nonlocal boundary condition in terms of an integral equation. We present an outline of the foundations of this new method.

1. Introduction

In a new numerical scheme for solving inverse boundary value problems for the Laplace equation in a doubly connected two-dimensional domain D via a conformal mapping technique introduced by Akduman, Haddar and Kress [1, 3, 5] the reconstruction of the non-accessible interior boundary curve Γ_0 from over determined Cauchy data on the accessible exterior boundary curve Γ_1 is based on a conformal map $\Psi : B \to D$ that takes an annulus B bounded by two concentric circles C_0 and C_1 onto D. The Cauchy–Riemann equations provide a nonlocal and nonlinear ordinary differential equation for the boundary values $\Psi|_{C_1}$ on the exterior circle that can be solved by successive approximations. Then a Cauchy problem for the holomorphic function Ψ has to be solved by a regularized Laurent expansion to retrieve the unknown interior boundary curve via $\Gamma_0 = \Psi(C_0)$. For the reconstruction of a perfectly conducting or a non-conducting inclusion, i.e., the inverse problem with a homogeneous Dirichlet or Neumann condition on Γ_0 this conformal mapping method separates the inverse problem into the nonlinear well-posed problem for the ordinary differential equation and the linear ill-posed Cauchy problem.

The inverse electrical impedance problem to reconstruct the shape of a conducting inclusion with a constant conductivity that is different from the constant background conductivity of D corresponds to an inverse transmission problem. For this case, when applying the conformal mapping idea, in principle, two conformal maps are required. In addition to the mapping $\Psi : B \to D$ also a map taking the interior of C_0 onto the interior of Γ_0 is needed. Furthermore, the homogeneous transmission condition on Γ_0 transforms into a more complicated transmission condition on C_0 containing the traces of the two conformal maps at different locations for both sides of C_0.

Restricting themselves to the case where the two conformal maps are extensions of each other, and consequently have to coincide with a Moebius transform, in a first attempt Dambrine and Kateb [2] were able to extend parts of the above methods to the inverse transmission problem. In a recent paper, Haddar and Kress [4] proposed a different approach that uses only the conformal map for the annulus and incorporates the transformed transmission condition on C_0 by a nonlocal boundary condition in terms of a boundary integral equation for the trace of the solution to the transmission problem on Γ_0.

2. The inverse algorithm

Let D_0 and D_1 be two simply connected bounded domains in $\mathbb{R}^2$ with C^2 smooth boundaries Γ_0 and Γ_1 such that $\overline{D}_0 \subset D_1$. Let $D := D_1 \setminus \overline{D}_0$ and assume the unit normal ν to Γ_0 and Γ_1 to be directed into the exterior of D_0 and D_1, respectively. For a given function $f \in H^{1/2}(\Gamma_1)$ and given positive constants σ_0 and σ_1 we consider the transmission problem for the Laplace equation

$$\Delta u_0 = 0 \quad \text{in } D_0, \quad \Delta u = 0 \quad \text{in } D \tag{2.1}$$

for $u_0 \in H^1(D_0)$ and $u \in H^1(D)$ with boundary condition

$$u = f \quad \text{on } \Gamma_1 \tag{2.2}$$

and transmission conditions

$$u_0 = u, \quad \sigma_0 \frac{\partial u_0}{\partial \nu} = \sigma_1 \frac{\partial u}{\partial \nu} \quad \text{on } \Gamma_0. \tag{2.3}$$

After denoting the normal derivative of u on Γ_1 by

$$g := \left. \frac{\partial u}{\partial \nu} \right|_{\Gamma_1}$$

the inverse problem under consideration is to determine the shape of the interior boundary curve Γ_0 from pairs of Cauchy data (f, g).

In the sequel we will identify $\mathbb{R}^2$ and $\mathbb{C}$. We introduce the annulus B bounded by the concentric circles C_0 with radius ρ and C_1 with radius one centered at the origin. By the conformal mapping theorem there exists a uniquely determined radius ρ and a holomorphic function Ψ that maps B bijectively onto D such that the boundaries C_0 and C_1 are mapped onto Γ_0 and Γ_1, respectively. Assuming for simplicity that the total length of Γ_1 is 2π, we denote by $\gamma : [0, 2\pi] \to \Gamma_1$ the parameterization of Γ_1 in terms of arc length.

We define a function $\varphi : [0, 2\pi] \to [0, 2\pi]$ by setting

$$\varphi(t) := \gamma^{-1}(\Psi(e^{it})). \tag{2.4}$$

Roughly speaking, φ describes how Ψ maps arc length on C_1 onto arc length on Γ_1. Clearly, the boundary values φ uniquely determine Ψ as the solution to the Cauchy problem with Ψ on C_1 given through $\Psi(e^{it}) = \gamma(\varphi(t))$. Hence, the operator

$$N_\rho : \varphi \mapsto \chi \tag{2.5}$$

where $\chi(t) := \Psi(\rho e^{it})$ is well defined. The function $\chi : [0, 2\pi] \to \mathbb{C}$ parameterizes the interior boundary curve Γ_0 and determining χ solves the inverse transmission problem.

We denote by $A_\rho : H^{1/2}[0, 2\pi] \times H^{1/2}[0, 2\pi] \to H^{-1/2}[0, 2\pi]$ the Dirichlet-to-Neumann operator for the annulus B that maps function pairs (F_1, F_2) onto the normal derivative

$$(A_\rho(F_1, F_2))(t) := \frac{\partial v}{\partial \nu}(e^{it}) - \frac{1}{2\pi} \int_0^{2\pi} \frac{\partial v}{\partial \nu}(e^{it})\, dt, \quad t \in [0, 2\pi],$$

of the harmonic function $v \in H^1(B)$ with boundary values on C_1 and C_0 given by

$$v(e^{it}) = F_1(t) \quad \text{and} \quad v(\rho e^{it}) = F_2(t) \quad \text{for} \quad t \in [0, 2\pi].$$

In terms of this operator A from the Cauchy–Riemann equations for u and $v = u \circ \Psi$ and their corresponding harmonic conjugates one can deduce the nonlocal differential equation

$$\frac{d\varphi}{dt} = \frac{A_\rho(f \circ \gamma \circ \varphi, u \circ N_\rho \varphi)}{g \circ \gamma \circ \varphi} \tag{2.6}$$

for the boundary map φ.

In order to eliminate the unknown trace of u on the interior boundary curve from (2.6) we introduce the double-layer operator $K : H^{1/2}(\Gamma_0) \to H^{1/2}(\Gamma_0)$ by

$$(K\beta)(x) := 2 \int_{\Gamma_0} \frac{\partial \Phi(x, y)}{\partial \nu(y)} \beta(y)\, ds(y), \quad x \in \Gamma_0,$$

with the fundamental solution

$$\Phi(x, y) := \frac{1}{2\pi} \ln \frac{1}{|x - y|}$$

to Laplace's equation in $\mathbb{R}^2$. In terms of the Cauchy data (f, g) we define the combined single- and double-layer potential

$$w(x) := \int_{\Gamma_1} \left\{ g(y)\Phi(x, y) - f(y) \frac{\partial \Phi(x, y)}{\partial \nu(y)} \right\} ds(y), \quad x \in \mathbb{R}^2 \setminus \Gamma_1. \tag{2.7}$$

Then, as a consequence of Green's representation theorem, the trace $\beta := u|_{\Gamma_0}$ is given as the unique solution to the integral equation

$$(1 + \mu)\beta + (1 - \mu)K\beta = 2\mu w|_{\Gamma_0}$$

where

$$\mu := \frac{\sigma_1}{\sigma_0}.$$

Via $\eta := u \circ \chi$ we transform this into

$$(1 + \mu)\eta + (1 - \mu)H_\chi \eta = 2\mu w \circ \chi \tag{2.8}$$

with the parameterized double-layer operator H_χ given by $H_\chi(\beta \circ \chi) := (K\beta) \circ \chi$. Since the double-layer operator is compact and has its spectrum contained in $[-1, 1]$, the operator on the left-hand side of (2.8) is bijective with bounded inverse

$$M_\chi := 2\mu[(1 + \mu)I + (1 - \mu)H_\chi]^{-1} : H^{1/2}[0, 2\pi] \to H^{1/2}[0, 2\pi].$$

Now we can write

$$u \circ N_\rho \varphi = M_{N_\rho \varphi}(w \circ N_\rho \varphi)$$

and, finally, the nonlocal differential equation for φ assumes the form

$$\frac{d\varphi}{dt} = \frac{A_\rho(f \circ \gamma \circ \varphi, M_{N_\rho \varphi}(w \circ N_\rho \varphi))}{g \circ \gamma \circ \varphi}. \tag{2.9}$$

The differential equation (2.9) has to be complemented by the boundary conditions $\varphi(0) = 0$ and $\varphi(2\pi) = 2\pi$.

To avoid difficulties in solving (2.9) arising from zeros of the function g occurring in the denominator, two pairs of Cauchy data can be used. If (f_1, g_1) and (f_2, g_2) are two pairs of Cauchy data on Γ_1, then

$$\frac{d\varphi}{dt} = \frac{\sum_{j=1}^{2}(g_j \circ \gamma \circ \varphi)\, A_\rho(f_j \circ \gamma \circ \varphi, M_{N_\rho \varphi}(w_j \circ N_\rho \varphi))}{\sum_{j=1}^{2}[g_1 \circ \gamma \circ \varphi]^2}, \tag{2.10}$$

where w_1 and w_2 denote the combined single- and double-layer potential (2.7) associated with the real-valued Cauchy pairs (f_1, g_1) and (f_2, g_2), respectively. To condense the notation, after introducing the complex valued functions

$$f = f_1 + if_2, \quad g = g_1 + ig_2 \quad \text{and} \quad w = w_1 + iw_2 \tag{2.11}$$

we rewrite (2.10) in the shorter form

$$\frac{d\varphi}{dt} = \frac{\Re\left[(\bar{g} \circ \gamma \circ \varphi)\, A_\rho(f \circ \gamma \circ \varphi, M_{N_\rho \varphi}(w \circ N_\rho \varphi))\right]}{|g \circ \gamma \circ \varphi|^2}.$$

After defining Fourier coefficients depending on the data (f, g) and on φ by setting

$$a_m(\varphi) := \int_0^{2\pi} f(\gamma(\varphi(t)))\, e^{-imt}\, dt,$$

$$b_m(\varphi) := \int_0^{2\pi} g(\gamma(\varphi(t)))\, \varphi'(t)\, e^{-imt}\, dt,$$

$$c_m(\varphi, \rho) := \int_0^{2\pi} (M_{N_\rho \varphi}(w \circ N_\rho \varphi))(t)\, e^{-imt}\, dt,$$

a straightforward application of Green's integral theorem yields

$$[|m|\, a_m(\varphi) + b_m(\varphi)]\rho^{2|m|} + |m|\, a_m(\varphi) - b_m(\varphi) = 2|m|\rho^{|m|} c_m(\varphi, \rho). \tag{2.12}$$

By this equation the radius is given in terms of φ and the data (f, g). Under appropriate assumptions, (2.12) can be solved iteratively via

$$\rho_{j+1} = \left|\frac{b_m(\varphi) - |m|\, a_m(\varphi) + 2|m|\rho_j^{|m|} c_m(\varphi, \rho_j)}{b_m(\varphi) + |m|\, a_m(\varphi)}\right|^{\frac{1}{2|m|}}. \tag{2.13}$$

Finally we define an operator V by setting $(V\psi)(t) = t + \psi(t)$ and introduce the operator $T : H_0^1[0, 2\pi] \to H_0^1[0, 2\pi]$ by

$$(T\psi)(t) := \int_0^t \left(U\psi - \frac{1}{2\pi}\int_0^{2\pi} U\psi\, d\theta\right) d\tau, \quad t \in [0, 2\pi],$$

where

$$U\psi := \frac{\Re\left[(\bar{g} \circ \gamma \circ V\psi)\, A_{\rho(V\psi)}(f \circ \gamma \circ V\psi, M_{N_{\rho(V\psi)} V\psi}(w \circ N_{\rho(V\psi)} V\psi))\right]}{|g \circ \gamma \circ V\psi|^2},$$

and $\rho(V\psi)$ indicates the solution of (2.12) for $\varphi = V\psi$ and an appropriately chosen $m \in \mathbb{N}$. Then we can summarize the above results into the following theorem.

Theorem 2.1 *Let (f, g) be a pair of Cauchy data of the form* (2.11) *for the transmission problem. Then, in terms of the holomorphic map $\Psi : B \to D$ and its boundary values φ the function $\psi = V^{-1}\varphi$ is a fixed point of T.*

Theorem 2.1 suggests the following iteration scheme: Given a current approximation ψ_0, we update it in two steps.

(i) For an appropriate choice of m we solve (2.12) with the boundary map $\varphi_0 = V\psi_0$ via iterations as indicated in (2.13) to obtain a radius ρ_0.

(ii) In view of Theorem 2.1 we update the boundary map by $\psi_1 = T(\psi_0)$ using the radius ρ_0 and a regularized version of the operator N_ρ for the Cauchy problem.

Of course, the whole scheme then consists in repeating these two steps iteratively. For a convergence result on this iteration scheme for sufficiently small transmission coefficients μ and numerical examples exhibiting the feasibility of the method we refer to [4].

Acknowledgement
This work was supported by the German Ministry of Education and Research in the BMBF-project Regularization techniques for electrical impedance tomography in medical science and geosciences.

References

[1] Akduman I and Kress, R 2002 Electrostatic imaging via conformal mapping. *Inverse Problems* **18** 1659
[2] Dambrine M and Kateb D 2007 Conformal mapping and inverse conductivity problem with one measurement. *ESAIM: Control, Optimisation and Calculus of Variations* **13** 163
[3] Haddar H and Kress R 2005 Conformal mappings and inverse boundary value problems. *Inverse Problems* **21** 935
[4] Haddar H and Kress R 2010 Conformal mapping and impedance tomography. *Inverse Problems* **26** 074002
[5] Kress R 2004 Inverse Dirichlet problem and conformal mapping. *Mathematics and Computers in Simulation* **66** 255

Journal of Physics: Conference Series **290** (2011) 012010 doi:10.1088/1742-6596/290/1/012010

On Carleman estimates with two large parameters

Jérôme Le Rousseau

Jérôme Le Rousseau. Université d'Orléans, Laboratoire Mathématiques et Applications, Physique Mathématique d'Orléans, CNRS UMR 6628, Fédération Denis-Poisson, FR CNRS 2964, B.P. 6759, 45067 Orléans cedex 2, France.

E-mail: jlr@univ-orleans.fr

Abstract. We provide a general framework for the analysis and the derivation of Carleman estimates with two large parameters. For an appropriate form of weight functions strong pseudo-convexity conditions are shown to be necessary and sufficient.

1. Introduction

Carleman estimates are an important tool in subjects in analysis of partial differential equations (PDEs) such as unique continuation, control theory, and inverse problems. They are weighted L^2-norm estimates of the solution of a PDE. Weights are of exponential form and involve a parameter that is taken large. The work of L. Hörmander in the late 50's-early 60's provided large classes of operators for which such estimates can be derived. The notions of principal normality for the operator and strong pseudo-convexity for the weight function were central in this seminal work.

If choosing a double exponential for weight function, sufficient conditions for the Carleman estimate to hold can be obtained from strong pseudo-convexity conditions. With this choice of weight function one introduces a second large parameter. Several authors have derived Carleman estimates for some operators in which the dependency upon the second large parameters is kept explicit. Such results can be very useful to address applications such as inverse problems. See for instance [1, 2, 3, 4].

We provide a general framework for the analysis and the derivation of Carleman estimates with two large parameters. Strong pseudo-convexity conditions are shown to be necessary and sufficient for the estimate to hold. Under further conditions on the operator and the weight function stronger Carleman estimates can also be achieved.

For an introduction to Carleman estimates for second-order elliptic and parabolic operator see [5].

Notation

Here Ω will always denote a bounded open subset of $\mathbb{R}^n$. When the constant C is used, it refers to a constant that is independent of the different parameter, e.g. τ and α. Its value may however change from one line to another. If we want to keep track of the value of a constant we shall use another letter.

Journal of Physics: Conference Series **290** (2011) 012010 doi:10.1088/1742-6596/290/1/012010

2. A pseudo-differential calculus with two large parameters

We set $W = \mathbb{R}^n \times \mathbb{R}^n$, often referred to as phase-space. A typical element of W will be $X = (x, \xi)$, with $x \in \mathbb{R}^n$ and $\xi \in \mathbb{R}^n$.

Let $\psi \in \mathscr{C}^\infty(\mathbb{R}^n; \mathbb{R})$ be such that

$$\psi \geq C > 0, \qquad |\psi'| > 0, \quad \text{and} \quad \|\psi'\|_\infty < \infty.$$

We then set

$$\varphi(x) = e^{\alpha\psi(x)}, \qquad \text{with} \ \ \alpha \geq 1.$$

We observe that $|\varphi'| > 0$. We make the following further assumption on the function ψ.

Assumption 2.1. *There exists $k \in \mathbb{N}$ such that $\sup_{\mathbb{R}^n} \psi \leq (k+1) \inf_{\mathbb{R}^n} \psi$.*

2.1. Metric and order function on phase-space

We consider the metric on phase-space:

$$g = \alpha^2 |dx|^2 + \frac{|d\xi|^2}{\mu^2}, \qquad \text{with} \ \ \mu^2 = \mu^2(x, \xi, \tau) = \big(\tau\alpha\varphi(x)\big)^2 + |\xi|^2, \ \ \text{and} \ \tau \geq 1, \ \alpha \geq 1.$$

The first result of this section shows that this metric on W defines a Weyl-Hörmander pseudo-differential calculus.

Proposition 2.2. *The metric g and the order function μ are admissible, in the sense that,*

(i) g satisfies the uncertainty principle, with $\lambda_h = h_g^{-1} = \alpha^{-1}\mu$.

(ii) μ and g are slowly varying;

(iii) μ and g are temperate.

For a presentation of the Weyl-Hörmander calculus we refer to [6], [7, Sections 18.4–6] and [8].

We may define Sobolev spaces associated with the calculus defined by the metric g as in [9]. For $k, s \in \mathbb{R}$, we set

$$\mathscr{H}_{k,s}(\mathbb{R}^n) = \Big\{ \operatorname{Op}^{\mathrm{w}}(\tilde{\tau}^{-s}\mu^{-k})v;\ v \in L^2 \Big\}, \quad \text{and} \ \|u\|_{k,s} = \|\operatorname{Op}^{\mathrm{w}}(\tilde{\tau}^{s}\mu^{k})u\|_{L^2(\mathbb{R}^n)}, \ \text{if} \ u \in \mathscr{H}_{k,s}(\mathbb{R}^n),$$

with $\tilde{\tau}(x) = \tau\alpha\varphi(x)$. Note that $\|.\|_{k,s}$ is a norm on $\mathscr{H}_{k,s}$.

Proposition 2.3. *Let $k, k', s, s' \in \mathbb{R}$. For $a \in S(\tilde{\tau}^s\mu^k, g)$ there exist $C > 0$ and $\tau_1 > 0$ such that for all $\tau \geq \tau_1$, we have*

$$\|\operatorname{Op}^{\mathrm{w}}(a)u\|_{k',s'} \leq C\|u\|_{k+k',s+s'}, \qquad u \in \mathscr{H}_{k+k',s+s'}(\mathbb{R}^n).$$

3. Carleman estimates under strong pseudo-convexity assumptions

Let $P(x, D_x)$ be a differential operator of order m, with homogeneous principal symbol $p(x, \xi)$.

Definition 3.1 (Principal normality [10, Definition 28.2.4]). The operator $P(x, D_x)$ is said to be principally normal on Ω if for some $C > 0$

$$\big|\{\overline{p}, p\}(x, \xi)\big| \leq C|p(x, \xi)|\,|\xi|^{m-1}, \quad x \in \overline{\Omega},\ \xi \in \mathbb{R}^n.$$

Elliptic operators and operators with real coefficients in the principal part are typical examples of principally normal operators.

We now revisit some consequences of pseudo-convexity and strong pseudo-convexity.

3.1. Pseudo-convexity properties and symbol estimates
Let $\psi \in \mathscr{C}^\infty(\overline{\Omega}, \mathbb{R})$. We recall the following definitions [11].

Definition 3.2 (pseudo-convexity). We say that ψ is pseudo-convex at $x \in \overline{\Omega}$ w.r.t. $P(x, D_x)$ if $\psi'(x) \neq 0$ and if

$$\forall \xi \in \mathbb{R}^n \backslash 0, \quad p(x,\xi) = 0 \text{ and } \big\{p, \psi\big\}(x,\xi) = 0 \quad \Rightarrow \quad \operatorname{Re}\big\{\overline{p}, \{p,\psi\}\big\}(x,\xi) > 0. \qquad (\Psi\mathrm{c})$$

The function ψ is said to be pseudo-convex w.r.t. Ω and p if $\psi' \neq 0$ in $\overline{\Omega}$ and $(\Psi\mathrm{c})$ is valid for all $x \in \overline{\Omega}$.

We note that

$$\big\{p, \psi\big\}(x,\xi) = \langle p'_\xi(x,\xi), \psi'(x)\rangle, \quad \text{and} \quad \operatorname{Re}\big\{\overline{p}, \{p,\psi\}\big\}(x,\xi) = \theta_{p,\psi}(x,\xi)$$

with $\theta_{p,\psi}(x,\xi) = \sum_{j,k} \partial^2_{x_j x_k}\psi(x)\, \partial_{\xi_k}\overline{p}(x,\xi)\, \partial_{\xi_j} p(x,\xi) + \operatorname{Re}\sum_j \partial_{x_j}\psi(x)\, \big\{\overline{p}, \partial_{\xi_j} p\big\}(x,\xi)$.

Definition 3.3 (strong pseudo-convexity). We say that ψ is strongly pseudo-convex at $x \in \overline{\Omega}$ w.r.t. p if

(i) ψ is pseudo-convex at x w.r.t. p;
(ii) if for all $\xi \in \mathbb{R}^n$ and $\tau > 0$,

$$\begin{aligned} &p(x, \xi + i\tau\psi'(x)) = 0 \text{ and } \big\{p,\psi\big\}(x,\xi + i\tau\psi'(x)) = 0 \\ &\qquad \Rightarrow \quad \frac{1}{2i}\big\{\overline{p}(x,\xi - i\tau\psi'(x)), p(x,\xi + i\tau\psi'(x))\big\} > 0. \qquad (\text{s-}\Psi\mathrm{c})\end{aligned}$$

The function ψ is said to be strongly pseudo-convex w.r.t. Ω and p if items (i) and (ii) are valid for all $x \in \overline{\Omega}$.

We note that

$$\begin{aligned} \frac{1}{2i}\big\{\overline{p}(x,\xi - i\tau\psi'(x)), p(x,\xi + i\tau\psi'(x))\big\} &= \big\{\operatorname{Re} p(x,\xi + i\tau\psi'(x)), \operatorname{Im} p(x,\xi + i\tau\psi'(x))\big\} \\ &= \Theta_{p,\psi}(x,\xi,\tau), \end{aligned}$$

with

$$\begin{aligned} \Theta_{p,\psi}(x,\xi,\tau) = \tau \sum_{j,k} &\partial^2_{x_j x_k}\psi(x)\, \partial_{\xi_j} p(x,\xi + i\tau\psi'(x))\, \partial_{\xi_k}\overline{p}(x,\xi - i\tau\psi'(x)) \\ &+ \operatorname{Im}\sum_j \partial_{x_j} p(x,\xi + i\tau\psi'(x))\, \partial_{\xi_j}\overline{p}(x,\xi - i\tau\psi'(x)). \end{aligned}$$

The following result sharpens the estimate of Proposition 28.3.3 in [10].

Proposition 3.4. *Let P be principally normal on Ω and let ψ be a strongly pseudo-convex function w.r.t. Ω and P. We set $\varphi = e^{\alpha\psi}$ and*

$$\zeta = \zeta(x,\xi,\tau) = \xi + i\tau\varphi'(x) = \xi + i\tilde{\tau}(x)\psi'(x), \quad \tilde{\tau}(x) = \tau\alpha\varphi(x).$$

There exist $C > 0$, $\tau_0 \geq 1$, $\alpha_0 > 0$ such that we have

$$\tilde{\tau}(x)\mu^{2(m-1)} \leq C\Big(|p(x,\zeta)|^2 + \frac{1}{2i}\big\{\overline{p(x,\zeta)}, p(x,\zeta)\big\}\Big), \quad \tau \geq \tau_0,\ \alpha \geq \alpha_0,\ (x,\xi) \in \overline{\Omega} \times \mathbb{R}^n. \quad (1)$$

The proof is based on a different treatment of the regions $\tau\alpha\varphi(x) \ll |\xi|$ and $\tau\alpha\varphi(x) \gtrsim |\xi|$.

3.2. Carleman estimate
We now state a Carleman estimate with two large parameters.

Theorem 3.5. *Let $P = P(x, D_x)$ be a principally normal operator in Ω of order m and $\psi \in \mathscr{C}^\infty(\overline{\Omega})$, $\psi \geq C > 0$ on Ω, be a strongly pseudo-convex function w.r.t. Ω and P. We set $\varphi = e^{\alpha\psi}$. If X is an open subset, $X \Subset \Omega$, there exist $C > 0$, $\alpha_1 > 0$, and $\tau_1 \geq 1$ such that*

$$\sum_{|\beta|<m} (\tau\alpha)^{2(m-|\beta|)-1} \|\varphi^{m-|\beta|-\frac{1}{2}} e^{\tau\varphi} D_x^\beta u\|_{L^2}^2 \leq C \|e^{\tau\varphi} P u\|_{L^2}^2, \tag{2}$$

for all $u \in \mathscr{C}_c^\infty(X)$, $\alpha \geq \alpha_1$, and $\tau \geq \tau_1$.

The proof is based on the pseudo-differential calculus introduced above, by mean of the Fefferman-Phong inequality (see [12] and [7, Theorem 18.6.8]).

4. Cases of stronger estimates

In this section we present classes of operators for which stronger Carleman estimates with two large parameters can be derived as compared to the result of Theorem 3.5.

4.1. Simple characteristics
We introduce the map

$$\rho : \mathbb{R}^+ \to \mathbb{C}, \qquad \hat{\tau} \mapsto p(x, \xi + i\hat{\tau}\psi'(x)),$$

We assume in this section that the operator satisfies the simple complex characteristic property:

$$\begin{cases} \hat{\tau} \mapsto \rho(\hat{\tau}) \text{ has simple roots} & \text{if } (x,\xi) \in \Omega \times (\mathbb{R}^n \setminus 0), \\ \hat{\tau} \mapsto \rho(\hat{\tau}) \text{ has simple nonzero roots} & \text{if } x \in \Omega \text{ and } \xi = 0. \end{cases} \tag{3}$$

Note that the case $\xi = 0$ is particular, as the root $\hat{\tau} = 0$ has of course multiplicity m. Note also that we have

$$\rho'(\hat{\tau}) = i\langle p'_\xi(x, \xi + i\hat{\tau}\psi'(x)), \psi'(x)\rangle = i\{p, \psi\}(x, \xi + i\hat{\tau}\psi'(x)). \tag{4}$$

With the simple-characteristic property we shall obtain a Carleman estimate with an additional power in the second large parameter α.

Examples 4.1. Evidently, first-order operators satisfy this property. Second-order elliptic operators with real coefficients also satisfy this property for any smooth function ψ such that $\psi' \neq 0$. Indeed, complex roots of $\sigma \mapsto p(x, \xi + \sigma\psi'(x))$ come in conjugated pairs. They can only be double if they are real. For the roots of $\hat{\tau} \mapsto \rho(\hat{\tau})$ this means $\hat{\tau} = 0$, that is $p(x, \xi) = 0$.

Note that elliptic operators of order higher than 2 may however not satisfy this property, e.g. the bilaplacian operator Δ^2. An example of an elliptic operator of order greater than two satisfying the simple-characteristic property (3) is $D_x^4 + D_y^4$ in $\Omega \subset \mathbb{R}^2$ for any function ψ whose gradient does not vanish in $\overline{\Omega}$.

An example of a second-order operator that is not elliptic satisfying the simple-characteristic property (3) is $\frac{1}{2}D_x^2 + D_x D_y$ in $\Omega \subset \mathbb{R}^2$, with $\psi(x) = \frac{1}{2}(x - x_0)^2$, where x_0 is chosen such that $|\psi'| \neq 0$ in $\overline{\Omega}$, i.e., $x_0 \notin \overline{\Omega}$.

Note that the examples we have just given are principally normal (see definition 3.1) as they have real coefficients.

Remark 4.2. The simple-characteristic property (3) implies that ψ is strongly pseudo-convex with respect to Ω and P.

Proposition 4.3. *Let P be principally normal on Ω and let ψ be such that the simple-characteristic property (3) is fulfilled. We set $\varphi = e^{\alpha\psi}$ and*

$$\zeta = \zeta(x,\xi,\tau) = \xi + i\tau\varphi'(x) = \xi + i\tilde{\tau}(x)\psi'(x), \quad \tilde{\tau}(x) = \tau\alpha\varphi(x).$$

There exist $C > 0$, $\tau_0 \geq 1$, $\alpha_0 > 0$, and ν_0 such that we have

$$\tilde{\tau}(x)^2\mu^{2(m-1)} \leq C\Big(\nu|p(x,\zeta)|^2 + \frac{\tau\varphi(x)}{2i}\{\overline{p(x,\zeta)}, p(x,\zeta)\}\Big),$$
$$\tau \geq \tau_0,\ \alpha \geq \alpha_0,\ \nu \geq \nu_0,\ (x,\xi) \in \overline{\Omega} \times \mathbb{R}^n. \quad (5)$$

With Proposition 4.3 we obtain the following Carleman estimate.

Theorem 4.4. *Let $P = P(x, D_x)$ be a principally normal operator in Ω of order m and $\psi \in \mathscr{C}^\infty(\overline{\Omega})$, $\psi \geq C > 0$ on Ω, a function such that the simple-characteristic property (3) is fulfilled. We set $\varphi = e^{\alpha\psi}$. If X is an open subset, $X \Subset \Omega$, there exist $C > 0$, $\alpha_1 > 0$, and $\tau_1 \geq 1$ such that*

$$\alpha \sum_{|\beta|<m} (\tau\alpha)^{2(m-|\beta|)-1} \|\varphi^{m-|\beta|-\frac{1}{2}} e^{\tau\varphi} D_x^\beta u\|_{L^2}^2 \leq C\|e^{\tau\varphi} Pu\|_{L^2}^2,$$

for all $u \in \mathscr{C}_c^\infty(X)$, $\alpha \geq \alpha_1$, and $\tau \geq \tau_1$.

We observe that we have gained a factor α on the l.h.s. in contrast to the Carleman estimate of Theorem 3.5.

4.2. Elliptic operators

For elliptic operators stronger results can also be achieved. First we shall consider general elliptic operators and second we shall consider elliptic operators with simple characteristics.

4.2.1. General elliptic operators under strong pseudo-convexity condition Let A be an elliptic operator of order m. We first note that if ψ is a smooth function such that $|\psi'| > 0$ on $\overline{\Omega}$, then it is a pseudo-convex function on Ω for the operator A.

Proposition 4.5. *Let A be elliptic on Ω and let ψ satisfy* (s-Ψc) *(point ii in Definition 3.3) for all $x \in \overline{\Omega}$. We set $\varphi = e^{\alpha\psi}$ and*

$$\zeta = \zeta(x,\xi,\tau) = \xi + i\tau\varphi'(x) = \xi + i\tilde{\tau}(x)\psi'(x), \qquad \tilde{\tau}(x) = \tau\alpha\varphi(x)$$

There exist $C > 0$, $\tau_0 \geq 1$, $\alpha_0 > 0$ such that we have

$$(\tau\alpha\varphi)^{-1}\mu^{2m} \leq C\Big(|p(x,\zeta)|^2 + \frac{1}{2i}\{\overline{p(x,\zeta)}, p(x,\zeta)\}\Big), \quad \tau \geq \tau_0,\ \alpha \geq \alpha_0,\ (x,\xi) \in \overline{\Omega} \times \mathbb{R}^n. \quad (6)$$

With Proposition 4.5 we then obtain the following Carleman estimate.

Theorem 4.6. *Let $P = P(x, D_x)$ be an elliptic operator in Ω of order m and $\psi \in \mathscr{C}^\infty(\overline{\Omega})$, $\psi \geq C > 0$ on Ω, be a strongly pseudo-convex function w.r.t. Ω and P. We set $\varphi = e^{\alpha\psi}$. If X is an open subset, $X \Subset \Omega$, there exist $C > 0$, $\alpha_1 > 0$, and $\tau_1 \geq 1$ such that*

$$\sum_{|\beta|\leq m} (\tau\alpha)^{2(m-|\beta|)-1} \|\varphi^{m-|\beta|-\frac{1}{2}} e^{\tau\varphi} D_x^\beta u\|_{L^2}^2 \leq C\|e^{\tau\varphi} Pu\|_{L^2}^2, \quad (7)$$

for all $u \in \mathscr{C}_c^\infty(X)$, $\alpha \geq \alpha_1$, and $\tau \geq \tau_1$.

We observe that we have gained an additional term in the sum on the l.h.s. in contrast to the Carleman estimate of Theorem 3.5.

Remark 4.7. From an estimate of the form of (7) by fixing the values of τ and α we obtain

$$\|u\|_{H^m(\mathbb{R}^n)} = \sum_{|\beta|\le m} \|D_x^\beta u\|_{L^2} \le C\|Pu\|_{L^2},$$

which implies that P is elliptic. The additional term we have obtained in the previous theorem is thus a privilege of elliptic operators.

4.2.2. Elliptic operators with simple characteristics Combining the simple characteristic and elliptic properties we obtain the following estimate.

Proposition 4.8. *Let $p(x,\xi)$ and $\psi(x)$ be such that* (3) *holds. Assume moreover that $p(x,\xi)$ is elliptic. We set $\varphi = e^{\alpha\psi}$ and*

$$\zeta = \zeta(x,\xi,\tau) = \xi + i\tau\varphi'(x) = \xi + i\tau\alpha\psi'(x)\varphi(x).$$

There exist $C > 0$, $\tau_0 \ge 1$, $\alpha_0 > 0$, $\nu_0 > 0$, such that we have

$$\mu^{2m} \le C\Big(\nu|p(x,\zeta)|^2 + \frac{\tau\varphi(x)}{2i}\{\overline{p(x,\zeta)}, p(x,\zeta)\}\Big), \quad \tau \ge \tau_0,\ \alpha \ge \alpha_0,\ \nu \ge \nu_0,\ (x,\xi) \in \overline{\Omega} \times \mathbb{R}^n. \tag{8}$$

With Proposition 4.8 we then obtain the following Carleman estimate.

Theorem 4.9. *Let $P = P(x, D_x)$ be an elliptic differential operator in Ω of order m and $\psi \in \mathscr{C}^\infty(\overline{\Omega})$, be such that the simple-characteristic property* (3) *holds. We set $\varphi = e^{\alpha\psi}$. If X is an open subset, $X \Subset \Omega$, there exist $C > 0$, $\alpha_1 > 0$, and $\tau_1 \ge 1$ such that*

$$\alpha \sum_{|\beta|\le m} (\tau\alpha)^{2(m-|\beta|)-1} \|\varphi^{m-|\beta|-\frac{1}{2}} e^{\tau\varphi} D_x^\beta u\|_{L^2}^2 \le C\|e^{\tau\varphi} Pu\|_{L^2}^2,$$

for all $u \in \mathscr{C}_c^\infty(X)$, $\alpha \ge \alpha_1$, and $\tau \ge \tau_1$.

We observe that we have gained a factor α and an additional term in the sum on the l.h.s. in contrast to the Carleman estimate of Theorem 3.5.

5. Necessary conditions on the weight function

Starting from Carleman estimates L. Hörmander derived necessary conditions on the weight function [11, 10]. We apply the same program in the case of Carleman estimates with two large parameters and weight functions of the form $\varphi = e^{\alpha\psi}$.

Lemma 5.1. *Let P be a differential operator of order m with smooth principal symbol $p(x,\xi)$ and let $\psi \in \mathscr{C}^\infty(\Omega)$ be such that the following estimate holds,*

$$\sum_{|\beta|<m} (\tau\alpha)^{2(m-|\beta|)-1} \|\varphi^{m-|\beta|-\frac{1}{2}} e^{\tau\varphi} D_x^\beta u\|_{L^2}^2 \le K\|e^{\tau\varphi} Pu\|_{L^2}^2, \quad \varphi = e^{\alpha\psi}, \tag{9}$$

for $\tau \ge \tau_0 > 0$ and $\alpha \ge \alpha_0 > 0$ and $u \in \mathscr{C}_c^\infty(\Omega)$. We then have

$$\sum_{|\beta|<m} \big(\tau\alpha\varphi(x)\big)^{2(m-|\beta|)-1} |\zeta^\beta|^2 \le \frac{K}{i}\{\overline{p}(x,\overline{\zeta}), p(x,\zeta)\}, \quad \zeta = \xi + i\tau\varphi'(x) = \xi + i\tau\alpha\psi'(x)\varphi(x), \tag{10}$$

for $\tau > 0$ and $\alpha \ge \alpha_0$, if $p(x,\zeta) = 0$. If $m \ge 2$ we have $\psi' \ne 0$ in Ω. Moreover $p(x,\xi)$ does not vanish at second order at any point of $T^*(\Omega) \setminus 0$.

This lemma is the counterpart with two large parameters of Theorem 28.2.1 in [10]. The proof is along the same lines.

Lemma 5.2. *If the sum in* (9) *is replaced by* $\sum_{|\beta|\leq m}$ *(resp.* $\alpha\sum_{|\beta|<m}$ *or* $\alpha\sum_{|\beta|\leq m}$*) then the same is true for* (10).

We next see that the strong pseudo-convexity condition on the function ψ is in fact *necessary* for the Carleman estimate (9) to hold.

Theorem 5.3. *Consider the same assumption as in Lemma 5.1 and assume further that P is principally normal. In the case of a first-order operator, further assume that $\psi' \neq 0$ in Ω. Then the function ψ is strongly pseudo-convex w.r.t. P and Ω.*

Along with Theorem 3.5 we then obtain that the strong pseudo-convexity of ψ is *necessary* and *sufficient* for the Carleman estimate (9) to hold. This result is in contrast with L. Hörmander's work where there is a gap between the necessary and the sufficient conditions on the weight function φ to have a Carleman estimate (compare Theorem 28.2.1 and Theorem 28.2.3 in [10]). Here, of course we impose a particular structure on φ, viz. $\varphi = e^{\alpha\psi}$.

We also obtain necessary conditions for a Carleman estimate as in Theorem 4.4 to hold.

Theorem 5.4. *Let P be a differential operator of order m with smooth principal symbol $p(x,\xi)$ and let $\psi \in \mathscr{C}^\infty(\Omega)$ be such that the following estimate holds,*

$$\alpha \sum_{|\beta|<m} (\tau\alpha)^{2(m-|\beta|)-1} \|\varphi^{m-|\beta|-\frac{1}{2}} e^{\tau\varphi} D_x^\beta u\|_{L^2}^2 \leq K \|e^{\tau\varphi} P u\|_{L^2}^2, \quad \varphi = e^{\alpha\psi}, \tag{11}$$

for $\tau \geq \tau_0 > 0$ and $\alpha \geq \alpha_0 > 0$ and $u \in \mathscr{C}_c^\infty(\Omega)$. Assume further that P is principally normal. Then the function ψ is such that $\psi'(x) \neq 0$ in Ω and the simple-characteristic property (3) *holds.*

Note that the case of a first-order operator, the stronger estimate (11) implies $\psi' \neq 0$ as opposed to the "regular" Carleman estimate with two large parameters (9).

With Theorem 4.4 we see that, in the case of principally normal operators, the simple-characteristic property (3) is a *necessary* and *sufficient* condition for the Carleman estimate (11) to hold.

Acknowledgments
The author wishes to thank L. Robbiano and N. Lerner for discussions on Carleman estimates, pseudo-convexity properties, and pseudo-differential calculus. The author was partially supported by l'Agence Nationale de la Recherche under grant ANR-07-JCJC-0139-01.

References

[1] Eller M 2000 *Journal of Mathematical Analysis and Applications* **249** 491–514
[2] Eller M and Isakov V 2000 *Contemporary Math.* **268** 117–136
[3] Isakov V and Kim N 2008 *Applicationes Mathematicae* **35** 447–465
[4] Bellassoued M and Yamamoto M 2010 *preprint*
[5] Le Rousseau J and Lebeau G 2009 *Preprint*
[6] Lerner N 2010 *Metrics on the Phase Space and Non-Selfadjoint Pseudo-Differential Operators* Pseudo-Differential Operators, Vol. 3 (Basel: Birkhäuser)
[7] Hörmander L 1985 *The Analysis of Linear Partial Differential Operators* vol III (Springer-Verlag) second printing 1994
[8] Hörmander L 1979 *Comm. Pure Appl. Math.* **32** 359–443
[9] Bony J M and Chemin J Y 1994 *Bull. Soc. Math. France* **122** 77–118
[10] Hörmander L 1985 *The Analysis of Linear Partial Differential Operators* vol IV (Springer-Verlag)
[11] Hörmander L 1963 *Linear Partial Differential Operators* (Berlin: Springer-Verlag)
[12] Fefferman C and Phong D H 1978 *Proc. Nat. Acad. Sci.* **75** 4673–4674

A numerical evaluation and regularization of Caputo fractional derivatives

Ming Li[1], X. T. Xiong[2] and Y. J. Wang[1]

[1] Department of Mathematics, Taiyuan University of Technology, Shanxi, China
[2] College of Mathematics and Information Science, Northwest Normal University, Gansu, China

Abstract. Numerical evaluations of Caputo fractional derivatives for scattered noisy data is an important problem in scientific research and practical applications. Fractional derivatives have been applied recently to the numerical solution of problems in fluid and continuum mechanics. The Caputo fractional derivative of order α is given as follows

$$f^{(\alpha)}(t) = \frac{1}{\Gamma(1-\alpha)} \int_0^t \frac{f^{(n)}(s)}{(t-s)^\alpha} ds, \qquad 0 < \alpha < 1$$

The above definition includes a Volterra integral equation with weakly singular kernels and difficult to calculate. In this paper, a faster convergence numerical schedule is given and applied to solve several fractional-diffusion heat conduction problems. Convergence rates of interest are also presented here. Several numerical results are given to show the effectiveness of the proposed numerical schedule.

Keywords: Numerical derivatives; fractional derivatives; ill-posed; regularization

1. Introduction

Fractional derivatives have been applied recently in fluid and continuum mechanics. Numerical derivative is a typical ill-posed in the sense that a small error in measurement data can induce a large error in the approximate derivatives and important in science and engineering [8, 9]. With the recent successful application of fractional derivatives in mechanics problems, an accurate numerical evaluation of fractional derivatives from scattered noisy data is important in scientific research and practical industries. Due to the ill-posedness of the numerical fractional derivative, both the convergence and stability of the numerical should be investigated.

In mathematics, the RiemannCLiouville integral is defined by

$$J^\alpha f(t) = \frac{1}{\Gamma(\alpha)} \int_0^t (s-t)^{\alpha-1} f(s) ds \tag{1}$$

Based on the the RiemannCLiouville integral, the Caputo fractional derivative was introduced by Caputo in 1967:

$$D^\alpha f(t) = J^{(n-\alpha)} D^n f(t), \qquad n = \lceil \alpha \rceil \tag{2}$$

where $\Gamma(\cdot)$ is the Gramma function. A number of computational methods have been researched in the past years: difference methods [2], Tikhonov regularization methods [3, 4], mollification methods [6, 7], fourier truncation method [1] and interpolation methods [5]. Most of these researches on numerical derivative are related to integer order derivative.

Although many numerical schemes are use to obtain the fractional order derivative, the difference method is still a simple and effective numerical method in one-dimensional case. In the present paper, we also give a new regularization difference method. Theoretical convergence and stability results and numerical examples will proposed and show the proposed regularization difference method is accurate and stable.

The structure of the paper is as follows. In Section 2, we briefly formulate the problem and propose the difference scheme. In Section 3, theoretical convergence and stability results are obtained respectively. In Section 4, numerical results illustrate the stability and accuracy of the proposed method.

2. Numerical Scheme

In this paper, let $\alpha \in (0,1)$ and n is a positive number. Let $f = f(x)$ be a function in $\mathbb{R}$ and $\{x_i = \frac{(b-a)i}{n}\}_{i=0}^{n}$ be a set of distinct points on [a,b]. Here, we consider the following problem: given the noisy measurement $\tilde{f}_i$ of the values $f(x_i)$ which satisfies

$$|\tilde{f}_i - f(x_i)| \leq \delta \tag{3}$$

where δ is called the noise level. The purpose of this paper is to obtain an approximate value of the Caputo derivative:

$$D_\alpha f(t) = \frac{1}{\Gamma(1-\alpha)} \int_0^t \frac{f'(s)}{(t-s)^\alpha} ds, \qquad 0 < \alpha < 1 \tag{4}$$

A difference scheme of Caputo fractional derivative

$$D_k^\alpha f(t_i) = \sigma_{\alpha,k} \sum_{j=0}^{i} \omega_j f_j \tag{5}$$

where

$$\begin{cases} \omega_0 &= (i-1)^{1-\alpha} - i^{1-\alpha} \\ \omega_j &= (i-j+1)^{1-\alpha} - 2(i-j)^{1-\alpha} + (i-j-1)^{1-\alpha} \\ & \qquad j = 1, \cdots, i-1 \\ \omega_i &= 1 \qquad \sigma_{\alpha,k} = \frac{1}{\Gamma(1-\alpha)} \frac{1}{(1-\alpha)k^\alpha} \end{cases} \tag{6}$$

3. Error Estimate

Lemma 3.1 *Suppose that* $\| f \|_\infty \leq E$. *For any* $t \in (t_i - \frac{k}{2}, t_i + \frac{k}{2}]$, $D_k^\alpha f(t) = D_k^\alpha f(t_i)$ *which defined by the formula (5). Then*

$$\| D^\alpha f(t) - D_k^\alpha f(t) \|_\infty \leq C k^{2-\alpha} \tag{7}$$

where C *is constants depending on* α *and* $\| f''(t) \|_\infty$

Journal of Physics: Conference Series **290** (2011) 012011 doi:10.1088/1742-6596/290/1/012011

Proof : First of all, we divide the integral (4) into two parts,

$$D^\alpha f(t_i) = \frac{1}{\Gamma(1-\alpha)}\int_0^{t_i}\frac{f'(s)}{(t-s)^\alpha}ds \tag{8}$$

$$= \underbrace{\tfrac{1}{\Gamma(1-\alpha)}\int_0^{t_{i-1}}\frac{f'(s)}{(t-s)^\alpha}ds}_{I} + \underbrace{\tfrac{1}{\Gamma(1-\alpha)}\int_{t_{i-1}}^{t_i}\frac{f'(s)}{(t-s)^\alpha}ds}_{II} \tag{9}$$

It's obvious that I part is a normal integral without singularity. Using the integration by parts, we obtain

$$I = f(s)(t_i-s)^{-\alpha}|_0^{t_{i-1}} - \alpha\int_0^{t_{i-1}} f(s)(t_i-s)^{-\alpha-1}ds \tag{10}$$

$$= f_{i-1}k^{-\alpha} - f_0 t_i^{-\alpha} - \alpha\sum_{j=1}^{i-1}\int_{t_{j-1}}^{t_j}\left(\frac{(t_{j-1}-s)f_j + (t_j-s)f_{j-1}}{k}\right. \tag{11}$$

$$\left.+f''(\xi_j)k^2\right)(t_i-s)^{-\alpha-1}ds \tag{12}$$

$$= \sigma_{\alpha,k} * \Gamma(1-\alpha)\left(\sum_{j=0}^{i-2}\omega_j f_j + f_{i-1}\right) - \alpha\sum_{j=1}^{i-1}\int_{t_{j-1}}^{t_j} f''(\xi_j)k^2(t_i-s)^{-\alpha-1}ds$$

And

$$II = \int_{t_{i-1}}^{t_i}\left(\frac{f_i - f_{i-1}}{k} + kf''(\xi_i)\right)(t_i-s)^{-\alpha}ds \tag{13}$$

$$= \frac{f_i - f_{i-1}}{(1-\alpha)k^\alpha} + f''(\xi_i)k^{2-\alpha} \tag{14}$$

Substituting (13)) and (14) into (4), we have

$$D^\alpha f(t) = D_k^\alpha f(t_i) - \alpha\sum_{j=1}^{i-1}\int_{t_{j-1}}^{t_j} f''(\xi_j)k^2(t_i-s)^{-\alpha-1}ds + f''(\xi_i)k^{2-\alpha} \tag{15}$$

Therefore,

$$\| D^\alpha f(t) - D_k^\alpha f(t) \|_\infty \leq \frac{\alpha+1}{\Gamma(1-\alpha)} \| f'' \|_\infty k^{2-\alpha} \tag{16}$$

□

Lemma 3.2 *Let* $D_k^\alpha \tilde{f}_i = \sigma_{\alpha,k}\sum_{j=0}^{i}\omega_j\tilde{f}_j$. *Then*

$$|D_k^\alpha f(t_i) - D_k^\alpha \tilde{f}_i| \leq \frac{\delta}{\Gamma(1-\alpha)(1-\alpha)k} \tag{17}$$

Proof : It's not difficult to show that

$$| D_k^\alpha f(t_i) - D_k^\alpha \tilde{f}_i | = \sigma_{\alpha,k} \sum_{j=0}^{i} \omega_j |f(t_j) - \tilde{f}_j| \tag{18}$$

$$\leq \sigma_{\alpha,k}\delta \sum_{j=0}^{i} \omega_j \leq \frac{\delta n^{1-\alpha}}{\Gamma(1-\alpha)(1-\alpha)k^\alpha} \tag{19}$$

$$= \frac{\delta}{\Gamma(1-\alpha)(1-\alpha)k} \tag{20}$$

□

As k tend to zero, the estimation of convergence (7) tent to zero. However, at the same time, the error (17) tent to infinity when the noise level $\delta > 0$. Hence, for balance them, the step size k should be chosen depending on the noise level δ. Then, combining the Lemma 1 and 2, the following theorem can be obtain.

Theorem 3.3 *Let* $k = \left(\frac{\delta}{(1-\alpha^2)\|f''\|_\infty}\right)^{\frac{1}{3-\alpha}}$. *Then*

$$\| D^\alpha f(t) - D_k^\alpha \tilde{f}_i \|_\infty \leq C' k^{\frac{2-\alpha}{3-\alpha}} \tag{21}$$

where $C' = \left(\frac{(\alpha+1)\|f''\|_\infty}{(1-\alpha)^{2-\alpha}}\right)^{\frac{1}{3-\alpha}}$

Remark 3.4 *In the real application,* $\| f'' \|_\infty$ *is usually not known, therefore we have no the exact step size* $k(\delta)$*. However, if we select* $k(\delta) = \left(\frac{\delta}{(1-\alpha^2)}\right)^{\frac{1}{3-\alpha}}$*, we can also obtain*

$$\| D^\alpha f(t) - D_k^\alpha \tilde{f}_i \|_\infty \leq C'' k^{\frac{2-\alpha}{3-\alpha}} \tag{22}$$

where the constant C'' *just only depends on* k *and* $\| f'' \|_\infty$*. The following numerical results show this choosing rule is still stable in the realistic computation.*

4. Numerical Examples

To demonstrate the effectiveness and stability of the proposed difference method, the following example is presented in this section. For simplicity, we assume that the interval $[a, b]$ always be $[0, 1]$. The accuracy of the numerical solution is measured by the the following root mean square error (RMSE):

$$RMSE = \sqrt{\frac{1}{n+1} \sum_{i=0}^{n} (D^\alpha f(t_i) - D_k^\alpha \tilde{f}_i)^2} \tag{23}$$

We consider a smooth function

$$f(t) = t^2$$

The exact α order fractional derivative is

$$D^\alpha f(t) = \frac{\Gamma(3)}{\Gamma(3-\alpha)} t^{2-\alpha} \tag{24}$$

Figure 1 shows the exact α order fractional derivative with different parameter α. For illustrate the convergence of the proposed difference method, first of all, we compute the fractional derivatives without noise by using the formula (5). From Figure 2, it's easy to see that the RMSEs by using our proposed algorithm converge to zero fast. And the numerical accuracies with small α are higher than large α. This is consistent with the theoretical result in Lemma 1.

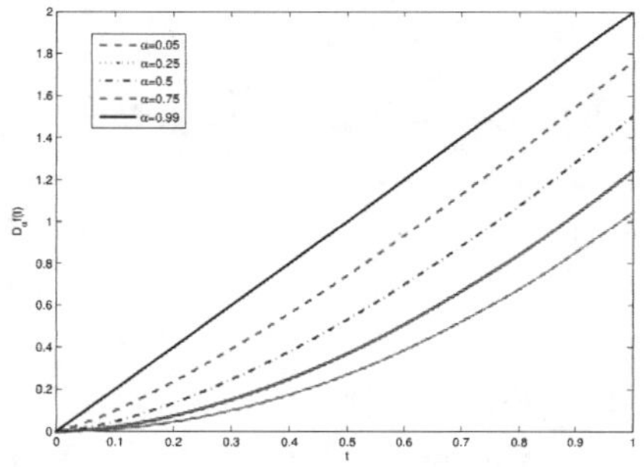

Figure 1. *The exact α order fractional derivative with different parameter α.*

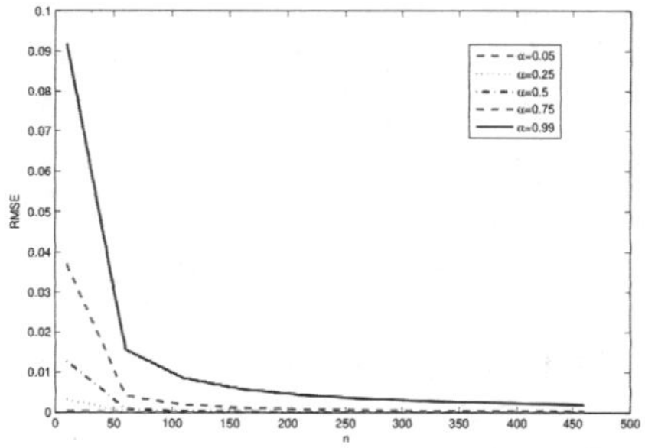

Figure 2. *The RMSEs without noise for different numbers of points n*

Then, for overcoming the ill-posedness, we choose the step size with the proposed rule $k = \left(\frac{\delta}{(1-\alpha^2)}\right)^{\frac{1}{3-\alpha}}$. Figure 3 shows that the proposed regularization difference method is stable for solving the numerical fractional derivative with $\alpha \in (0, 1)$.

5. Acknowledgement

The work described in this paper was supported by a grant from the Research Grants Council of the Hong Kong Special Administrative Region, China (Project No. CityU 101310).

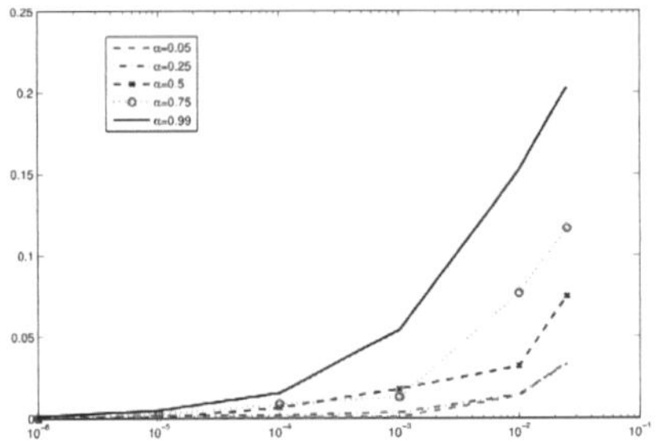

Figure 3. *The RMSEs with different noise level δ.*

6. Conclusion

In this paper, a regularization difference method is proposed to solve numerical fractional derivatives. The convergence rate and the error estimation of the proposed method are obtained. Numerical results indicate that the proposed method gives an accurate and reliable scheme and performances.

[1] Qian Z., Fu C. L., Xiong X. T. and Wei T., Fourier truncation method for high order numerical derivatives, Applied Mathematics and Computation, Vol. 181, pp. 940-948, 2006.

[2] Qu R., A new approach to numerical differentiation and integration, Mathematical and Computer Modelling Vol. 24, pp. 55-68, 1996.

[3] Cullum J., Numerical differentiation and regularization, SIAM Journal on Numerical Analysis, Vol. 8, pp. 254-265, 1971.

[4] Wang Y. B., Jia X. Z. and Cheng J., Anumerical differentiation method and its application to reconstruction of discontinuity, Inverse Problems, Vol. 18, pp. 1461-1476, 2002.

[5] Wei T. and Li M., High order numerical derivatives for one-dimensional scattered noisy data, Applied Mathematics and Computation, Vol. 175, pp. 1744-1759, 2006.

[6] Hào D. N., a mollification method for ill-posed problems, Numerical Mathematics, Vol. 68, pp. 469-506, 1994.

[7] Murio D. A., Mejia C. E. and Zhan S., Discrete mollification and automatic numerical differentiation, Computers and Mathematics with Applications, Vol. 35, pp. 1-16, 1998.

[8] Liu F., Anh V. and Turner I., Numerical solution of the space fractional Fokker-Planck equation, Journal of Computational and Applied Mathematics - Special issue: Proceedings of the international conference on boundary and interior layers, Vol. 166, pp. 209-219, 2004.

[9] Yuste S. B. and Acedo L., An explicite finite difference method and a new von Neumann-type stability analysis for fractional diffusion equations, SIAM Journal on Numerical Analysis, Vol. 42, pp. 1862-1874, 2005.

International Conference on Inverse Problems 2010 IOP Publishing
Journal of Physics: Conference Series **290** (2011) 012012 doi:10.1088/1742-6596/290/1/012012

Determination of welding residual stresses by inverse approach with eigenstrain formulations of boundary integral equation

Hang Ma[1,3], Ying Wang[1] and Qing-Hua Qin[2]

[1] College of Sciences, Shanghai University, Shanghai 200444, China
[2] School of Engineering, Australian National University, Canberra, ACT 0200, Australia
[3] Corresponding author: hangma@staff.shu.edu.cn

Abstract. Based on the concept of eigenstrain, a straightforward computational model of the inverse approach is proposed for determining the residual stress field induced by welding using the eigenstrain formulations of boundary integral equations. The eigenstrains are approximately expressed in terms of low-order polynomials in the local area around welded zones. The domain integrals with polynomial eigenstrains are transformed into the boundary integrals to preserve the favourable features of the boundary-only discretization in the process of numerical solutions. The sensitivity matrices in the inverse approach for evaluating the eigenstrain fields are constructed by either the measured deformations (displacements) on the boundary or the measured stresses in the domain after welding over a number of selected measuring points, or by both the measured information. It shows from the numerical examples that the results of residual stresses from deformation measurements are always better than those from stress measurements but they are sensitive to the noises from experiments. The results from stress measurements can be improved by introducing a few deformation measuring points while reducing the number of points for stress measuring to reduce the cost since the measurement of deformation is easier than that of stresses in practice.

1. Introduction

The residual stresses, induced during the process of manufacturing or technological treatment of engineering structures, have a significant influence on the performance of related components in service [1]. When such a component is in service, the associated residual stresses may superimpose on the applied stress to influence the deformation behavior of components. Therefore, the residual stresses are generally regarded as detrimental. Whatever the impact of residual stresses on the component, their detection and measurement is an important technical issue. However, since the nature of residual stress is in self-equilibrium, the determination of it is not an easy task.

There are a great number of techniques to detect the residual stresses in a solid which can be classified as three major groups: the physical, the mechanical and the numerical techniques. In the mechanical techniques, since the direct detection is impossible, parts of the material have to be removed from the solid to disturb the stress balance while the response of the specimen is measured in terms of either strains or shape changes at some other locations on the surface of the body. Obviously, in addition to the great cost of the mechanical methods, all of them are more or less destructive to the measured component by the material removal. It is evident that the mechanical techniques always provide a limited level of detail, due to the finite number of discrete data points that restricts the possibility of reconstructing full-field stress distributions.

Numerical methods present a supplementary but effective means for determining residual stresses. As is widely accepted, residual stresses in engineering components are caused by incompatible internal permanent strains, named originally as the inherent strains [2-3] and lately as the eigenstrains [4], induced by any inhomogeneous inelastic deformation, temperature gradients or phase transformations during the manufacturing and processing of components. On the other hand, the theory of inverse methods has been actively developed over the past decades. In particular, inverse problems in elasticity have gained great attention including that of determining residual stresses based on the concept of eigenstrain which makes use of the information observed from experiments [5-7]. With the aid of numerical methods, generally the finite element method (FEM) [3,5-7], the unknown eigenstrain distributions can be retrieved following the mathematical framework of inverse problem with eigenstrain theory to obtain the whole field of residual stresses. In spite of the inelastic origin of eigenstrains, the inherent state of residual stress fields falls really into elastic regime so that the boundary element method (BEM) would be the most efficient numerical means to deal with the residual stress problems [8-9]. With the aid of the BEM it is also possible to determine the residual stresses nondestructively utilizing the deformations measured after welding with a forward approach [10].

In the present work, based on the concept of eigenstrain, a straightforward computational model of inverse approach is proposed with the eigenstrain formulations of boundary integral equations to determine the welding residual stresses, together with the numerical examples verifying the proposed computational model.

2. Eigenstrain Formulations of Boundary Integral Equation

The boundary integral equations for displacements u_i and stresses σ_{ij}, respectively [11], can be written as follows for an elastic domain Ω with the boundary Γ containing eigenstrains ε^0_{ij} in local area Ω_I ($\Omega_I \in \Omega$):

$$Cu_i(p)+\int_\Gamma u_j(q)\tau^*_{ij}(p,q)\mathrm{d}\Gamma(q)=\int_\Gamma \tau_j(q)u^*_{ij}(p,q)\mathrm{d}\Gamma(q)+\int_{\Omega_I}\varepsilon^0_{jk}(q)\sigma^*_{ijk}(p,q)\mathrm{d}\Omega(q) \quad (1)$$

$$C\sigma_{ij}(p)=\int_\Gamma \tau_k(q)u^*_{ijk}(p,q)\mathrm{d}\Gamma(q)-\int_\Gamma u_k(q)\tau^*_{ijk}(p,q)\mathrm{d}\Gamma(q)$$
$$+\int_{\Omega_I-\Omega_\varepsilon}\varepsilon^0_{kl}(q)\sigma^*_{ijkl}(p,q)\mathrm{d}\Omega(q)+\varepsilon^0_{kl}(p)O^*_{ijkl} \quad (2)$$

where p and q are the source and the field points, u^*_{ij}, τ^*_{ij}, and σ^*_{ij} represent the fundamental solutions for displacement, traction and stress, respectively. u^*_{ijk}, τ^*_{ijk}, and σ^*_{ijk} are the related derivatives. C is a conventional boundary shape coefficient. Ω_ε is a tiny ε region around p when $p\in\Omega_I$ and O^*_{ijkl} is the corresponding free term from the domain integral in (2).

It is obvious from (1) and (2) that once the distributions of eigenstrains ε^0_{ij} in domain integrals are known, the unknown boundary displacements can be solved using (1) and the total fields of stresses can be computed using (2). Considering the features of thermal cycles in welding, suppose that the distributions of eigenstrains can be approximately expressed in terms of low-order polynomials in the local area Ω_I as follows:

$$\varepsilon^0_{ij}=\sum_{m=0,n=0}^{m+n=M}\alpha^{mn}_{ij}x^m_1(q)x^n_2(q) \quad (3)$$

where M is the number of terms of polynomials and α_{ij}^{mn} the coefficients to be identified. m and n are integers. The above equation gives a smooth constraint on the eigenstrain field. The domain integrals with polynomial eigenstrains in (1) and (2) can be transformed into the boundary integrals by introducing the two-point variables as follows

$$x_i=x_i(q)-x_i(p) \quad (4)$$

which are in fact the projection of the two-point distance, r_i. With this definition, the domain integrals with the certain term of polynomials in (1) and (2), respectively, can be expressed in the form of two-point polynomials and then the domain integrals at the right hand sides in (1) and (2) with eigenstrains can be transformed into the boundary integrals [12]. In this way, the favorable features of the boundary-only discretization are preserved.

3. Inverse Approach

In the inverse approach, the information from experiments is required to identify the unknown coefficients α_{ij}^{mn} in (3), including selected M_U points for displacements measured on boundary and selected M_S points for stresses measured in domain after welding. In the present work, since the residual stresses of welded plates are two-dimensional, the displacements are to be measured only in one direction, x_1 or x_2 for u_1 or u_2, at one point. The stresses are to be measured only for two normal components, σ_{11} and σ_{22}, at one point. Therefore the numbers of known information from experiments are M_U and $2M_S$, for displacements and stresses, respectively. Owing to the same reason, only normal components of eigenstrains, ε^0_{11} and ε^0_{22}, are considered so that the number of unknown coefficients α_{ij}^{mn} in (3) is $2M$.

Using the boundary point method (BPM) [13] and noticed the traction-free condition of residual problems, the displacement equation (1) can be discretized into a system of algebraic equation in matrix form as

$$\mathbf{Hu} = \mathbf{B}\boldsymbol{\alpha} \tag{5}$$

where **u** is the vector of displacements at all the N nodal points on the boundary Γ, **α** the vector of unknown coefficients in (8). After **α** is identified, the boundary displacements can be computed:

$$\mathbf{u} = \mathbf{H}^{-1}\mathbf{B}\boldsymbol{\alpha} \tag{6}$$

3.1. Residual stresses determined using measured displacements

Making use of the proposed algorithm with eigenstrain formulations, it is possible to determine the residual stresses nondestructively utilizing the deformations measured after welding [10]. Suppose that the displacements are measured on boundary Γ at selected M_U points from the experiment, the required sensitivity matrix $\mathbf{S}_u$ can be obtained by reducing the matrix product $\mathbf{H}^{-1}\mathbf{B}$ according to the displacement measuring points, that is

$$\mathbf{H}^{-1}\mathbf{B} \xrightarrow{reduce} \mathbf{S}_u \tag{7}$$

Then the unknown coefficients of eigenstrains can be obtained using the least squire method by minimizing the object function, **Φ**, defined as follows

$$\boldsymbol{\Phi} = \frac{1}{2}\left\|\mathbf{S}_u\boldsymbol{\alpha} - \bar{\mathbf{u}}\right\|^2 \tag{8}$$

where $\bar{\mathbf{u}}$ represent the vector of measured displacements. The minimizing condition of (8) is

$$\mathbf{S}_u^{\mathrm{T}}\left(\mathbf{S}_u\boldsymbol{\alpha} - \bar{\mathbf{u}}\right) = 0 \tag{9}$$

Therefore the unknown eigenstrain coefficients can be computed by

$$\boldsymbol{\alpha} = \left(\mathbf{S}_u^{\mathrm{T}}\mathbf{S}_u\right)^{-1}\mathbf{S}_u^{\mathrm{T}}\bar{\mathbf{u}} \tag{10}$$

It is obvious in practice that the number of measured displacement points is far less than that of the boundary nodal points ($M_U<N$). However, the number of measured points should not be less than that of eigenstrain coefficients, i.e. $M_U \geq 2M$.

3.2. Residual stresses determined using measured stresses

Suppose that the stresses are measured in domain Ω at selected M_S points from the experiment, the sensitivity matrix $\mathbf{S}_\sigma$ can be obtained by expressing the stresses using the stress equation (2) in matrix form and then by substituting (6) into it as follows

$$\boldsymbol{\sigma} = \mathbf{Fu} + \mathbf{D}\boldsymbol{\alpha} = \left(\mathbf{FH}^{-1}\mathbf{B} + \mathbf{D}\right)\boldsymbol{\alpha} = \mathbf{S}_\sigma\boldsymbol{\alpha} \tag{11}$$

where **F** and **D** stands for the corresponding coefficient matrices in the stress equation (2). Similarly, the unknown coefficients of eigenstrains can be obtained using the least squire method by minimizing the object function, $\mathbf{\Phi}$, defined as follows

$$\mathbf{\Phi} = \frac{1}{2}\left\|\mathbf{S}_u\boldsymbol{\alpha} - \bar{\boldsymbol{\sigma}}\right\|^2 \tag{12}$$

where $\bar{\boldsymbol{\sigma}}$ represent the vector of measured stresses. The unknown eigenstrain coefficients can be computed by the minimizing condition of (12) and the unknown α can be obtained

$$\mathbf{S}_\sigma^{\mathrm{T}}\left(\mathbf{S}_\sigma\boldsymbol{\alpha} - \bar{\boldsymbol{\sigma}}\right) = 0 \tag{13}$$

$$\boldsymbol{\alpha} = \left(\mathbf{S}_\sigma^{\mathrm{T}}\mathbf{S}_\sigma\right)^{-1}\mathbf{S}_\sigma^{\mathrm{T}}\bar{\boldsymbol{\sigma}} \tag{14}$$

The number of measured points for stresses should obey the inequality $2M_S \geq 2M$.

3.3. Residual stresses determined using both of the measured displacements and stresses

In order to cut down the cost of stress measurements by reducing the number of measuring points, the hybrid procedure can be used by combining the measured displacement and stress data. Suppose that the displacements be measured on boundary Γ at selected M_U points and the stresses be measured in domain Ω at selected M_S points from the experiment. The object function $\mathbf{\Phi}$ can be defined as follows using the combined sensitivity matrix:

$$\mathbf{\Phi} = \frac{1}{2}\left\|\begin{Bmatrix}\mathbf{S}_u \\ \mathbf{S}_\sigma\end{Bmatrix}\boldsymbol{\alpha} - \begin{Bmatrix}\bar{\mathbf{u}} \\ \bar{\boldsymbol{\sigma}}\end{Bmatrix}\right\|^2 \tag{15}$$

The unknown eigenstrain coefficients α can then be computed in a way similar to those described above by minimizing (15):

$$\left(\mathbf{S}_u^{\mathrm{T}}, \mathbf{S}_\sigma^{\mathrm{T}}\right)\left(\begin{Bmatrix}\mathbf{S}_u \\ \mathbf{S}_\sigma\end{Bmatrix}\boldsymbol{\alpha} - \begin{Bmatrix}\bar{\mathbf{u}} \\ \bar{\boldsymbol{\sigma}}\end{Bmatrix}\right) = \mathbf{0} \tag{16}$$

$$\boldsymbol{\alpha} = \left(\mathbf{S}_u^{\mathrm{T}}\mathbf{S}_u + \mathbf{S}_\sigma^{\mathrm{T}}\mathbf{S}_\sigma\right)^{-1}\left(\mathbf{S}_u^{\mathrm{T}}, \mathbf{S}_\sigma^{\mathrm{T}}\right)\begin{Bmatrix}\bar{\mathbf{u}} \\ \bar{\boldsymbol{\sigma}}\end{Bmatrix} \tag{17}$$

The number of measured points in the hybrid procedure should satisfy the inequality $M_U + 2M_S \geq 2M$. With the known eigenstrains, the total stress fields in the domain can be computed using the stress equation (2), after the boundary displacements are solved using (1).

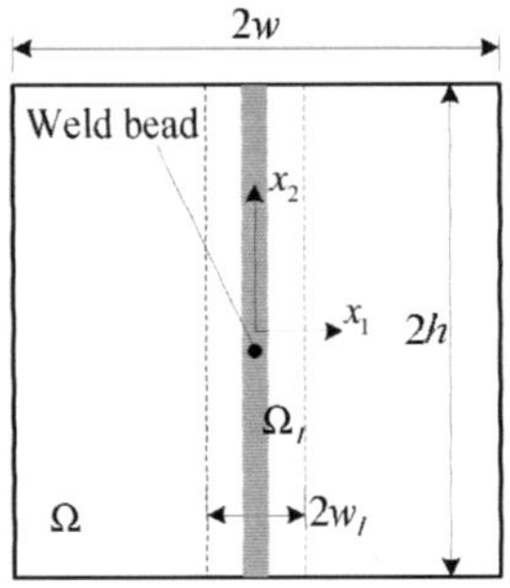

Figure 1. Weld plate with centre weld bead

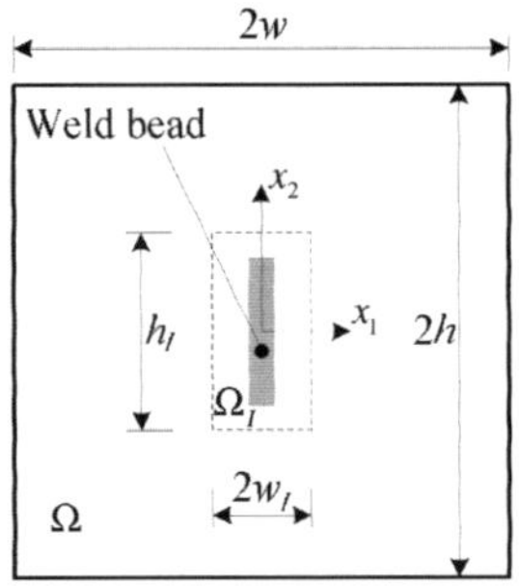

Figure 2. Weld plate with constrained weld bead

4. Numerical examples

In the numerical examples, two square weld plates (h=w) are considered with a center weld bead (figure 1) and a restrained weld bead (figure 2), respectively. The latter is usually used as the test plate for weld cracking. The width of local area Ω_l where the eigenstrains distribute is set as w_l=0.2w for the two plates, which is somewhat wider than that of the HAZ according to the parameters of material and welding [10]. The length of local area Ω_I is set as h_I=0.4h for the plate with restrained weld bead (figure 2). The distribution of eigenstrains in local region is obviously incompatible.

Table 1. The coefficients of eigenstrains employed

α^0_{11}	α^1_{11}	α^2_{11}	α^0_{22}	α^1_{22}	α^2_{22}
$-c_1$	$4c_1w_I^4$	$3c_1h_I^2$	$-c_2$	$3c_2w_I^2$	$4c_2h_I^4$

Table 2. The parameters in table 1

Plate	Centre bead	Constrained bead
c_1	4	4
c_2	1.1	0.8

The 'exact' solution of residual stresses can be obtained as controls using the BPM with the values of eigenstrains listed in tables 1 and 2. The measuring points for stresses and displacements are shown in figure 3.

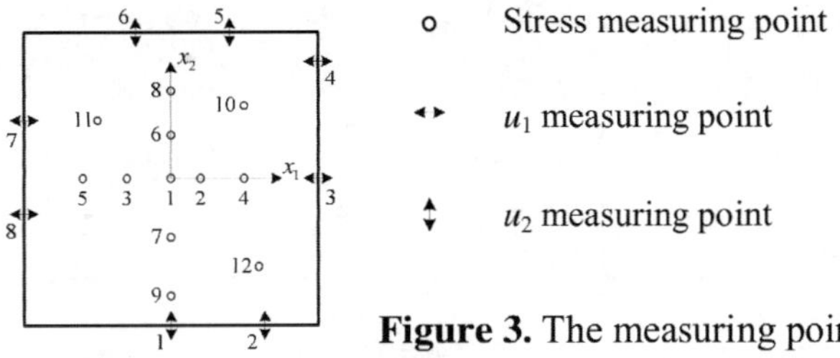

Figure 3. The measuring points

The idealized measuring data for stress and displacement are computed using the BPM at the measuring points with the values of eigenstrains listed in tables 1 and 2. With these idealized data, the residual stress can be reconstructed after solving the eigenstrains using the inverse approaches stated above. However, as there are always errors in the experimental measurements, 10% noises are introduced into the idealized data as follows

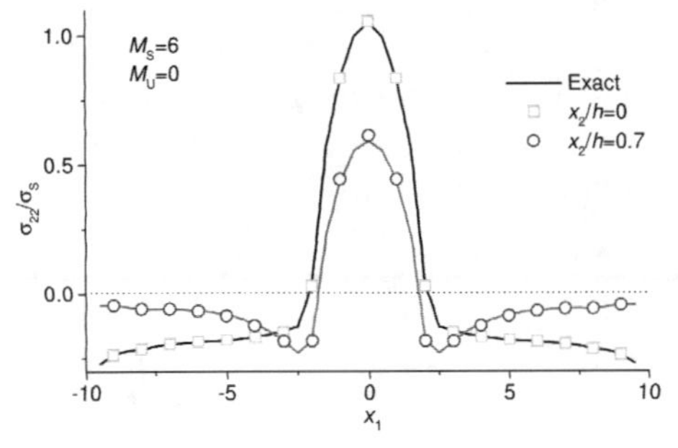

Figure 4. Longitudinal stresses in the plate with centre weld bead

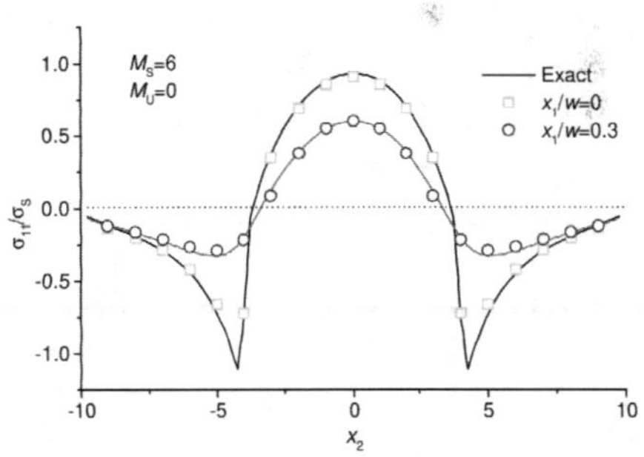

Figure 5. Transverse stresses in the plate with constrained weld bead

$$\bar{\mathbf{u}}_{noise} = (1 \pm 0.1ran)\bar{\mathbf{u}} \tag{20}$$

$$\bar{\boldsymbol{\sigma}}_{noise} = (1 \pm 0.1ran)\bar{\boldsymbol{\sigma}} \tag{21}$$

where *ran* is the random function varying between 0 and 1. With these noise introduced data, the residual stress can also be reconstructed after solving the eigenstrains using inverse approaches.

The normalized longitudinal and the transverse stresses, $\sigma_{22}\cdot\sigma_S^{-1}$ and $\sigma_{11}\cdot\sigma_S^{-1}$, for the two plates are compared with the 'exact' solutions in figures 4 and 5, respectively, with the measured stresses at a few points (M_S=6, M_U=0), indicating the feasibility and effectiveness of the present algorithm where σ_S is the yield strength of material. The present eigenstrain formulations of the BIE are considered to be in the most consistent with the concept of eigenstrain.

Table 3. Errors for the plate with centre weld bead using idealized measuring data

M_S	M_U	$\varepsilon_{MAX}(\sigma_{11})$	$\varepsilon_{MAX}(\sigma_{22})$	$\varepsilon_{RMS}(\sigma_{11})$	$\varepsilon_{RMS}(\sigma_{22})$
0	8	0	0	0	0
6	0	0.0516	0.0194	0.008590	0.000066
9	0	0.0153	0.0146	0.002480	0.000667
12	0	0.0153	0.0143	0.002482	0.000650
2	8	0.0019	0.0086	0.000320	0.000050
4	8	0.0038	0.0096	0.000637	0.000010

Table 4. Errors for the plate with centre weld bead using noise introduced measuring data

M_S	M_U	$\varepsilon_{MAX}(\sigma_{11})$	$\varepsilon_{MAX}(\sigma_{22})$	$\varepsilon_{RMS}(\sigma_{11})$	$\varepsilon_{RMS}(\sigma_{22})$
0	8	0.0835	0.6114	0.015592	0.003765
6	0	0.0793	0.0492	0.013604	0.000183
9	0	0.0253	0.0325	0.003887	0.000714
12	0	0.0254	0.0328	0.003882	0.000698
2	8	0.0312	0.0430	0.006563	0.004127
4	8	0.0309	0.0409	0.006051	0.004006

The maximum errors, ε_{MAX}, and the root mean square errors, ε_{RMS}, for the plate with the center weld bead using the idealized data and the noise introduced data are compared in tables 3 and 4, respectively, showing that the results of the inverse approach with measured displacements are satisfactory with the idealized data but very sensitive to the noises. In contrast, the results of the inverse approach with measured stresses are relatively insensitive to the noises, which can be improved further by introducing a few deformation measuring points while reducing the number of stress measuring points.

It is demonstrated in practice that the welding residual stresses can be determined nondestructively with the measuring deformations [10] using a forward approach with iterations. With the inverse approach, however, it cannot be employed independently due to its high sensitivity to the experimental noises. The inverse approach with measured displacements should be used in combination with the measured stresses for the purpose of cutting down the cost in practice.

5. Conclusions

(1) A straightforward computational model of the inverse approach is proposed for determining the welding residual stress field with the eigenstrain formulations of boundary integral equations, which are considered to be in most consistent with the concept of eigenstrain.

(2) The sensitivity matrices in the inverse approach for evaluating the eigenstrain fields can be constructed by either the measured deformations or the measured stresses or by both the measured information over a number of selected measuring points.

(3) Using the boundary point method, the numerical examples verify the feasibility and the effectiveness of the present algorithm. The results from stress measurements can be improved by introducing a few deformation measuring points while reducing the number of stress measuring points to cut down the cost.

Acknowledgments

The work was supported by the National Natural Science Foundation of China (Grant No. 10972131) and the Graduate Innovative Foundation of Shanghai University (Grant No. SHUCX102351).

References

[1] Masubuchi K 1980 *Analysis of Welded Structures-Residual Stresses, Distortion and Their Consequences* (New York: Pergamon Press)
[2] Ueda Y, Fukuda K, Nakacho K and Endo S 1975 *Transactions of JWRI* **4** 123–31
[3] Ueda Y, Jirn Y and Yuan M 1986 *Transactions of the Japan Welding Society* **6** 59–64
[4] Mura T, Shodja H M and Hirose Y 1996 *Applied Mechanics Review* **49** S118–S127
[5] Korsunsky A M, Regino G M and Nowell D 2007 *International Journal of Solids and Structures* **44** 4574–91
[6] Jun T-S and Korsunsky A M 2010 *International Journal of Solids and Structures* **47** 1678–86
[7] Cao Y P, Hu N, Lu J, Fukunaga H and Yao Z H 2002 *Journal of Strain Analysis for Engineering Design* **37** 345–59
[8] Qian X Q, Yao Z H, Cao Y P and Lu J 2004 *Engineering Analysis with Boundary Elements* **28** 205–11
[9] Qian X Q, Yao Z H, Cao Y P and Lu J 2005 *Engineering Analysis with Boundary Elements* **29** 986–99
[10] Ma H and Deng H L 1998 *Advances in Engineering Software* **29** 89–95
[11] Ma H, Yan C and Qin Q H 2009 *Engineering Analysis with Boundary Elements* **33** 410–19
[12] Ma H, Kamiya N and Xu S Q 1998 *Engineering Analysis with Boundary Elements* **21** 271–75
[13] Ma H, Zhou J and Qin Q H 2010 *Advances in Engineering Software* **41** 480–8

Journal of Physics: Conference Series **290** (2011) 012013 doi:10.1088/1742-6596/290/1/012013

Experimental studies on illposed singularly perturbed boundary value problems for parabolic differential equations

Hans-Jürgen Reinhardt

Department of Mathematics, University of Siegen, Emmy-Noether-Campus, Walter-Flex-Str. 3, D-57072 Siegen, Germany

E-mail: `reinhardt@mathematik.uni-siegen.de`

Abstract. In this paper singularly perturbed parabolic initial-boundary value problems are considered which, in addition, are illposed. The latter means that at one end of the 1-d spatial domain two conditions (for the solution and its spatial derivative) are given while on the other end the corresponding quantities are to be determined. It is well-known that such problems are illposed in the mathematical sense. Here, in addition, boundary layers may occur which make the problems more difficult. For relatively simple examples numerical experiments have been carried out and numerical results are shown. The Conjugate Gradient Methods is used to find the desired quantities iteratively. It will be explained what has to be done in any iteration step. A regularisation is performed by means of discretization and by determining an optimal final iteration step via a stopping rule.

1. Introduction

Singularly perturbed differential equations have the essential feature that a small parameter multiplies the highest derivative. As a consequence, boundary layers may appear at parts of the boundary or even at inner locations. We restrict ourselves to a one-dimensional setting. Typical examples of singularly perturbed ordinary differential equations are displayed in the following Figures 1 and 2, (see the books [1], [2], [3] and the references therein). In parabolic equations, the boundary layers can move (in time). To make it as simple as possible, we do not consider moving boundary layers, and we assume that the solution of the parabolic equation exits and has the form $v_\varepsilon(x,t) = w(t)u_\varepsilon(x)$, where u_ε may be the solution of a singularly perturbed ordinary differential equation having boundary layers - or not.

In parabolic equations one type of inverse problems is the 'Inverse Heat Conduction Problem' or the 'noncharacteristic Cauchy problem' (see [4], [5], [6], [7] and many others). Here, the problem consists in determining the boundary data - i.e. temperature and heat flux - at inaccessible parts of the boundary from corresponding data at accessible parts. The latter are called the 'Cauchy data'. In a one-dimensional setting, the Cauchy data are given at one side of the spatial interval while the unknown quantities have to be determined at the other side. Such problems are known to be ill-posed in the mathematical sense (see, e.g. [8]).

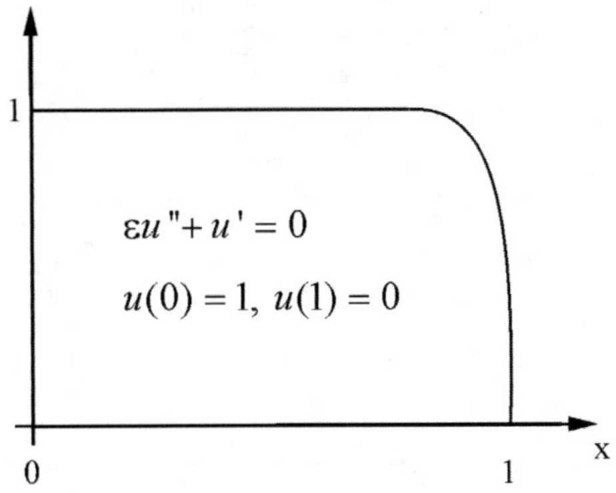

Figure 1. Example of singularly perturbed ode

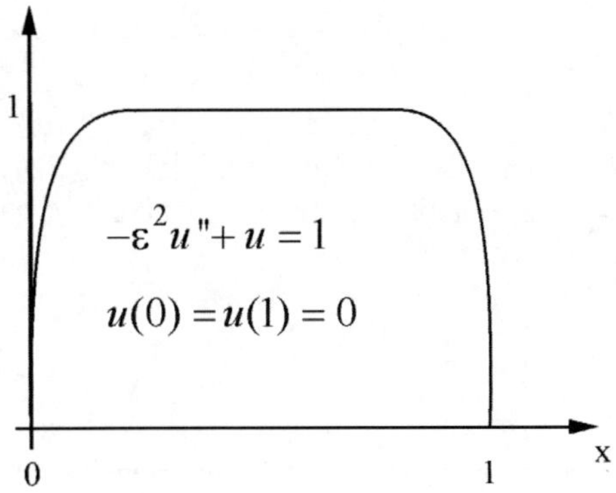

Figure 2. Example of singularly perturbed ode

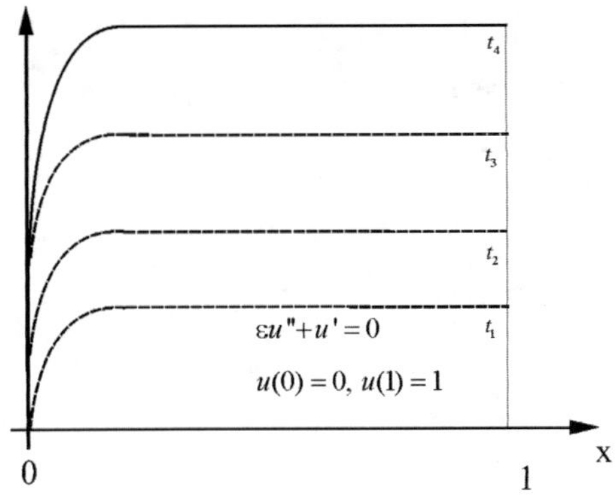

Figure 3. 2 Example of singularly perturbed parabolic equation, $v_\varepsilon(x,t) = w(t)u_\varepsilon(x)$

In this contribution we pose the question, whether in an inverse heat conduction problem the boundary data at the inaccessible part of the boundary can be determined if the underlying equation is singularly perturbed. More precisely, we consider two situations. At first, the Cauchy data are given at the side of the spatial interval where a boundary appears; at the other side no boundary layer exists. Secondly, at the side of the Cauchy data there is no boundary layer while at the other (inaccessible) part of the spatial interval a boundary layer appears. One may consider further situations - like boundary layers of both sides of the spatial interval or equations with inner layers - which will be the subjects of further studies.

The present study tries to answer the question of identifiability of boundary data in case of boundary layers by means of numerical methods. For the iterative solution of the illposed problem we use the Conjugate Gradient Method and solve the underlying parabolic problems by the Crank-Nicolson-Galerkin Method.

For the Conjugate Gradient Method solving Inverse Heat Conduction Problems we have a lot of experience (see [9], [10], [6], [11], [12], [13], [14]). Concerning the other aspect of solving singularly perturbed ordinary and partial differential equations we have used adaptive finite element methods to solve such problems effectively in [Rei81a], [Rei81b], [Rei82a], [Rei82b], [Rei85], [Rei91]. The adaptive and automatic construction of grids is no subject of this paper but can be exploited in further studies.

This paper is organized as follows. After this introduction, illposed singularly perturbed parabolic equations will be introduced. Then the variational approach for solving such problems is explained and the Conjugate Gradient Methods is presented in detail. Finally, experimental results for three examples are displayed and discussed. Conclusions and a list of references finalize this work.

2. Illposed Singularly Perturbed Parabolic Equations

We consider singularly perturbed parabolic partial differential equations of the form

$$v_t(x,t) = \varepsilon^2 v_{xx}(x,t) + c(x)v(x,t) + F(x,t),\ x \in [0,1],\ t \in (0,T]\,, \tag{1}$$

together with boundary conditions at $x = 0$ and $x = 1$ and an inital condition at $t = 0$,

$$v(x,0) = v_0(x) \tag{2}$$

Here, ε denotes a small positive parameter. We assume that $c = c(x)$ only depends on the spatial variable. The solution $v = v_\varepsilon(x,t)$ is assumed to be of the form $v_\varepsilon(x,t) = w(t)u_\varepsilon(x)$ where w can be of the form

$$w(t) = 1 - \exp(-t/\beta) \text{ or } w(t) = t\,. \tag{3}$$

The 'source term' $F = F_\varepsilon(x,t)$ in equation (1) is chosen appropriately.

With this framework we have in mind that $u_\varepsilon = u_\varepsilon(x)$ is some solution of a singularly perturbed ordinary differential equation having boundary layers at $x = 0$ or $x = 1$. We do not consider inner boundary layers. Also, we exclude moving boundary layers (in time).

In the associated *inverse problem*, for given c, F and

$$v_0 = v(x,0),\ \varphi(t) = v|_{x=0},\ g(t) = -\varepsilon^2 v_x|_{x=0} \tag{4}$$

one wants to find the heat flux

$$q(t) = \varepsilon^2 v_x|_{x=1}$$

where v is the solution of equation (1), (2) (see Fig. 4).

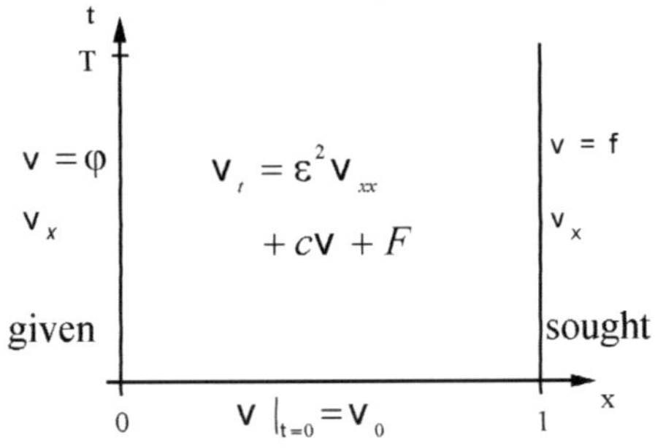

Figure 4. Inverse heat conduction problem

When q is available, also $f = v|_{x=1}$ can be determined. Alternatively, one can first try to find f and then determine q. Such a problem is also called *noncharacteristic Cauchy problem for a parabolic equation*; φ and g are called the *Cauchy data.*

In the following, we start with an initial guess for q and try to improve this iteratively until an appropriate stopping criterion is fulfilled. For this, we also use the notation

$$v = v_\varepsilon(x,t;q)\,.$$

We assume that the above inverse problem (1), (2), (4) has a unique solution.

3. Variational Approach and Conjugate Gradient Method

In the variational approach, instead of solving (1), (2), (4) one tries to minimize the functional

$$J(q) = \frac{1}{2}\|v(0,\cdot;q) - \varphi(\cdot)\|^2_{L^2(0,T)} \tag{5}$$

where v is the solution of the following *direct problem* (DP)

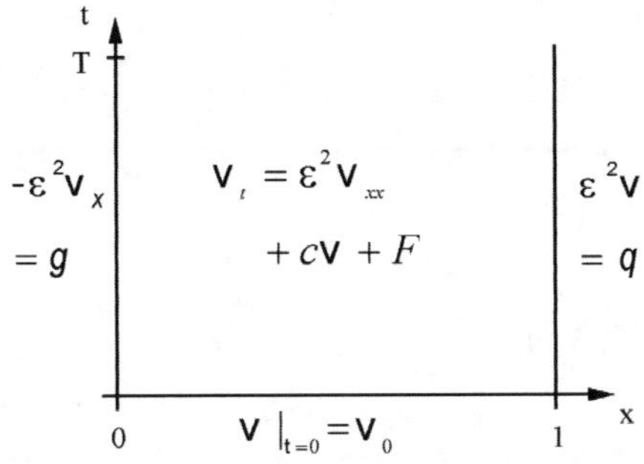

$$\textbf{(DP)} \begin{cases} v_t = \varepsilon^2 v_{xx} + cv + F \\ -\varepsilon^2 v_x|_{x=0} = g,\ \varepsilon^2 v_x|_{x=1} = q \\ v(\cdot,0) = v_0 \end{cases} \tag{6}$$

Figure 5. Direct problem

In (5) one can also add a penalty term, and minimize the functional.

$$J_\gamma(q) = \frac{1}{2}\|v(0,\cdot;q) - \varphi(\cdot)\|^2_{L^2(0,T)} + \gamma\|q\|^2_{L^2(0,T)}\,. \tag{7}$$

This is the well-known *Tikkonov regularization*, and $\gamma > 0$ is the *regularization parameter* which has to be chosen in an optimal way.

The Conjugate Gradient Method (abr.: CGM) is a suitable method for solving such minimisation problems. For this one needs the gradient of the functional (5) w.r.t. q. One can obtain the gradient by means of the solution of the associated *adjoint problem* (AP),

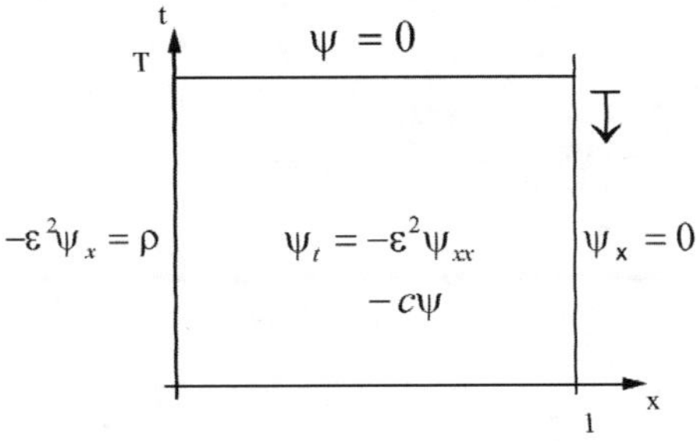

$$\textbf{(AP)} \begin{cases} \psi_t = -\varepsilon^2\psi_{xx} - c\psi,\ 0 \le x \le 1, \quad 0 \le t < T \\ \psi(x,T) = 0 \\ -\varepsilon^2\psi_x|_{x=0} = \underbrace{v(0,t;q) - \varphi(t)}_{=:\rho(t)=\text{ defect/residual}} \\ \psi_x|_{x=1} = 0 \end{cases} \tag{8}$$

Figure 6. Adjoint problem

The gradient is then obtained by

$$J'(q) = \psi(x,t;q)|_{x=1}\,. \tag{9}$$

Note, that the parabolic equation (8) is backward in time but wellposed. Indeed the transformation

$$\tau = T - t,\ \hat{\psi} = \hat{\psi}(x,\tau) = \psi(x, T - \tau)$$

yields the following equations equivalent to (8),

$$(\text{AP}) \Longleftrightarrow \begin{cases} \hat{\psi}_\tau = \varepsilon^2 \hat{\psi}_{xx} + c\hat{\psi},\ 0 \le x \le 1,\ 0 < \tau \le T \\ \hat{\psi}(x,0) = 0(\text{ initial cond.}) \\ -\varepsilon^2 \hat{\psi}_x|_{x=0} = \rho(T-\tau),\ \hat{\psi}_x|_{x=1} = 0 \end{cases} \tag{10}$$

The CGM for solving the minimisation problem (5), (6) proceeds as follows:

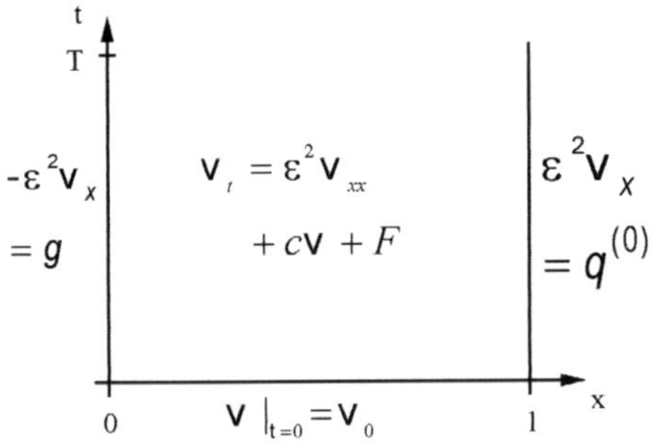

Figure 7. Step 0 of CGM

Step 0: Guess $q = q(t) = q^{(0)}(t)$ (e.g. $q^{(0)} = 0$) and solve

$$(\text{DP}) \begin{cases} v_t = \varepsilon^2 v_{xx} + cv + F,\ 0 \le x \le 1,\ 0 < t \le T \\ -\varepsilon^2 v_x|_{x=0} = g,\ \varepsilon^2 v_x|_{x=1} = q^{(0)}, \\ v|_{t=0} = v_0\,; \end{cases}$$

with solution $v^{(0)} = v(x,t;q^{(0)})$; determine *defect* $\rho^{(0)} = v^{(0)}(0,\cdot) - \varphi(\cdot)$ and set $k = 0$.

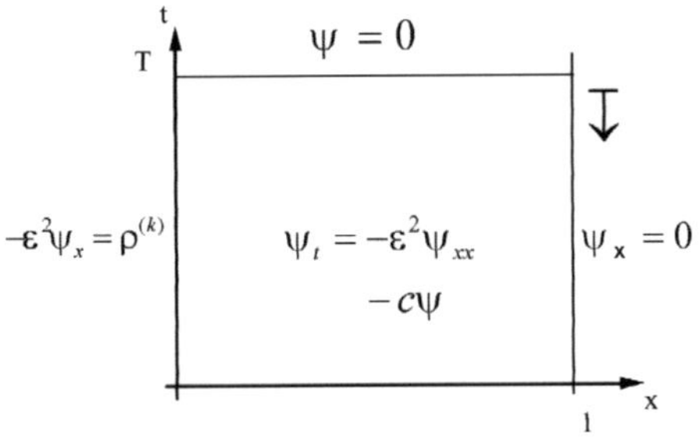

Figure 8. Step 1 of CGM

Step 1: Solve (AP) with $q^{(k)}$ ($\to$ solution $\psi^{(k)}$) i.e.

$$\begin{aligned} &\psi_t^{(k)} = -\varepsilon^2 \psi_{xx}^{(k)} - c\psi^{(k)}, \quad 0 \le x \le 1, \quad 0 \le t < T, \\ &\psi^{(k)}(x,T) = 0, \quad 0 \le x \le 1, \\ &-\varepsilon^2 \psi_x^{(k)}|_{x=0} = \underbrace{v^{(k)}(0,\cdot) - \varphi(\cdot)}_{\rho^{(k)}}, \quad \psi_x^{(k)}|_{x=1} = 0, \\ &\qquad\qquad 0 \le t < T; \end{aligned}$$

set $d^{(0)} = -r^{(0)} (k=0)$, $d^{(k)} = -r^{(k)} + \beta_k d^{(k-1)}$ $(k \ge 1)$ where $\beta_k = \|r^{(k)}\|_{L^2}^2 / \|r^{(k-1)}\|_{L^2}^2 (k \ge 1)$, $r^{(k)} = \psi^{(k)}|_{x=1}$.

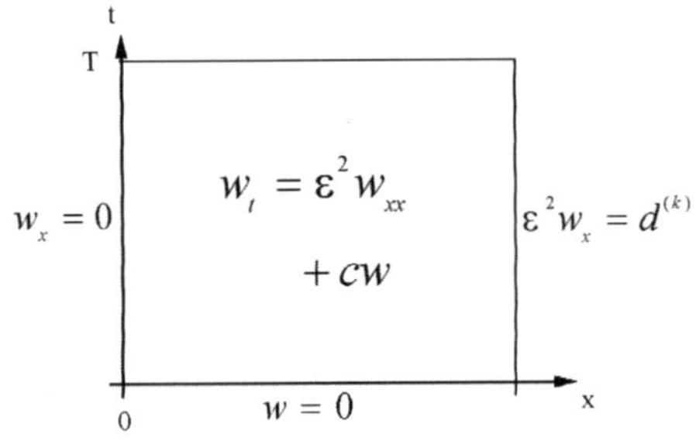

Figure 9. Step 2 of CGM

Step 2: Solve (DP) with $g = 0$, $F = 0$ with initial-boundary conditions

$$\begin{aligned} &w_x|_{x=0} = 0,\ \varepsilon^2 w_x\,|_{x=1} = d^{(k)},\ 0 < t \le T \\ &w|_{t=0} = 0 \end{aligned}$$

($\longrightarrow$ solution $w = w^{(k)} = w^{(k)}(x,t;d^{(k)})$) and set
$\alpha_k = \|r^{(k)}\|_{L^2}^2 \,/\, \|w^{(k)}(0,\cdot;d^{(k)}))\|_{L^2}^2$,
$q^{(k+1)} = q^{(k)} + \alpha_k d^{(k)}$

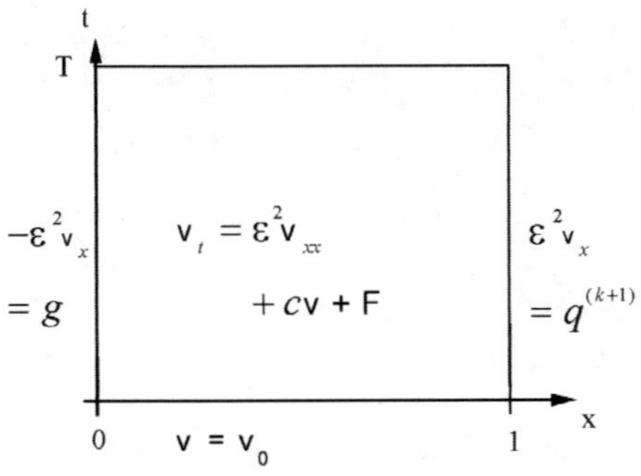

Step 3: Solve (DP) with
$-\varepsilon^2 v_x|_{x=0} = g, \quad \varepsilon^2 v_x|_{x=1} = q^{(k+1)}$
($\longrightarrow$ solution $v^{(k+1)}$) and determine *defect*
$\rho^{(k+1)} = v^{(k+1)}(0, \cdot) - \varphi(\cdot)$.
Go to Step 1 with $k + 1 \longrightarrow k$.
Remark: Only 2 direct solutions are needed in every iteration step!

Figure 10. Step 3 of CGM

Note that perturbations of the Cauchy data are allowed. Here, we consider only perturbations of φ,

$$\|\varphi - \varphi_\delta\|_{L^2(0,T)} \leq \delta\,. \tag{11}$$

The stopping rule due to Morozov [21] stops at the first index $k = k(\delta)$ where the defect is of the magnitude of the data perturbation; more precisely, when

$$\|v^{(k)}(0, \cdot; q^{(k)}) - \varphi_\delta(\cdot)\|_{L^2(0,T)} \leq \gamma_1 \delta\,. \tag{12}$$

Here, γ_1 is a constant close to one, $\gamma_1 \approx 1$. If we use discretization or projection methods of accuracy $O(h)$ to solve the underlying parabolic equations numerically, the stopping rule has to be extended to

$$\|v^{(k)}(0, \cdot; q^{(k)}) - \varphi_\delta(\cdot)\|_{L^2(0,T)} \leq \gamma_1(\delta + h\|q^{(k)}\|_{L^2})\,. \tag{13}$$

(s. Nemirowski [22]).

The following results are well-known for the CGM:

- The CGM is a 'regularization method', i.e. the stopping rule gives final $k's$; $k = k\,(\delta, h)\; s.t.$, $q = q^{\left(k(\delta,h)\right)} \longrightarrow q(\delta \longrightarrow 0,\, h \longrightarrow 0)$
- There may be orders of convergence under strong assumptions.
- Big errors can occur near the final time T (explanation known; improvements possible).
- In (7), γ can be also determined in an optimal way by a stopping rule.

Note that everything depends on the small parameter $\varepsilon > 0$ which is assumed to be fixed.

4. Experimental Results

In our experimental studies, we have considered three examples with the following solutions of (1):

Example 1: $v(x,t) = \left(1 - \exp(-\frac{t}{0.4})\right) \frac{1-\exp\left(-(1-x)/\varepsilon\right)}{1-\exp(-1/\varepsilon)}$

Example 2: $v(x,t) = t\left\{x + \frac{\exp\left(-(1+x)/\varepsilon\right) - \exp\left(-(1-x)/\varepsilon\right)}{1-\exp(-2/\varepsilon)}\right\}$

Example 3: $v(x,t) = \left(1 - \exp(-\frac{t}{0.4})\right) \frac{1-\exp(-x/\varepsilon)}{1-\exp(-1/\varepsilon)}$

The functions F are chosen accordingly; c is set to be -1. In Example 1 a boundary layer appears at $x = 1$ (see Fig. 12) while in the remaining part of the spatial interval the solution is nearly constant. Because of the exponential factor $w(t) = 1 - \exp(-t/0.4)$, for $t \approx 1$ and larger the solution is nearly a stationary one.

In Example 2, again a boundary layer appears at $x = 1$ while in the remaining x-interval the solution is nearly linear. The factor w is here chosen to be $w(t) = t$. Example 3 is obtained from Example 1 by using the transformation $x \longrightarrow 1 - x$. Hence, the boundary layer appears at $x = 0$.

As we have said in the beginning, in Example 1 and 2 we try to identify the boundary data at the side of the x-interval where the boundary layer appears; the Cauchy data are given at $x = 0$ where no boundary layer exists. In Example 3 the situation is just the converse: The boundary layer appears at the side where Cauchy data are given while at the inaccessible part of the boundary no boundary layer shows up.

All examples are computed with the Crank–Nicolson–Galerkin Method (abr. CNG) with continuous, piecewise linear basis functions for the spatial approximation. If nothing else is said we have used

$$\triangle x = 0.125,\ \triangle t = 0.05,\ T = 1.0,\ \delta = 10^{-8}\,.$$

For the ε's we have taken $\varepsilon^2 = 0.1$ and $\varepsilon^2 = 0.005$. In the first case, because of the relatively large ε, no boundary layer appears.

In Figure 11 very good numerical results can be seen for Example 1 at two different times. Here, $\varepsilon^2 = 0.1$ which is the case of no boundary layers. Figure 12 shows the exact solution for three different times for the same example but with boundary layer at $x = 1$ - due to $\varepsilon^2 = 0.005$. In the following Figure 13 the numerical solutions (together with the exact solution) are displayed for two different times. Near the boundary layer (at $x = 1$) the numerical results become worse. However, the results are not completely bad. Here, the CGM has stopped after two iterations while in Figure 11 twenty iterations have been performed. In Figure 14 the desired heat flux and its numerical approximation at $x = 1$ is seen for Example 1 and $\varepsilon^2 = 0.005$. The exact heat flux in this example is given by $\varepsilon^2 v_x(1,t) = -\varepsilon\frac{1-\exp(-t/0.4)}{1-\exp(-1/\varepsilon)}$. Due to the factor ε, the heat flux at $x = 1$ is nearly zero. The absolute error for the numerical heat flux is moderate.

Numerical results for Example 2 are shown in Figure 15 to Figure 17. In Figure 15 the results for a relatively large ε, $\varepsilon^2 = 0.1$, are again very good. Also, the results for $\varepsilon^2 = 0.005$ in Figures 16 and 17 are good. The exact heat flux at $x = 1$ (s. Figure 17) is again nearly zero,

$$\varepsilon^2 v_x(1,t) = \varepsilon t \left\{ \varepsilon - \frac{1 + \exp(-2/\varepsilon)}{1 - \exp(-2/\varepsilon)} \right\} ;$$

the numerical heat flux is nearly zero, as well.

One possible explanation for the relatively good results in the above mentioned figures may be that the CGM is started with initial guess $q^{(0)} = 0$. This is already a good approximation for the desired heat flux at $x = 1$ as explained above. In Figure 18 for Example 1 one can see the initial approximation $v^{(0)}(x,t)$, for $t = 1$ and with $\Delta x = 1/16$, calculated with a 'bad' initial guess $q^{(0)} = 0.5$ for the heat flux at $x = 1$; here again $\varepsilon^2 = 0.005$. It can be observed that the defect at $x = 0$, $\rho^{(0)}(1) = v^{(0)}(0,1;q^{(0)}) - \varphi(1)$, nearly vanishes. Hence, in Step 1 of the CGM the gradient obtained by the solution of the Adjoint Problem nearly vanishes, and further iterations of the CNG yield no improvements. It can be even observed that further iterations produce useless results with large errors. One may conjecture that the boundary layer at $x = 1$ is responsible for the fact that boundary data at the boundary layer location have no influence on the solution outside the boundary layer.

In Example 3 the situation is just the converse, namely the Cauchy data are at the boundary layer location and no boundary layer is at $x = 1$. Again, the initial guess for the heat flux at $x = 1$ is $q^{(0)} = 0.5$. In Figure 19 it can be seen that, similarly to the last figure, the boundary values at $x = 1$ have nearly no influence on the boundary values at $x = 0$ of the numerical solution and, hence, the defect and the gradient nearly vanishes. Here, again $\Delta x = 1/16$. One may say that a 'numerical boundary layer' appears at $x = 1$ while the (analytical) boundary layer is at $x = 0$. Here, as in the previous figure, further iterations will not improve the numerical approximations of the heat flux at $x = 1$ – they even produce larger and larger errors.

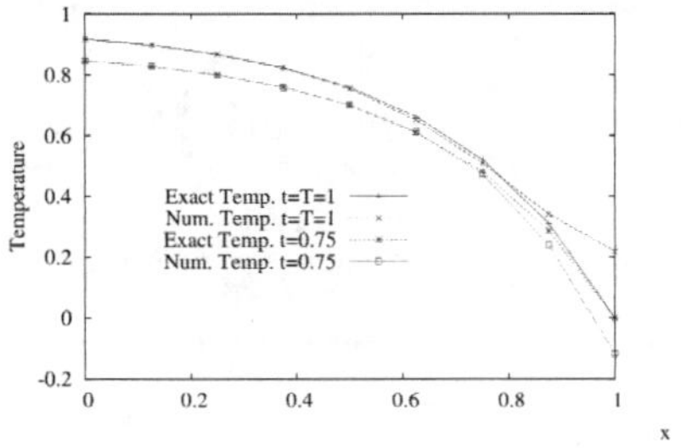

Figure 11. Ex. 1: Temp. (in x) for 2 times, $\varepsilon^2 = 0.1$.

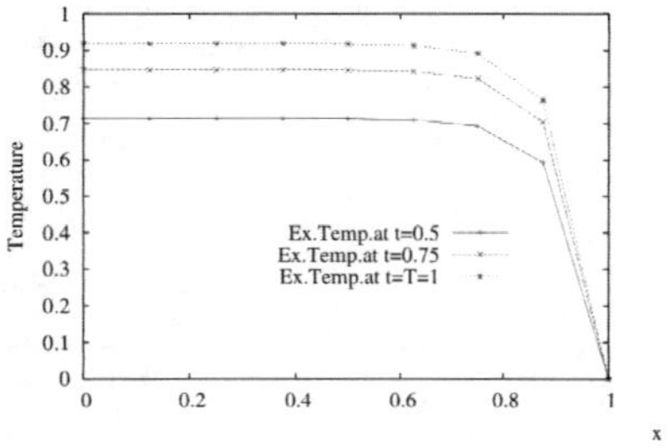

Figure 12. Ex. 1: Solution (in x) at 3 times, $\varepsilon^2 = 0.005$.

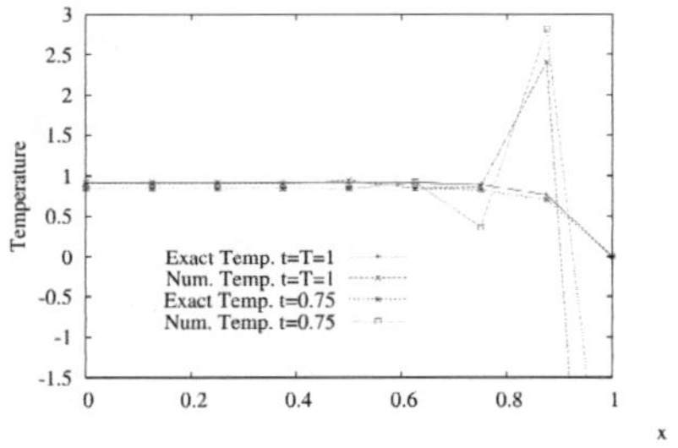

Figure 13. Ex. 1: Temp. (in x) for 2 times, $\varepsilon^2 = 0.005$.

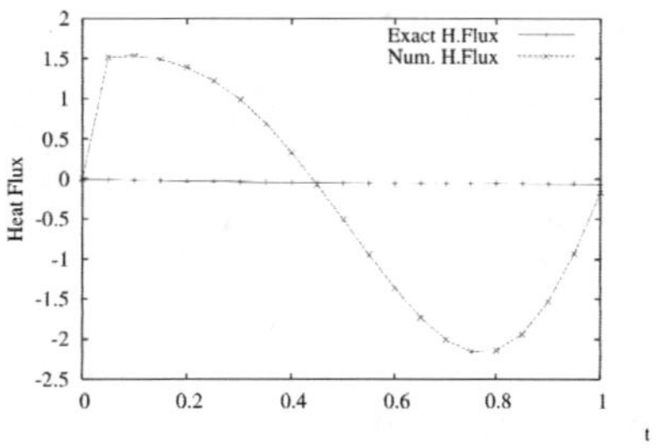

Figure 14. Ex. 1: Heat Flux at $x = 1$ for $\varepsilon^2 = 0.005$.

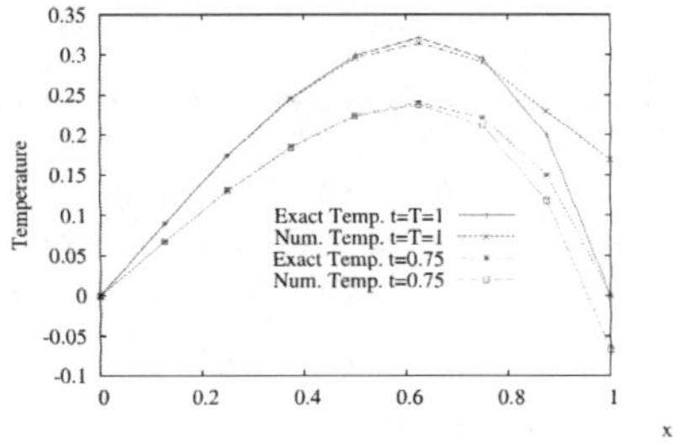

Figure 15. Ex. 2: Temp. (in x) for 2 times, $\varepsilon^2 = 0.1$.

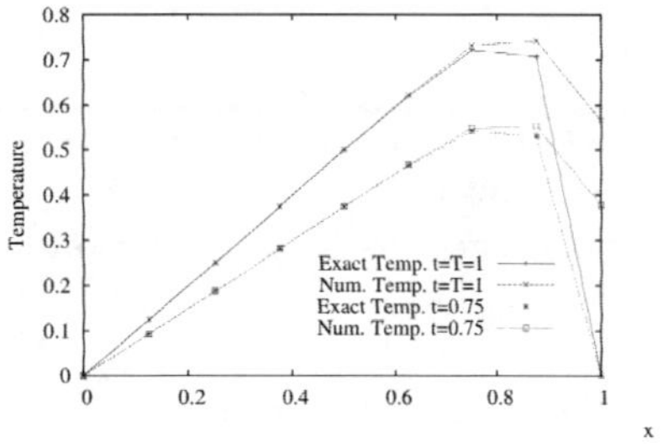

Figure 16. Ex. 2: Temp. (in x) for 2 times, $\varepsilon^2 = 0.005$.

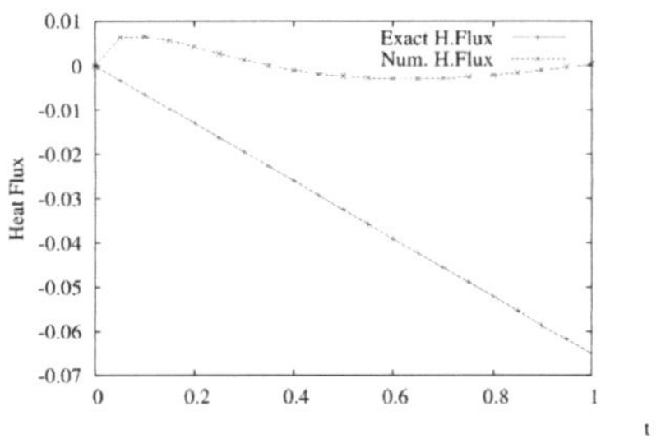

Figure 17. Ex. 2: Heat Flux at $x = 1$ for $\varepsilon^2 = 0.005$.

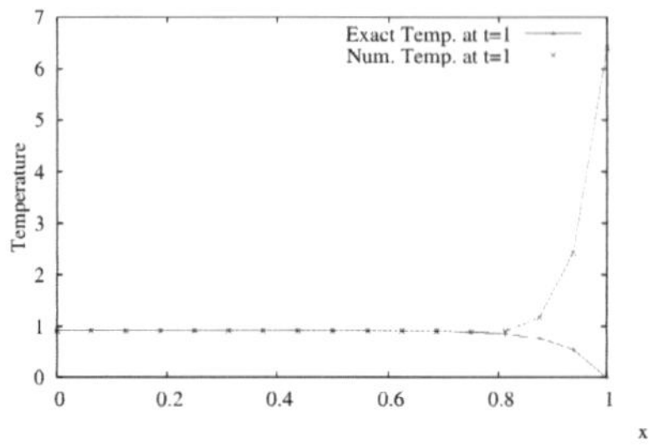

Figure 18. Ex. 1: Temp. (in x), $\varepsilon^2 = 0.005$, for $q^{(k)} = 0.5, k = 0$.

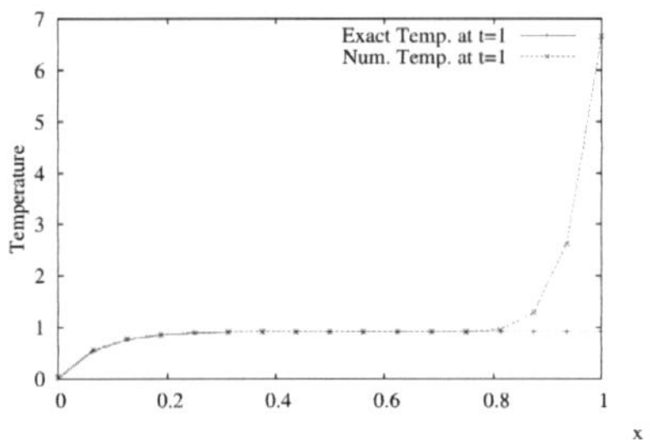

Figure 19. Ex. 3: Temp. (in x), $\varepsilon^2 = 0.005$, for $q^{(k)} = 0.5, k = 0$.

5. Questions and Conclusions

Our numerical experiments have left a series of problems open. First of all, for small ε it is not clear whether the boundary values (heat flux and temperature) can be identified at the inaccessible part of the boundary. This is an open question in cases when the Cauchy data are at the boundary layer location or when the boundary layer appears at the inaccessible part of the boundary.

The above examples are given in dimensionless form – in space and time. It is well-known that boundary data cannot be identified if the inaccessible parts are to far away from the accessible parts where the Cauchy data are given. Here, the influence of small ε's should be analysed with regard to 'far' and 'close'.

There are numerical methods for noncharacteristic Cauchy problems for parabolic equations, like the Beck method (see [4], [11], [12]) which may be more stable in cases of small ε's.

Singularly perturbed equations should be numerically solved with adapted grids. The choice of the grid points could be done a-prioriy – like the Shishkin mesh (see [3], [23]) – or a-posteriorily with automatically chosen grids (see e.g. [15], [16], [18], [19], [20]). Finally, the influence of the small ε on iterative methods for solving the illposed problems is completely open.

References

[1] Eckhaus W 1979 *Asymptotic analysis of singular perturbations* (Amsterdam: North Holland)

[2] O'Malley R E Jr 1974 *Introduction to singular perturbations* (New York: Academic Press)

[3] Roos H-G, Stynes M and Tobiska L 2008 *Robust Numerical Methods for Singularly Perturbed Differential Equations – Convection-Diffusion-Reaction and Flow Problems* Springer Series in Computational Mathematics Vol. 24

[4] Beck J V, Blackwell B and St Clair C R Jr 1985 *Inverse heat conduction problems* (New York: Wiley)

[5] Dinh Nho Hào 1998 *Methods for inverse heat conduction problems* Methoden und Verfahren der mathem. Physik Vol 4 (P Lang)

[6] Dinh Nho Hào and Reinhardt H-J 1998 *A variational method for multi-dimensional linear inverse heat conduction problems* Matimyas Matematika p 48 - 56.

[7] Dinh Nho Hào and Gorenflo R 1991 *A noncharacteristic Cauchy problem for the heat equation* Acta Appl. Math. **24** p 1-27

[8] Cannon J R 1984 *The One-Dimensional Heat Equation* vol 23 Encyclomedia of Mathematics and Its Applications (Addison-Wesley)

[9] Dinh Nho Hào and Reinhardt H-J 1996 *Stable Numerical Solution to Linear Inverse Heat Conduction by the Conjugate Gradient Method* J. Inverse and Ill-posed Problems **3** p 447-467

[10] Dinh Nho Hào and Reinhardt H-J 1998 *Gradient Methods for Inverse Heat Conduction problems* Inverse Problems in Engineering 6 **3** p 177-211

[11] Reinhardt H-J 1994 *Analysis of the discrete Beck method solving illposed parabolic equations* Inverse Problems **10** p 1-16.

[12] Reinhardt H-J 1994 *Analysis of the Discrete Beck Method Solving Linear Illposed Parabolic Equations* Inverse Problems **10** No. 6 p 1345-1360.

[13] Reinhardt H-J and Dinh Nho Hào 1996 *A Sequential Conjugate Gradient Method for the Stable Numerical Solution to Inverse Heat Conduction Problems* Inverse Problems in Engineering **2** No. 4 p 267-272.

[14] Reinhardt H-J and Dinh Nho Hào 2004 *A Generalization of Beck's Method for Inverse Heat Conduction Problems* In: Chuong N M et al. (eds.): ICAAA (International Conference on Abstract and Applied Analysis, Hanoi 2002), p 271-303 (Singapore, World Scientific)

[15] Reinhardt H-J 1981 *A-posteriori error estimates for the finite element solution of a singularly perturbed linear ordinary differential equation* SIAM J. Numer. Anal. **18** p 406-430.

[16] Reinhardt H-J 1981 *A-posteriori error estimation and adaptive finite element computations for singularly perturbed one space dimensional parabolic equations* In: Axelsson, O., Frank, L. S., van der Sluis, A. (eds.), Analytical and Num. Approaches to Asymptotic Probl. in Analysis, p 213-233 (Nijmegen, 1980), (Amsterdam: North-Holland)

[17] Reinhardt H-J *Analysis of adaptive finite element methods for $-\varepsilon u'' + u' = f$ based on a-posteriori error estimates* In: Eckhaus, W., de Jager, E. M. (eds.), Theory and applications of singular perturbations, p 207-227, (Oberwolfach, 1981), LNM, Vol. 942. (Berlin: Springer) 1982

[18] Reinhardt H-J 1982 *A-posteriori error analysis and adaptive finite element methods for singularly perturbed convection-diffusion equations* Math. Methods Appl. Sci. **4** p 529-548

[19] Reinhardt H-J 1985 *A-posteriori error estimation for finte element modifications of line methods applied to singulary perturbed partial differential equations* Appl. Numer. Math. **1** p 145-176.

[20] Reinhardt H-J 1991 *On adaptive numerical methods for singular perturbation problems* In: Roos, H.-G., Felgenhauer, A., Angermann, L. (eds.), Numerical Methods in Singularly Perturbed Problems, ISAM'91, p 111-112, (Dresden, 1991). TU Dresden

[21] Morozov V A 1984 *Methods for Solving Incorrectly Posed Problems* (New York: Springer)

[22] Nemirovskii A S 1986 *The regularizing properties of the adjoint gradient method in ill-posed problems* U.S.S.R. Comput. Math. and Math. Phys. **26/2** p 7-16

[23] Roos H-G and Teofanov L 2008 *A singularly perturbed problem with two small parameters II: The Galerkin finite element method on a Shishkin mesh* J. Comput. Appl. Math. **212** p 374-389

SVD analysis in application to full waveform inversion of multicomponent seismic data

Ilya Silvestrov and Vladimir Tcheverda
Institute of Petroleum Geology and Geophysics SB RAS, Novosibirsk, 630090, Russia
E-mail: SilvestrovIY@ipgg.nsc.ru

Abstract. An inverse problem of recovery the Earth's interior by multi-shot/multi-offset multicomponent seismic data is considered in this work. This problem may be considered as a nonlinear operational equation, and local derivative-based techniques are commonly used for its solution. Such method is known in seismic precessing as "full-waveform inversion". The major properties of the inversion process are governed by a Frechet derivative of the forward map. We show and study these properties by means of singular value decomposition(SVD) truncation. This decomposition depends strongly on the acquisition system and on the parameterization of the problem. We show, that it is very important to study the inverse problem in each particular case, otherwise unreliable results may be obtained. Surface and cross-well acquisition systems are considered in this work. Appropriate parameterizations for them are determined, and typical behavior of the inverse problem solution is studied.

1. Introduction

An inverse problem of recovery the Earth's interior by multi-shot/multi-offset multicomponent seismic data is considered in this work. The data are treated as an image of some nonlinear operator (forward map) which is implicitly introduced by initial-boundary value problem for elastic wave equations. In order to recover elastic parameters, this nonlinear operator equation should be resolved. The usual way to do this numerically is implementation of some local derivative-based minimization technique like conjugate gradient or Gauss-Newton methods and various their modifications. In seismic processing this method is known as "full-waveform inversion".

This approach was proposed in early 1980's [1] and was intensively studied and developed by many researches. The method seemed to be very promising at first, because of its apparent ability to recover elastic parameters under a few preprocessing of data. However its application to realistic synthetic and field data exposed its major drawbacks. The most significant of them were troubles in reconstruction of the smooth velocity constituent, the so-called macrovelocity or migration model. Certain difficulties were found in connection with coupling of elastic parameters[2] and different sensitivity of the method to their perturbations[3].

Our belief is that all peculiarities of the inversion procedure are controlled mainly by the Frechet derivative of the forward map, because this derivative determines the gradient and Hessian of the L2 data misfit functional. Singular Value Decomposition of this derivative is a powerful tool for analyzing its key features. In current work this analysis is applied to 2D elastic full-waveform inversion. We show that all of the mentioned above drawbacks of the method are explained and predicted by the structure of right singular vectors. The notion of stable subspace in model space is introduced and discussed and its geometry is analyzed. The results of synthetic data processing for surface and for cross-well acquisition systems are presented and discussed.

2. Statement of the problem

The seismic inverse problem, which we are targeted to solve, formally may be written as a following equation:

$$B\vec{m} = \vec{u}^{obs},$$

where $\vec{u}^{obs}$ defines recorded seismic data(seismograms), $\vec{m}$ defines unknown elastic parameters of the subsurface, and B is a nonlinear operator, which is defined implicitly by some boundary-value problem for elastic wave equations. A common way to solve such kinds of problems is an application of Newton method to this nonlinear equation. This leads us to the following well-known iterative process:

$$L[\vec{m}_k](\vec{m}_{k+1} - \vec{m}_k) = \vec{u}^{obs} - B(\vec{m}_k),$$

where operator L is a formal Frechet derivative of the considered operator B and k defines a step number of the iterative process. Certainly, the question of differentiability of this operator is a non-trivial in each particular case, however it is out of scope of this work. On each step of the iteration process we are required to solve the following linear equation:

$$L\delta\vec{m} = \delta\vec{u}, \tag{1}$$

where $\delta\vec{u}$ is a small perturbation in the recorded data, and $\delta\vec{m}$ is a small unknown update to the elastic parameters of the subsurface. Two-dimensional isotropic elastic medium is defined by three parameters (e.g Lame parameters and density) and the recorded data in thas case may have two components (e.g. horizontal and vertical displacements). Therefore equation (1) may be written in the form:

$$\begin{pmatrix} L_{11} & L_{12} & L_{13} \\ L_{21} & L_{22} & L_{23} \end{pmatrix} \begin{pmatrix} \delta m_1 \\ \delta m_2 \\ \delta m_3 \end{pmatrix} = \begin{pmatrix} \delta u_1 \\ \delta u_2 \end{pmatrix},$$

where L_{ij} defines partial derivative with respect to corresponding parameter. A common way to obtain these derivatives is to apply a single-scattering(Born's) approximation, according to which we obtain the following integral equations of first kind:

$$\int_\Omega K_{ij}(\vec{x_s}, \vec{x_r}, \vec{x}, \omega)\delta m_j(\vec{x})d\vec{x} = \delta u_j(\vec{x_s}, \vec{x_r}, \omega),$$

where the seismic data are considered as functions of frequency ω after applying Fourier transform. $\vec{x}_s$, $\vec{x}_r$ stand for sources and receivers position correspondingly, and Ω defines the subdomain in the subsurface, where the elastic parameters are targeted to be inverted. The kernels K_{ij} of these integral operators are defined by Green's function of the considered problem.

From the above relations we see, that on each step of Newton method we have to solve system of integral equations of first kind, which is a classical ill-posed problem, and it requires regularization for its solution. In this work we will focus on a truncated SVD regularization method, particularly on its generalization for compact operators in Hilbert spaces that were developed in works of Cheverda and Kostin[4, 5]. In this technique the so-called r-solution is constructed as a projection of a true solution onto a "stable subspace", which is defined as a linear span of singular vectors, which correspond to largest singular values. The number of the involved singular vectors is controlled by the acceptable condition number and therefore by the noise level in the data and by the error due to numerical approximating of the operator. Using this technique in the rest part of the paper we will construct the regularized solutions for two problems of seismic prospecting, and will analyze the stable subspaces for both of them. We will study only the first step of the non-linear iterative process. However, this step usually brings main contribution into solution, and its consideration allows us to determine main features of the problem.

One of the most important step in solution of the inverse problem is a proper choice of the unknown parameters. For example, in case of isotropic elastic medium we may consider following sets of unknown elastic properties: pressure(V_p) and shear(V_s) waves velocities and density(ρ); or Lame parameters (λ, μ) and density; or pressure(IP) and shear(IS) impedances and density. These parameters are connected with each other by following relations:

$$V_p = \sqrt{\frac{\lambda + 2\mu}{\rho}},\ V_s = \sqrt{\frac{\mu}{\rho}},\ IP = \rho V_p,\ IS = \rho V_s.$$

First of all we note, that the parameters inside these sets have different physical dimensions, and therefore a compulsory nondimensionalization is required. For example, instead of V_p, V_s, ρ one may use following dimensionless parameters for inversion: $\frac{\delta V_p}{V_p^0}, \frac{\delta V_s}{V_s^0}, \frac{\delta \rho}{\rho^0}$, where V_p^0, V_s^0, ρ^0 define elastic properties in some reference medium. However, even after such normalization not every set of parameter is suitable for inversion[3]. As will be shown in next section it requires very careful study in each particular case.

3. Numerical results of SVD analysis

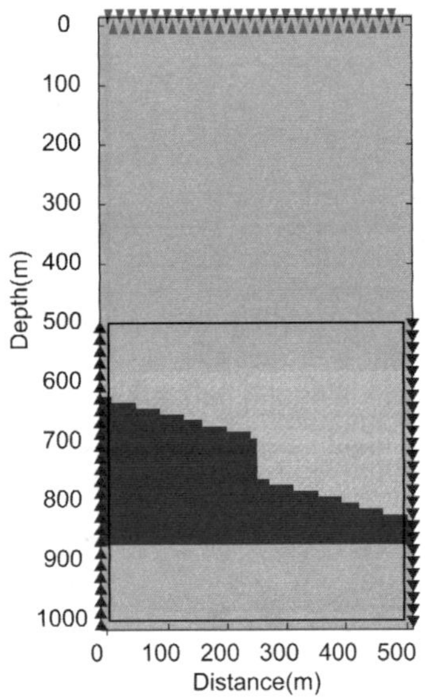

Figure 1. A model and acquisition systems used for numerical experiments. Sources are marked by (∇) character, and receivers by ($\triangle$). Red color stands for surface acquisition, and blue color for cross-well experiment.

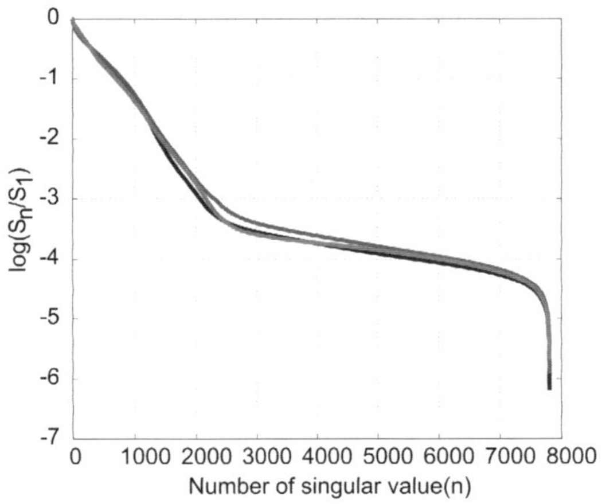

Figure 2. Singular values of linearized forward modeling operator in case of surface acquisition for different sets of parameters: velocities(blue line), impedances(red line), Lame parameters(green line).

Let us consider a model illustrated in Figure 1. We assume that the structure inside the area marked by black square is unknown, and our aim is to determine elastic parameters there. We will consider two cases. The first one is inversion of surface data, in which sources and receivers of seismic waves are placed uniformly at the Earth's surface. The second one is so-called cross-well data inversion, in which we assume, that there are two wells at the vertical edges of the model, and the sources are placed into one well, while receivers are place into another. This acquisition system is marked by blue rectangles.

The most time and numerically consuming part of the utilized technique is a construction of numerical approximation of operator DB and its singular value decomposition. To reduce this cost in current work we use rather simple approach, which however allows us to catch main features of the inverse problems. We assume that the reference medium, with respect to which we are going to obtain perturbations of elastic parameters, is homogeneous. It allows us to use explicit well-known form of Green's function for such medium. It is also assumed that source generates only pressure wave. The next simplifications are made by assuming only far field approximation to the Green's function, and by ignoring the free surface. The first of them is common for seismic prospecting because only far field effects usually contain the desired information about the subsurface, and the second one is usually achieved by additional preprocessing of the recorded data and suppressing the surface waves. All these simplifications allow us to use explicit formulas for constructing the matrix approximation of operator DB.

We consider following parameters of homogeneous background medium: $V_p = 3000m/s$, $V_s = 1500m/s$, $\rho = 1000kg/m^3$. To construct the matrix approximation, elementary step-like finite-dimensional bases were introduced into the model and into the data spaces. In the model space a grid with cell size of 10m in both directions is used, and in the data space the frequency range from 0.1Hz to 60Hz is uniformly sampled by interval of 0.5Hz. A source wavelet is Ricker and the central frequency is 30Hz.

3.1. Surface data inversion

The singular values of the constructed matrices for different parameterizations are illustrated in Figure 2 in logarithmic scale. The problem is overdetermined and the total number of the shown values is equal to number of the unknown parameters. We see that the singular values rapidly tends to zero, and it is explained by compactness of the considered integral operator. The graphic inflection near 2500 singular value corresponds to loss of machine accuracy, because a normal matrix form is used for calculating the SVD. Therefore, the rest singular values are incorrect.

We observe, that the singular values are very similar for different parameterizations. It means that from point of view of conditioning of the problem, all considered parameters sets are equivalent for inversion. The next step is to understand what part of the model may be inverted, when different parameterizations are used. Let us consider at first the model, in which there is inhomogeneity with respect to only one parameter Lame λ. The another Lame parameter μ and density are homogeneous. However, when we are solving the inverse problem we do not usually have such information, and we would like to invert for all parameters simultaneously. To predict what kind of the model may be inverted, we truncate the SVD of the corresponded matrix and construct the projection of the true model onto largest singular vectors. We consider a condition number of 10 to be acceptable and use 800 largest singular vectors. Figure 3 illustrates the true perturbation in the target area and its constructed projection that can be treated as a result of inversion. First of all we see, that we do not invert any interior of the layers, but we invert only positions of interfaces in the model. Also we see, that initial perturbation with respect to λ leaks into density and slightly into μ. Thus using such result of inversion we may come to a wrong conclusion, that the model has interfaces with respect to all parameters. Certainly, this is a numerical artifact and it should be eliminated by using for example some another

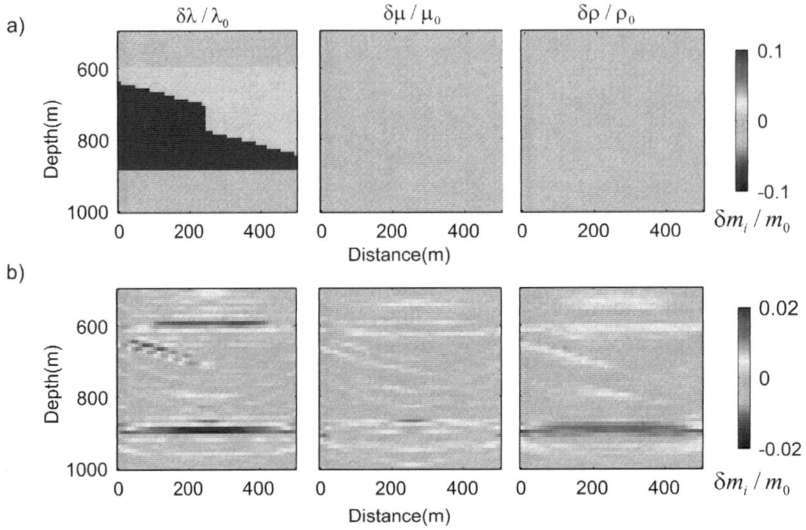

Figure 3. Inversion of Lame parameters in case of surface acquisition: a) the true normalized model perturbation with respect to homogeneous background; b) the projection of the true perturbation onto linear span of 800 largest right singular vectors.

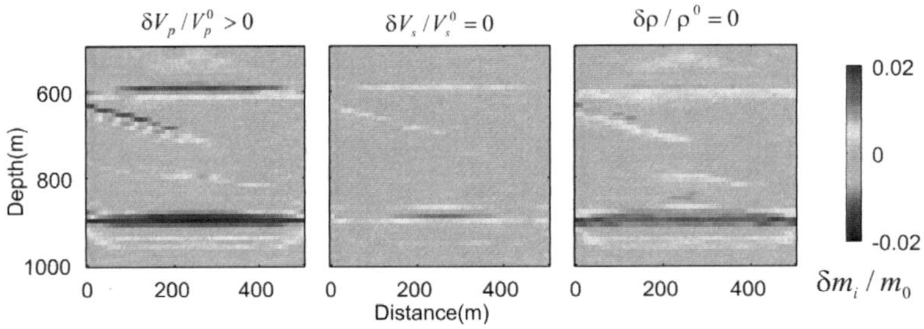

Figure 4. Inversion of pressure wave velocity in case of surface acquisition. The projection of P wave velocity perturbation onto 800 largest right singular vectors is shown.

parameterization. Inversion for velocities gives us very common result, as shown in Figure 4. In this example only pressure wave velocity was perturbed in the true model, but on a result we see interfaces with respect to shear wave velocity and density as well. Therefore, such parameterization is also not suitable for inversion in that case. Another behavior of the inverse problem solution we may observe using the impedances. The projections onto largest singular vectors in that case are shown in Figure 5. Each row corresponds to an inversion test, in which only one parameter is perturbed, and the others are zero. We observe that pressure and shear impedances can be inverted rather well, and they are not coupled with other parameters. However, when such parameterization is used, density can not be inverted at all. So, we may conclude, that for inversion using surface data the most appropriate parameters are pressure and shear impedances, and density can not be inverted in that case.

3.2. Cross-well data inversion

In this example we will use the same model as in previous one, but another positions of sources and receivers will be considered, as shown in Figure 1. The singular values for matrix approximation of operator DB in case of the cross-well acquisition system are shown

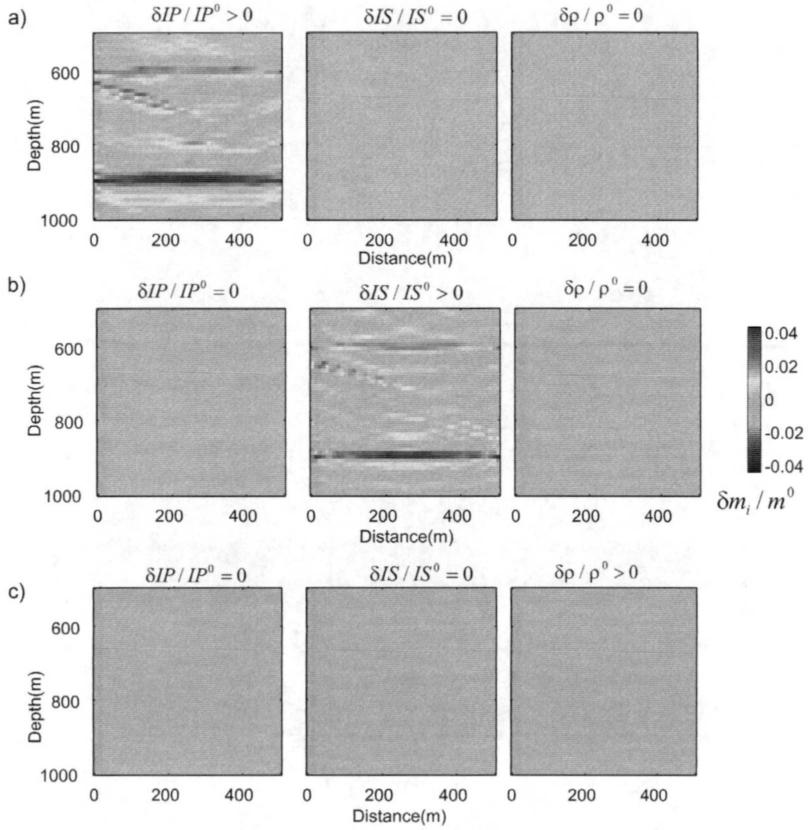

Figure 5. Inversion of elastic impedances and density in case of surface acquisition. The projections of single parameter perturbation onto 800 largest right singular vectors are shown for: a) pressure impedance; b) shear impedance; c) density.

in Figure 6, compared with values for surface acquisition. We see that singular values for these two experiments are different. As in previous example we again fix condition number to be 10, and it corresponds to 1000 largest right singular vectors that is approximately equal to the number of singular vectors used for surface acquisition.

Previously we obtained, that impedances are the best parameters for inversion using surface data. Now we repeat for cross-well data the numerical experiment shown in Figure 5(b), where single P impedance perturbations were recovered. Using the same technique we construct the projection of such perturbation onto largest singular vectors and obtain the result shown in Figure 7. We see, that while the true model is the same, the projection in case of cross-well acquisition is totaly different. We observe a strong low density anomaly in the result, that means a strong coupling between pressure impedance and density. Therefore this parameterization can not be used for inversion in case of cross-well acquisition. Figure 8 illustrates, that velocities are the most appropriate parameters in this case. We note that P wave velocity is recovered almost ideally. The inversion process is able in that case to detect not only the interfaces, but interior of the layers. No coupling between different parameters is observed. The fail to obtain the shear wave velocity is explained because of sources type, which generate only P wave, and apparently that pressure waves propagation is not sensitive to shear waves velocity perturbations. Therefore, for shear velocity we may obtain only interfaces, where waves conversion occurs. The density is also almost not recovered, however some week interfaces are observed.

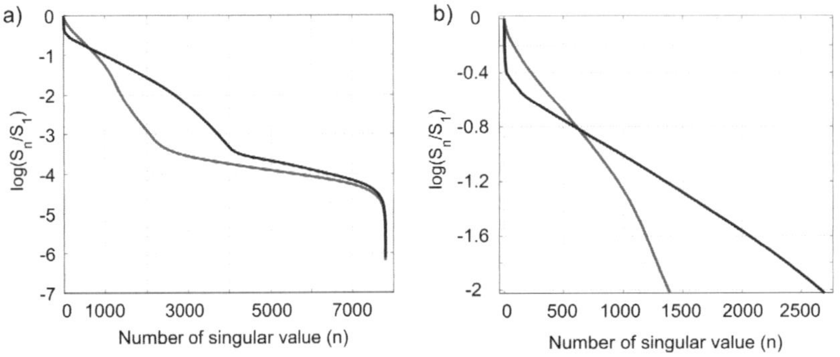

Figure 6. Singular values of linearized forward modeling operator in case of cross-well acquisition(blue line) and their comparison with values for surface acquisition(red line): a) singular values overview; b) zoom view of largest singular values.

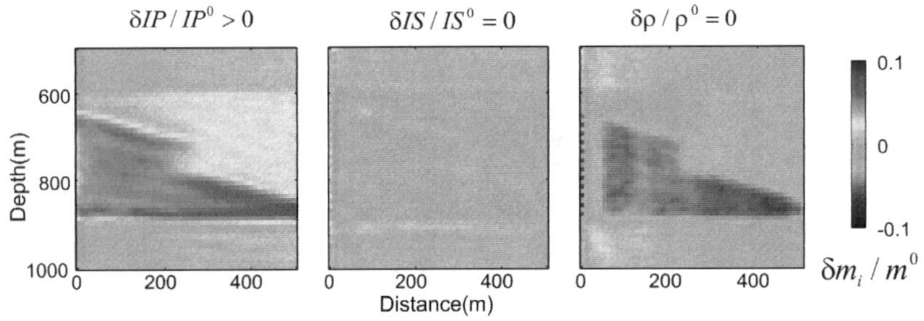

Figure 7. Inversion of pressure impedance perturbation in case of cross-well acquisition. The projection onto 1000 largest right singular vectors is shown.

4. Conclusion

In this work we use singular value decomposition to study the properties of multiparameter seismic inverse problem. As was expected, a Frechet derivative of the non-linear forward modeling operator determines major properties of the inverse problem, and truncating its SVD allows us to analyze them in detail. We show that these properties strongly depend on the considered arrangement of sources and receivers, and they should be studied in each particular case. In this paper we were focused on a correct parameterization of the problem that is one of the important step for successful solution. As was shown, a wrong choice of parameters may lead to totally incorrect results of inversion. It was proved, that pressure and shear impedances are the best choice for inversion using surface acquisition system. However density can not be inverted in that case. Such acquisition allows us to determine only interfaces of reflectors. In case of cross-well experiment waves velocities are the most appropriate parameters for inversion. We obtain, that pressure wave velocity may be inverted almost ideally, and this result is very stable. The results were obtained by analyzing projections of the true solution onto largest right singular vectors. However, we do not analyze here these vectors themselves, that also may provide us additional useful information about the inverse problem. We hope that it will be done in future works. We would like also to note, that though only the SVD truncation regularization technique is considered in this work, the obtained results make sense for others regularization methods such as Tikhonov regularization, or regularization by iterative procedures, because there is a strong relation between all of them.

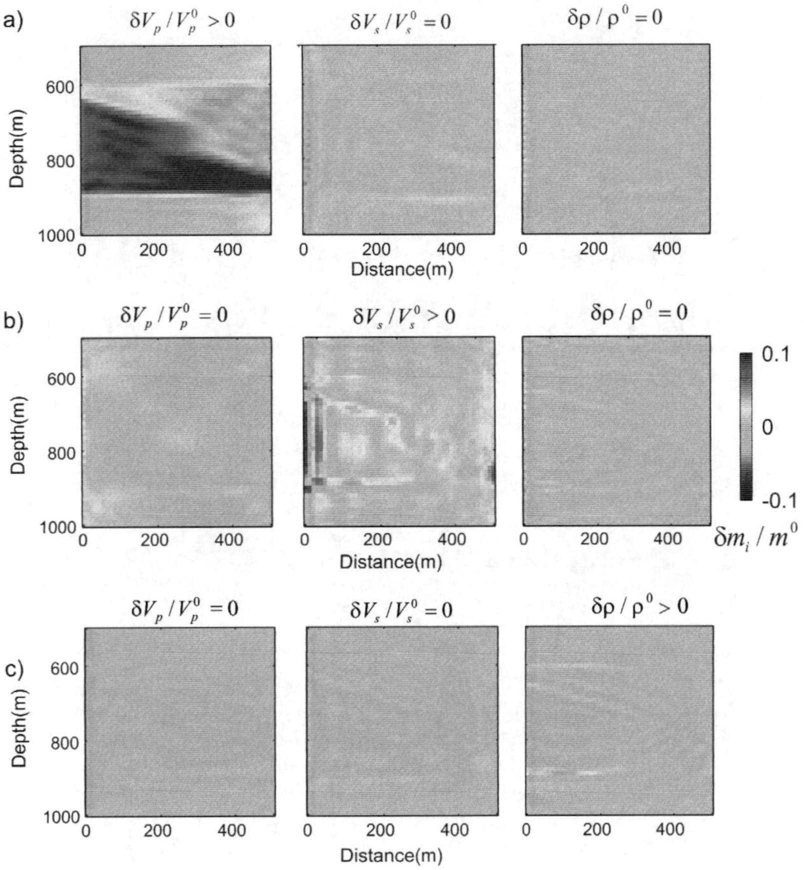

Figure 8. Inversion of elastic waves velocities and density in case of cross-well acquisition. The projections of single parameter perturbation onto 1000 largest right singular vectors are shown for: a) pressure velocity; b) shear velocity; c) density.

Acknowledgments
This research was done in cooperation with Schlumberger Moscow Research and was partially supported by Russian Foundation of Basic Research, projects 10-05-00233, 11-05-00238, 11-05-00947.

References
[1] Tarantola A 1984 Inversion of seismic reflection data in the acoustic approximation *Geophysics* **49** 1259–1266
[2] Assous F and Collino F 1990 A numerical method for the explanation of sensitivity: the case of the identification of the 2D stratified elastic medium *Inverse Prob.* **6** 487–514
[3] Tarantola A 1986 A strategy for nonlinear elastic inversion of seismic reflection data *Geophysics* **51** 1893–1903
[4] Cheverda V A and Kostin V I 1995 r-pseudoinverses for compact operators in hilbert spaces: existence and stability *J. of Inverse and Ill-posed Prob.* **3** 131–148
[5] Cheverda V A, Clement F, Khaidukov V G and Kostin V I 1998 Linearized inversion of data of multi-offset data for vertically inhomogeneous background *J. of Inverse and Ill-posed Prob.* **6** 453–484

Journal of Physics: Conference Series **290** (2011) 012015 doi:10.1088/1742-6596/290/1/012015

Bi-quadratic polynomial approach for global convergent algorithm in high dimensions coefficient inverse problems

Quan-Fang Wang

Mechanical and Automation Engineering, Chinese University of Hong Kong, Shatin, N.T., Hong Kong

E-mail: quanfangwang@hotmail.com

Abstract. Sequential minimization algorithm in convexification approach established a stable approximate solution via minimizing a finite sequence of strictly convex objective function. Its application to 1D/2D dimension cases is reported in a great deal literatures. An interesting question should be asked is how about the development of the convex approach to high dimension case. That is to say, whether the approach could be applied to realistic three dimensions issue by a routing way. It becomes a much more challenging topic in present research. In this work, a newly modified global convergent algorithm for recovering the profile of coefficient inverse problems will be constructed by using so-called bi-quadratic polynomial (BQP) in 2D/3D cases. Hope the developed paradigm would be applied in a broad class application.

1. Introduction

Global convergent algorithm for coefficient inverse problems (CIP) has been used in solving a lot of problems. Our referred publications included [1],[2], [3],[4] and [8]. But for high dimension case, the approach is quite limited to be used for dealing with a variety of issues arising in real world. The motivation of this paper is to systematically develop a bi-quadratic polynomial (BQP) approach for constructing approximate solution in global convergent scheme for two/three dimensional cases. The methodology was originally proposed in [11] and relative works see [12] and [13]. Particularly, this work is to address a updated global convergent algorithm for solutions of coefficient inverse problems by using BQP approximate, see [14]. Fortunately, the later papers [5, 6, 7, 9] have used the method to give the sufficient evidences in advance.

The final goal of this paper is to apply the BQP approximate methodology for solving coefficient inverse problems in high dimension.

Let $(\mathbf{x}, t) \in \mathbf{R}^3 \times (0, \infty)$ for $\mathbf{x} = (x_1, x_2, x_3)$. In order to propose BQP approach (cf.[5]), we consider the Cauchy problem given by parabolic partial differential equation in the form of

$$\begin{cases} \dfrac{\partial u(\mathbf{x}, t)}{\partial t} - \nabla^2 u(\mathbf{x}, t) + \rho(\mathbf{x}) u(\mathbf{x}, t) = -s(\mathbf{x}, t), \\ \dfrac{\partial u(\mathbf{x}, t)}{\partial \eta}\Big|_{t=0} + k u(\mathbf{x}, t)|_{t=0} = u_0(\mathbf{x}), \end{cases} \tag{1}$$

where $\rho(\mathbf{x}) \geq 0$ is an unknown coefficient, k is a constant and $\dfrac{\partial}{\partial \eta}$ represents the unit outer normal derivative in the Robin initial condition. Let $\mathbf{R}^3_- = \{\mathbf{x} \in \mathbf{R}^3 \mid x_2 > 0 \text{ or } x_3 > 0\}$ and $\mathbf{R}^3_+ = \{\mathbf{x} \in \mathbf{R}^3 \mid x_2 > 0 \text{ and } x_3 > 0\}$. $S = \{(x_2, x_3) \mid x_1 = 0\}$ denotes a plane. Assume that the source function $s(\mathbf{x}, t) = \delta(\mathbf{x} - \mathbf{x}_0)\delta(t) \geq 0$ is compactly supported in $\mathbf{R}^3$. Let $\Omega \subset \mathbf{R}^3_+$ be a prism/quadrangle defined by $\Omega = \{\mathbf{x} \in \mathbf{R}^3_+ : -\mathbf{R} < x_1 < \mathbf{R},\ \ x_2 \in (0, L),\ x_3 \in (0, L)\}$. The support of $s(\mathbf{x}, t)$ belongs to the set $\mathbf{R}^3_+ \setminus \bar{\Omega}$. Using the convexification approach (cf. [4]) to denote the Laplace transform of $u(\mathbf{x}, t)$ by $w(\mathbf{x}, s) = \int_0^\infty e^{-st} u(\mathbf{x}, t)dt$ for $s > 0$. Applying Laplace transform in Cauchy problem (1) to obtain

$$\begin{cases} [s + \rho(\mathbf{x})]w - \nabla^2 w = S_0(\mathbf{x}, s), \\ \lim\limits_{|\mathbf{x}| \to \infty} w = 0, \\ \dfrac{\partial w(\mathbf{x}, s)}{\partial \eta} + kw(\mathbf{x}, s) = \tilde{u}_0(\mathbf{x}, s), \quad \mathbf{x} \in S. \end{cases} \tag{2}$$

Here $\tilde{u}_0(\mathbf{x}, s)$ and $S_0(\mathbf{x}, s)$ represents Laplace transforms of u_0 and source function $s(x, t)$, respectively. For simplicity, assume that $S_0(\mathbf{x}, s) = \delta(\mathbf{x} - \mathbf{x}_0)$, where $\mathbf{x}_0 = (0, x_2, L - a) \in \mathbf{R}^3_+ \setminus \bar{\Omega}$ is a fixed source position, and a is a small positive number. $\dfrac{\partial w}{\partial \eta}$ is the outer normal derivative of w at the boundary, which corresponding to $\dfrac{\partial u}{\partial \eta}$ in (1). Refer Lemma 1 and 2 in [4] for the positivity of $w(\mathbf{x}, s)$.

CIP for system (1): Given data on $w(\mathbf{x}, s), \mathbf{x} \in S$ for a fixed source position and sufficiently large T. The coefficient inverse problem (CIP) is to identify the unknown coefficient $\rho(\mathbf{x})$ in Ω.

2. Bi-quadratic polynomial approach

Owing to the source function $S_0(\mathbf{x}, t)$ can be existed or not, so it's convenient to neglect this term in following deduces. For $w > 0$ in Ω, let $v = \ln w$ and transform (2) as

$$\nabla^2 v + (\nabla v)^2 = s + \rho(\mathbf{x}), \quad \mathbf{x} \in \Omega. \tag{3}$$

If introducing $f(\mathbf{x}, s) = v + s \cdot l(\mathbf{x}, \mathbf{x}_0), q(\mathbf{x}, s) = \dfrac{\partial f}{\partial s}$, then by formulation (6) in Lemma 2 of [4] to imply that $f(\mathbf{x}, s) = O(s^{-1}), q(\mathbf{x}, s) = O(s^{-2})$ as $s \to \infty$. Hence to have approximate

$$f(\mathbf{x}, s) \approx -\int_s^{\bar{s}} q(\mathbf{x}, \nu)d\nu,$$

where $\bar{s}$ is a sufficiently large number. For $\mathbf{x} \in \Omega, \Gamma = \partial\Omega$ and $s \in (s_0, \bar{s})$, the boundary problem (2) is converted to

$$\begin{cases} \displaystyle\int_s^{\bar{s}} \nabla^2 q(\mathbf{x}, \nu)d\nu + \left(\int_s^{\bar{s}} \nabla q(\mathbf{x}, \nu)d\nu\right)^2 + s^2 - s = 0, \\ \dfrac{\partial q(\mathbf{x}, s)}{\partial \eta} + kq(\mathbf{x}, s) = q_0(\mathbf{x}, s) \quad \mathbf{x} \in \Gamma. \end{cases} \tag{4}$$

Here q_0 correspond to $\tilde{u}_0$. For numerical computation, q_0 is calculated by using some regularizing scheme of numerical differential on s. Obviously (4) does not contain the unknown coefficient $\rho(\mathbf{x})$. Set $H(m') = \left\{q(\mathbf{x}, s) \mid \dfrac{\partial q}{\partial \eta} + kq = q_0, \mathbf{x} \in \Gamma,\ \max\limits_{s \in (s_0, \bar{s})} \|q(\mathbf{x}, s)\|_{C^3(\bar{\Omega})} \leq m'\right\}$, where $m' > 0$ is a known number.

Existence: For the problem (4), then there exists at most one solution $q(\mathbf{x}, s)$ such that $\max\limits_{s \in (s_0, \bar{s})} \|q(\mathbf{x}, s)\|_{C^3(\bar{\Omega})} < \infty$.

Let $\Omega = \Omega_1 \times [0, L] \times [0, L]$ for $L > 0$, where $\Omega_1 = \{x_1 | -\mathbf{R} \le x_1 \le \mathbf{R}\}$. Suppose that q_0 is observed on $\Gamma = \Omega_1 = \bar{\Omega} \cap S \subset \{x_2 = 0, x_3 = 0\}$. Take $\{\phi_k(x_1)\}\Big|_{k=1}^{K} \subset C^3(\bar{\Omega}_1)$ as a set of linear (or quadratic, cubic) basis functions that approximate $q(\mathbf{x}, s)$ and its $\mathbf{x}$-derivatives up to the third order (cf. [10]), i.e.,

$$D^{\alpha}_{\mathbf{x}} q(\mathbf{x}, s) \approx D^{\alpha}_{\mathbf{x}} \sum_{k=1}^{K} \eta_k(x_2, x_3, s)\phi_k(x_1), \quad (\mathbf{x}, s) \in \bar{\Omega} \times [s_0, \bar{s}], \ |\alpha| \le 3. \tag{5}$$

Here, $K = n \times n$ is the total number of sub-rectangles. Then, for $q \in H(m')$, a number n can be chosen corresponding to $K(\varepsilon, m')$ for sufficiently small $\varepsilon > 0$ and $\eta_k(x_2, x_3, s)$, such that $\max\limits_{s \in (s_0, \bar{s})} \Big\| q - \sum\limits_{k=1}^{K(\varepsilon, m')} \eta_k(x_2, x_3, s)\phi_k(x_1) \Big\|_{C^3(\bar{\Omega})} < \varepsilon$. Denote $\mathbf{p}(x_2, x_3, s) = (\eta_1(x_2, x_3, s), \cdots, \eta_k(x_2, x_3, s), ..., \eta_K(x_2, x_3, s))^T$ and substitute (5) into (4). The result is quoted as in [4] to parabolic coefficient inverse problems. Let $q(\mathbf{x}, s) \approx \sum\limits_{k=1}^{K} \eta_k(x_2, x_3, s)\phi_k(x_1)$ for $(\mathbf{x}, s) \in \bar{\Omega} \times [s_0, \bar{s}]$. Then we have

$$\begin{aligned} &\mathcal{L}(\mathbf{p}) = \mathbf{p}_1'' + \mathbf{p}_2'' \\ &+F\left(\mathbf{p}_1', \mathbf{p}_2', \mathbf{p}, \int_s^{\bar{s}} \mathbf{p}_1'(x_2, x_3, \nu)d\nu, \int_s^{\bar{s}} \mathbf{p}_2'(x_2, x_3, \nu)d\nu, \int_s^{\bar{s}} \mathbf{p}(x_2, x_3, \nu)d\nu, x_2, x_3\right). \end{aligned}$$

From (4), it is easy to deduce that

$$\begin{cases} \mathcal{L}(\mathbf{p}) = 0, \ \mathbf{p} = \mathbf{p}(x_2, x_3, s), x_2 \in (0, L), \ x_3 \in (0, L), \ s \in (s_0, \bar{s}), \\ \dfrac{\partial \mathbf{p}(x_2, x_3, s)}{\partial \eta} + k\mathbf{p}(x_2, x_3, s) = \mathbf{p}_0(x_2, x_3, s), \quad (x_2, x_3) \in \Gamma. \end{cases} \tag{6}$$

Here $\mathbf{p}_0$ corresponds to the initial value q_0 in (4).

Third kind boundary condition can be converted to Dirichelet boundary condition without describing the extrapolated boundary condition. In numerical approximate of finite element method (FEM) as in [10], for constructing the bi-quadratic polynomials as approximate solutions at each element, we discretize the coefficient inverse problems (6) on a rectangle $[0, L] \times [0, L]$ of two dimension domain. Denote $y = x_2, z = x_3$ with the grid

$$\begin{aligned} 0 = y_0 < y_1 < ... < y_{n-1} < y_n = L, \quad h^i = y_i - y_{i-1}, \\ 0 = z_0 < z_1 < ... < z_{n-1} < z_n = L, \quad h^j = z_j - z_{j-1}. \end{aligned}$$

The indexes $k = 1, 2, \cdots, K$ of $\eta_k(x_2, x_3, s)$ to the indexes ij of $p_{ij}(y, z, s)$ have the corresponding relationship: $k = (i-1)n + j$. See detailed array in Table 1. Using the Taylor expansion for three arguments of the function $f^T(y, z, s)$ at point (y_0, z_0, s), then we obtain the polynomial in the form of

$$\begin{aligned} f^T(y, z, s) = &\frac{\partial^2 f^T(y_0, z_0, s)}{\partial y^2}\frac{(y - y_0)^2}{2} + \frac{\partial^2 f^T(y_0, z_0, s)}{\partial z^2}\frac{(z - z_0)^2}{2} \\ &+\frac{\partial^2 f^T(y_0, z_0, s)}{\partial y \partial z}(y - y_0)(z - z_0) + \frac{\partial f^T(y_0, z_0, s)}{\partial y}(y - y_0) \\ &+\frac{\partial f^T(y_0, z_0, s)}{\partial y}(z - z_0) + f^T(y_0, z_0, s). \end{aligned} \tag{7}$$

Table 1. Bi-quadratic polynomials at each element

p_{11}	p_{12}	p_{1j}	p_{1n}
p_{21}	p_{22}	p_{2j}	p_{2n}
p_{i1}	p_{i2}	p_{ij}	p_{in}
p_{n1}	p_{n2}	p_{nj}	p_{nn}

Considering the Taylor expansion of p_{ij} at point (y_0, z_0, s), it is easy from (7) to guess the leading coefficients in the expansion of p_{ij} as:

$$\frac{\partial^2 p_{ij}(y_0, z_0, s)}{\partial y^2} = a_i(s), \frac{\partial^2 p_{ij}(y_0, z_0, s)}{\partial z^2} = b_j(s), \frac{\partial^2 p_{ij}(y_0, z_0, s)}{\partial y \partial z} = c_{ij}(s),$$

where $a_i(s), b_j(s), c_{ij}(s)$ need to be determined by calculating at each element:

$$a_i(s) = \alpha_i s^2 + s, \;\; b_j(s) = \beta_j s^2 + s, \;\; c_{ij}(s) = \gamma_{ij} s^2 + s,$$

for constants α_i, β_j and γ_{ij} in $\mathbf{R}^1$.

Therefore, we can approximate the component of the vector function $\mathbf{p}(x_2, x_3, s)$ via a quadratic polynomial in each sub-rectangle as follow:

$$\begin{aligned}
&\mathbf{p}(y, z, s) \approx p_{ij}(y, z, s) \\
&= a_i(s)\frac{(y - y_{i-1})^2}{2} + b_j(s)\frac{(z - z_{j-1})^2}{2} + \\
&c_{ij}(s)(y - y_{i-1})(z - z_{j-1}) + p'_{ij}(y_{i-1}, z_{j-1}, s)(y - y_{i-1}) \\
&+p'_{ij}(y_{i-1}, z_{j-1}, s)(z - z_{j-1}) + p_{ij}(y_{i-1}, z_{j-1}, s)
\end{aligned}$$

for $i = 1, 2, \cdots, n$ and $j = 1, 2, \cdots, n$. For more detailed configuration, see reference [14].

By considering the minimization problem for the criteria function, we obtain the resulting functionals with respect to the (i, j)-th leading coefficients α_i, β_i, γ_{ij} of the quadratic polynomial as:

$$J_{\lambda,i,j}(\mathbf{p}) = \int_{s_0}^{\bar{s}} \int_{z_{j-1}}^{z_j} \int_{y_{i-1}}^{y_i} \left|\mathcal{L}(p_{ij}(y, z, \nu))\right|^2 C^2_{\lambda,i,j}(y, z) \; dydzd\nu, \tag{8}$$

where $C_{\lambda,i,j}(y, z) = \exp[-\lambda(y - y_{i-1}) - \lambda(z - z_{j-1})]$ ($i = 1, 2, \cdots, n$ and $j = 1, 2, \cdots, n$) are Carleman's weighted functions, which appear in Carleman estimates for the operator $\frac{d^2}{dy^2}, \frac{d^2}{dz^2}$. The calculation of $J_{\lambda,i,j}$ in (8) can be executed by using Gauss Legendre quadrature. By taking norms as $\|a_i\| = \max\limits_{s \in [s_0, \bar{s}]} |a_i(s)|$, $\|b_j\| = \max\limits_{s \in [s_0, \bar{s}]} |b_j(s)|$, $\|c_{ij}\| = \max\limits_{s \in [s_0, \bar{s}]} |c_{ij}(s)|$, the sequential minimizers of $J_{\lambda,i,j}(p_{ij})$ will be searched on the bounded set $G_{abc}(m) = \Big\{(a_i(s), b_j(s), c_{ij}(s)) : \|a_i\| \le m, \|b_j\| \le m, \|c_{ij}\| \le m\Big\}$. Let the corresponding set of p_{ij} be denoted by $G(m) = \Big\{g(s) : \|g\| \le m\Big\}$. By citing the result given in [4] and the convexity of $J_{\lambda,i,j}(\mathbf{p})$, the uniqueness of minimizers for these functionals and the convergence result can be proven certainly (cf. [14]).

Exact solution: Assume that there exists a vector function $\mathbf{p}^*(y,z,s)$, such that

$$\begin{cases} \hat{\mathcal{L}}(\mathbf{p}^*) = 0, \quad y \in (0,L), \ z \in (0,L), s \in (s_0, \bar{s}), \\ \dfrac{\partial \mathbf{p}^*(y,z,s)}{\partial \eta} + k\mathbf{p}^*(y,z,s) = \mathbf{p}_0^*(s), \quad (y,z) \in \Gamma. \end{cases}$$

where $\hat{\mathcal{L}}(\mathbf{p}^*) \equiv \mathbf{p}^{*\prime\prime} + \mathbf{p}^{*\prime} + F^*(\mathbf{p}^{*\prime}, \mathbf{p}^*, \int_s^{\bar{s}} \mathbf{p}^{*\prime}(y,z,\nu)d\nu, \int_s^{\bar{s}} \mathbf{p}^*(y,z,\nu)d\nu, y, z)$. Then $\mathbf{p}^*(y,z,s)$ is called the exact solution to (6).

Set $\Omega_0 = [0,L] \times [0,L]$, and assume that $\mathbf{p}^* \in C^3(\Omega_0)$ satisfy $\max_{s \in (s_0, \bar{s})} \|\mathbf{p}^*(y,z,s)\|_{C^3(\Omega_0)} \le c_0 m$ for constant c_0. The function F is known approximately by

$$\begin{aligned} & F(\mathbf{p}', \mathbf{p}, \int_s^{\bar{s}} \mathbf{p}'(y,z,\nu)d\nu, \int_s^{\bar{s}} \mathbf{p}(y,z,\nu)d\nu, y, z) \\ = \ & F^*(\mathbf{p}^{*\prime}, \mathbf{p}^*, \int_s^{\bar{s}} \mathbf{p}^{*\prime}(y,z,\nu)d\nu, \int_s^{\bar{s}} \mathbf{p}^*(y,z,\nu)d\nu, y, z) + \hat{\varepsilon}(y,z,s), \end{aligned}$$

where $\max_{s \in [s_0, \bar{s}]} \|\hat{\varepsilon}(y,z,s)\|_{C^3(\Omega_0)} \le \varepsilon$ and $\max_{s \in [s_0, \bar{s}]} |\mathbf{p}_0(s) - \mathbf{p}_0^*(s)| \le \varepsilon$. Denote that

$$F_{ij} = F\Big(p'_{ij}, p_{ij}, \int_s^{\bar{s}} p'_{ij}(y,z,\nu)d\nu, \int_s^{\bar{s}} p_{ij}(y,z,\nu)d\nu, y, z\Big)\Big|_{\substack{y = y_{i-1}, \\ z = z_{j-1}}}$$

$$F^*_{ij} = F\Big(p^{*\prime}_{ij}, p^*_{ij}, \int_s^{\bar{s}} p^{*\prime}_{ij}(y,z,\nu)d\nu, \int_s^{\bar{s}} p^*_{ij}(y,z,\nu)d\nu, y, z\Big)\Big|_{\substack{y = y_{i-1}, \\ z = z_{j-1}}}$$

Existence of minimizer: Assume that all above assumptions are satisfied. Let (a_0, b_0, c_{00}),(a_i, b_j, c_{ij}) and (a'_i, b'_j, c'_{ij}) be the corresponding coefficients of p_0, p_{ij} and p'_{ij}, respectively. Let h_i,h_j be the discrete lengths of the subintervals at y, z directions, respectively. Assume that $(a_0, b_0, c_{00}) \in G_{abc}(c_0 m)$. Then there exists sufficiently small positive numbers $h_{i0}(m)$, $h_{j0}(m)$ such that for all $h_i \in (0, h_{i0}), h_j \in (0, h_{j0})$

$$\begin{aligned} & (a_i, b_j, c_{ij}), (a'_i, b'_j, c'_{ij}) \in G_{abc}(m), \\ & p_{ij}(y_{i-1}, z_{j-1}, s), p'_{ij}(y_{i-1}, z_{j-1}, s) \in G(m), \\ & F_{ij} \in G(c'_0 m), \quad (i = 1, 2, ..., n, \ j = 1, 2, ..., n) \end{aligned}$$

All functionals $J_{\lambda,i,j}(\mathbf{p})$ are strictly convex on $G(m)$. The unique minimizer pairing $(\bar{a}_i, \bar{b}_j, \bar{c}_{ij})$ of $J_{\lambda,i,j}(\mathbf{p})$ is an interior point of $G_{abc}(m)$. It can also be regarded as the solution of operation equation $(a_i, b_j, c_{ij}) = \Phi_{\lambda,i,j}(a_i, b_j, c_{ij})$ with the relationship: $\Phi_{\lambda,i,j} : G(m) \to G(m)$.

The proof in [14] can be obtained by citing [4], [5], [11] and [12].

Furthermore, it illustrated that the convergence of the bi-quadratic polynomials algorithm to the exact solution is guaranteed for initial guess (a_0, b_0, c_{00}) in the set $G_{abc}(m)$ without the lower bounded condition. It implies that the global convergence of the BQP paradigm in two-dimension.

Inversion: If the approximating field $\tilde{f}(\mathbf{x}, \nu)$ is solved, the unknown coefficient $\rho(\mathbf{x})$ can be determined by (3) approximately. The inversion formula for $\tilde{\rho}(\mathbf{x})$ can be deduced as

$$\tilde{\rho}(\mathbf{x}) = \nabla^2 \tilde{f} + (\nabla \tilde{f})^2 - s, \quad \forall s \in [s_0, \bar{s}]. \tag{9}$$

Hence, without needing to compute the solutions $u(\mathbf{x}, t)$ for getting $\tilde{\rho}(\mathbf{x})$ (cf. [4]).

Due to the limit of paper length, we omit the discussion on convergence and error estimation. We also neglect the computational procedure for simplification (cf. [14]).

Notice: For the construction of bi-quadratic polynomial in three dimensional case, we render to our further publication in details.

3. Conclusions and future work

This work solved the coefficient inverse problems of nonlinear parabolic partial differential equations (PDE) for two dimensional spaces using developed globally convergent algorithm (i.e., convexification approach). The newly proposed bi-quadratic polynomial (BQP) approach is presented for constructing approximate solution in high dimensions architecture.

Although we have worked out a great deal problems in the determination of unknown coefficient inverse problem by current level global convergent algorithm using the convexification approach as well as adaptive techniques in recovering property (i.e. profile) of materials, mechanical and medicine process. However, it still has fatal limition for realistic issues, which needs to be solved in future advantage global convergent algorithm. Some questions and answers is posed for referring.

- Q1: How about the theory and numerical evidence in 3 dimension?
  A1: In the case of unknown coefficient only dependent to spatial variable, current BQP, CIP and globally convergent algorithm would be extended and valid.
- Q2: How about the coefficient depended on the time t?
  A2: In that case, current methodology will not valid. It needs to establish new theory for various case arising in real world.
- Q3: Other solution for Q2?
  A3: Mathematically, one can develop a third path to convert the issue in Q2 to Q1 if any. It is also difficult obviously.

Acknowledgments

The author would like to express full gratitude to the conference committee of International Conference on Inverse Problems (ICIP2010) held at City University of Hong Kong for kind invitation and financial support to participate the conference.

References

[1] Klibanov M 1997 *SIAM J. Math. Aanl.* **28** 1371
[2] Klibanov M and Timonov A 2003 *Inverse Problems* **19** 331
[3] Klibanov M and Timonov A 2003 *Numerical Methods and Programming* **4** 52
[4] Klibanov M and Timonov A 2004 *Applicable Analysis* **83** 993
[5] Klibanov M and Timonov A 2007 *Inverse Problems* **23** 123
[6] Klibanov M and Timonov A 2007 *Journal Inverse Ill-Posed Problems* **15** 167
[7] Beilina L and Klibanov M 2008 *SIAM Journal on Scientific Computing* **31** 478
[8] Nachman A 1996 *Ann. Math.* **143** 71
[9] Shan H Klibanov M Liu H Pantong N and Su J 2008 *Inverse Problem* **24** 1
[10] Wang Q F 2011 *Optimal Control for Nonlinear Parabolic Distributed Parameter Systems: with numerical analysis*, LAMBERT Academic Publishing, Germany
[11] Wang Q F 2005 *Proceeding of Conference on Differential and Difference Equations and Applications* Florida Institute of Technology USA 1109
[12] Wang Q F 2006 *Inverse Problems in Applied Sciences - towards breakthrough* Hokkaido University Japan 91
[13] Wang Q F 2007 *International Conference of Inverse and Ill-Posed Problems of Mathematical Physics* Novosibirsk Russia
[14] Wang Q F 2010 *International Conference of Inverse Problems*, City University of Hong Kong, Invited lecture Hong Kong 22

Journal of Physics: Conference Series **290** (2011) 012016 doi:10.1088/1742-6596/290/1/012016

An alternative approach of Kirsch-Kress Method for reconstructing the shape of a sound-soft obstacle from several incident fields

Zewen Wang and Bin Hu

School of Mathematics and Information Science, East China Institute of Technology, Fuzhou, Jiangxi, 344000, P.R.China.

E-mail: zwwang@ecit.cn

Abstract. The inverse problem we consider is to reconstruct the shape of a plane acoustically sound-soft obstacle from the knowledge of the far-field patterns for several incident fields. An alternatively iterative procedure is proposed based on the numerical method which was developed by Kirsch and Kress. For several incident fields, a computational technique with its corresponding iterative method is presented to reduce the amount of computation by the superposition of the single-layer potential and of the far-field equation. Numerical examples show that the methods give accurate numerical approximations in relatively few iterations for several incoming waves.

1. Introduction

We consider an inverse scattering problem where the shape of a plane acoustically sound-soft obstacle is to be determined from measurements of the far-field patterns for one or several incident fields with wave number $k > 0$. This problem has many important applications, such as remote sensing, ultrasound tomography and seismic imaging. It is difficult to solve since they are nonlinear and also ill-posed, since the solution does not depend continuously on the data. Most of the Newton type methods for approximately solving such inverse scattering problems require the solution of a corresponding direct problem at each iteration step [1, 2, 3].

The aim of this paper is to provide an alterative approach to the Kirsch-Kress method [4, 5, 6] by a boundary operator. Compared with Newton type methods, the method which we describe in this paper belongs to a group of schemes which do not require the forward solver. However, a disadvantage of the Kirsch-Kress method is the increase of computing time when the number n of incoming plane waves is increased. This is due to the fact that the Kirsch-Kress method requires the determination of $n + 1$ unknown functions. On the other hand, we hope to take several incoming plane waves from different directions (so that there is no shadow region) for an accurate reconstruction of the shape. By the superposition of the single-layer potential and of the far-field equation we only need to determine two unknown function for n incoming plane waves, which reduces the amount of computation. Our idea which is very simple and is feasible from the reconstructions of three different shapes.

After precisely formulated the inverse scattering problem in this introduction, the outline of Kirsch-Kress method is presented in section 2. An alternative approach of Kirsch-Kress method with two iterative procedures is proposed in section 3. For numerical implementation, a parametrization of the boundary curve of the obstacle is found in section 4. Finally, numerical experiments for some two-dimensional examples are given in section 5. Although we can extend our analysis to the Neumann or the impedance boundary condition, for the sake of exposition and simplicity, we confine our presentation to the Dirichlet boundary condition.

Let $D \subset \mathbb{R}^2$ be a simply connected domain with C^2-boundary ∂D. The incident field u^i is given by $u^i(x) = e^{ikx\cdot d}$, where k is the wave number with $k > 0$ and d is the direction of the incident plane wave. The scattering of time-harmonic acoustic plane wave for the obstacle D with sound-soft boundary condition is modeled by an exterior boundary value problem of the Helmholtz equation. That is, for a given incident plane wave $u^i(x) = e^{ikx\cdot d}$, the direct scattering problem for the sound-soft obstacle is to find the total field $u = u^i + u^s \in C^2(\mathbb{R}^2\backslash\bar{D}) \cap C(\mathbb{R}^2\backslash D)$ which satisfies

$$\begin{cases} \Delta u + k^2 u = 0, & \text{in } \mathbb{R}^2 \setminus \bar{D}; \\ u = 0; & \text{on } \partial D; \\ \lim\limits_{r\to\infty} \sqrt{r}\left(\frac{\partial u^s}{\partial r} - iku^s\right) = 0, & r = |x|. \end{cases} \tag{1}$$

where ν is the unit normal vector of ∂D directed into the exterior of D. u^s is the scattered field corresponding to the incident field $u^i(x)$. As we know, there exists a unique solution for the direct scattering problem (1) [7]. Furthermore, the forward scattering problem is well posed, i.e., the solution u depend continuously the data k, d and the boundary ∂D. It is straightforward to show that the scattered field u^s has the following asymptotic behavior

$$u^s(x) = \frac{e^{ik|x|}}{\sqrt{|x|}}\left\{u_\infty(\hat{x}; d) + O\left(\frac{1}{|x|}\right)\right\}, \; |x| \to \infty. \tag{2}$$

Here, $u_\infty(\hat{x}; d)$ is the far-field pattern or the scattering amplitude, and $\hat{x} = \frac{x}{|x|}$ belongs to Ω that is a unit circle.

Now we give the inverse scattering problem: *Determine the shape of the scatterer D by the given the far-field pattern $u_\infty(\hat{};d)$ on the unit circle for one or several incident waves with different directions d and the fixed wave number k.* This type of inverse scattering problem is well studied. For an overview of this inverse scattering problem, we refer to the monograph by Colton & Kress [6]. And it is both severely ill-posed and nonlinear, thus difficult to be solved. The question of uniqueness for the inverse scattering problem for sound-soft obstacle is not completely solved. According to results of [6] we have the following uniqueness.

Lemma 1 *Let D_1 and D_2 be two planar scatterers which are contained in a disk of radius R such that $kR < \pi$. If the far patterns coincide for one incident plane wave with wave number k, then $D_1 = D_2$.*

2. Outline of the Kirsch-Kress method

We now describe a method of approximately solving the inverse scattering problem which was suggested by Kirsch and Kress [4, 5]. The method is divided into two steps. In the first step (Sl) the scattered field u^s is determined from the given far-field pattern. Then in the second step (S2) the boundary of the unknown scatterer is found as a zero-level curve of the total field $u^i + u^s$.

Step(S1). By a weak *a priori* information about D, we can choose an auxiliary closed curve Γ contained in the unknown scatterer D, so that k^2 is not a Dirichlet eigenvalue of the Laplacian in the interior of Γ. We try to find the scattered field u^s as a single-layer potential

$$u^s(x) = \int_\Gamma \Phi(x, y)\varphi(y)ds(y), \tag{3}$$

with some unknown density $\varphi \in L^2(\Gamma)$. Here $\Phi(x, y) = \frac{i}{4}H_0^{(1)}(k|x-y|)$ denotes the fundamental solution to the Helmholtz equation and $H_0^{(1)}(k|x|)$ is the first kind Hankel function of order zero. From the asymptotic behavior of $H_0^{(1)}(k|x|)$ [6, 7], the far-field pattern of the single-layer potential is given by

$$u_\infty(\hat{x}) = \sigma \int_\Gamma \varphi(y) e^{-ik\hat{x}\cdot y} ds(y),\ \hat{x} \in \Omega. \tag{4}$$

where $\sigma = \frac{e^{i\pi/4}}{\sqrt{8\pi k}}$. Now, we define the integral operator $F : L^2(\Gamma) \to L^2(\Omega)$ by

$$(F\varphi)(\hat{x}) = \sigma \int_\Gamma \varphi(y) e^{-ik\hat{x}\cdot y} ds(y),\ \hat{x} \in \Omega. \tag{5}$$

Hence, given the far-field pattern u_∞, we have to solve the integral equation of the first kind

$$F\varphi = u_\infty. \tag{6}$$

The integral operator F has an analytic kernel and therefore equation (6) is a severely ill-posed equation of the first kind. Provided k^2 is not a Dirichlet eigenvalue of the Laplacian in the interior of Γ, F is injective and has dense range in $L^2(\Omega)$ [6]. Applying the Tikhonov regularization method to (6) [5, 6] leads to the minimization of

$$\|F\varphi - u_\infty\|^2_{L^2(\Omega)} + \alpha\|\varphi\|^2_{L^2(\Gamma)}, \tag{7}$$

where α is the regularization parameter.

Step(S2). Through the solution φ_α (depends on α) of (7) we obtain a corresponding approximation u^s_α for the scattered field by the single-layer potential (3) with density φ_α. Then, given the approximation u^s_α, we can now seek the boundary of the obstacle D as the location of the zeros of $u^i + u^s_\alpha$ in a minimum norm sense, i.e., we can approximate ∂D by minimizing the defect

$$\|u^i + u^s_\alpha\|_{L^2(\Lambda)} \tag{8}$$

over some suitable class U of admissible surface Λ.

3. An alternative approach to Kirsch-Kress method

From section 2, we know the step(S2) of the Kirsch-Kress method is to seek the unknown boundary ∂D as the location where the total field $u = u^i + u^s$ satisfies the boundary condition $u = 0$. For a closed C^2-surface Λ in the domain of definition of u, we define the operator

$$G : \Lambda \to u|_\Lambda, \tag{9}$$

which maps Λ onto the trace of the total field u on Λ. Assuming that the boundary ∂D of the obstacle D is analytic, the total wave u can be extended as a solution to the Helmholtz equation into a neighborhood of ∂D. Therefore, the mapping G is well defined for surfaces Λ in a neighborhood of ∂D. Clearly, finding the boundary ∂D of the obstacle is equivalent to solving the nonlinear operator equation

$$G(\Lambda) = 0. \tag{10}$$

Again, we only consider that G maps C^2-surface Λ into $C(\Lambda)$. From results of [8], the Fréchet derivative of G is given by

$$G'(\Lambda)h = \operatorname{grad} u \cdot h, \text{ on } \Lambda. \tag{11}$$

Consequently, the Newton update reads

$$G(\Lambda) + G'(\Lambda)h = 0. \tag{12}$$

To sum up, the first iterative procedure of this paper is the following:

Step(A1). Given the far-field pattern u_∞ corresponding to the incident wave u^i and the auxiliary closed curve Γ contained in the unknown boundary ∂D. We approximately solve the ill-posed linear integral equation (6) for the density φ_α.

Step(A2). Given the initial guess Λ of ∂D and Λ contains the auxiliary curve Γ, we evaluate the scattered field u^s on Λ by the single-layer potential (3).

Step(A3). Solve the nonlinear equation (10) via the linear equation (12) by the Newton iteration method.

Step(A4). If the tolerance is not satisfied or the maximum iteration step is not reached, update Λ by $\Lambda := \Lambda + h$, Goto *Step(A2)*.

The convergence [6] of the above algorithm can be obtained by combining the minimization of Tikhonov functional (7) and the defect minimization (8) into one cost functional

$$\|F\varphi - u_\infty\|^2_{L^2(\Omega)} + \alpha\|\varphi\|^2_{L^2(\Gamma)} + \|u^i + u^s_\alpha\|_{L^2(\Lambda)}. \tag{13}$$

If we have the a priori information that the diameter of the unknown obstacle is less than $\frac{2\pi}{k}$, then by Lemma 1 the obstacle can be uniquely determined by the far-field pattern that coincides one incident plane wave with wave number k. Hence, due to the lack of a uniqueness result for one wave number and one incoming plane wave, we can not assure that we always have convergence towards the boundary of the unknown scatterer.

We can try to achieve more accurate reconstructions by using more incident waves $u^i_1, u^i_2, \cdots, u^i_n$ with different directions $d_1, d_2, \cdots, d_n$ and corresponding far field patterns $u^1_\infty, u^2_\infty, \cdots, u^n_\infty$. If we solve the ill-posed linear integral equation (6) and calculate the scattered field (3) for each far-field pattern u^j_∞, the computational effort will be large. Due to the linear properties of the far-field operator and of the single-layer potential, we note that

$$\sum_{j=1}^{n} F\varphi_j = \sigma \int_\Gamma e^{-ik\hat{x}\cdot y} \sum_{j=1}^{n} \varphi_j(y)ds(y), \; \hat{x} \in \Omega. \tag{14}$$

and

$$\sum_{j=1}^{n} S\varphi_j = \int_\Gamma \Phi(x,y) \sum_{j=1}^{n} \varphi_j(y)ds(y), \; x \in \Lambda. \tag{15}$$

Denoting that $\psi(y) := \sum_{j=1}^{n} \varphi_j(y)$, we obtain the superposition field equation

$$(F\psi)(\hat{x}) = U_\infty := \sum_{j=1}^{n} u^j_\infty, \; \hat{x} \in \Omega. \tag{16}$$

and its approximately scattered field

$$U^s = (S\psi)(x) = \int_\Gamma \Phi(x,y)\psi(y)ds(y), \; x \in \Lambda. \tag{17}$$

Therefore, the superposition total field

$$U = U^i + U^s. \tag{18}$$

where $U^i = \sum_{j=1}^{n} e^{ikx\cdot d_j}$.

We get immediately the second iterative procedure for reconstructing the unknown obstacle D from several incoming wave u^i when we embed (16), (17) and (18) into the first iterative procedure. The second iterative procedure is

Step(B1). Given the far-field pattern u_∞^j corresponding to the incident wave $u_j^i, j = 1, 2, \cdots, n$ and the auxiliary closed curve Γ contained in the unknown boundary ∂D. We approximately solve the ill-posed linear integral equation (16) for the density ψ_α.

Step(B2). Given the initial guess Λ of ∂D and Λ contains the auxiliary curve Γ, we evaluate the scattered field U^s on Λ by the single-layer potential (17).

Step(B3). Solve the nonlinear equation (10) via the linear equation (12) with the superposition total field U.

Step(B4). If the tolerance is not satisfied or the maximum iteration step is not reached, update Λ by $\Lambda := \Lambda + h$, Goto *Step(B2)*.

4. Numerical implementation

For the practical computation, we represent the curves Γ and Λ by a parametrization

$$\Lambda = \{x(t) \mid t \in [0, 2\pi]\}, \tag{19}$$

where $x(t) : [0, 2\pi] \to \mathbb{R}^2$ is a two times continuously differential 2π-periodic vector. Looking for the location where the boundary condition $u = 0$ is satisfied, we solve the linear equation (12) of (10) with respect to the normal direction at Λ. To this end, we try to update Λ in the following form

$$\Lambda := \Lambda + h = \{x(t) = x(t) + h(t)\nu(t) \mid t \in [0, 2\pi]\}, \tag{20}$$

where $h(t) : [0, 2\pi] \to \mathbb{R}$ is a 2π-periodic vector and $\nu(t)$ is the outward normal vector of the former curve Λ. The normal vector can be evaluated through the parametrization (19) via

$$\nu(t) = \frac{(x_2'(t), -x_1'(t))}{|x'(t)|}, \; t \in [0, 2\pi], \tag{21}$$

where $x_1(t)$ and $x_2(t)$ are two components of the vector $x(t)$, $|x'(t)|$ is the Euclid norm of tangent vector $x'(t)$. The update $h(t)$ is taken as by trigonometric polynomials of degree less than or equal to m, i.e.,

$$h(t) = a_0 + \sum_{j=1}^{m} (a_j \cos jt + b_j \sin jt). \tag{22}$$

For the above representations of curves Γ and Λ, it is easy to compute the integrals in terms of the parameter t. Hence,

$$(F\varphi)(\hat{x}) = \sigma \int_0^{2\pi} \varphi(y(t)) e^{-ik\hat{x}\cdot y(t)} |y'(t)| dt, \; \hat{x} \in \Omega, \tag{23}$$

and

$$u^s(x(t)) = \int_0^{2\pi} \Phi(x(t), y(\tau)) \varphi(y(\tau)) |y'(\tau)| d\tau, \; t \in [0, 2\pi], \tag{24}$$

where $x(t) \in \Lambda$ and $y(t) \in \Gamma$. Since the auxiliary closed curve Γ is contained in the approximate curve Λ, there is no singular in the latter integrals. Also, we do not ask the jump relation of the single-layer potential to compute the normal derivative of the scattered field u^s.

Next, we show how to get the update $h(t)$ via the linear equation (12). Here, we solve (12) in a least squares sense, that is, the coefficients $a_0, a_j, b_j, j = 1, 2, \cdots, m$ are chosen such that for a set of collocation points $t_1, t_2, \cdots, t_M$ in $[0, 2\pi)$ the least squares sum

$$\sum_{l=1}^{M} \left| u(x(t_l)) + \frac{\partial u(x(t_l))}{\partial \nu} \left(a_0 + \sum_{j=1}^{m} (a_j \cos jt_l + b_j \sin jt_l) \right) \right|^2 . \tag{25}$$

is minimized. According to results of [8, 9, 10], we can also take a penalized least squares sum for the sake of stability. But we find that we do not need the penalized term in the following example to obtain more accurate reconstructions. Therefore we always take the common form (25) of least squares sum in this paper.

5. Numerical examples

In this section, we present the reconstructions of three different obstacles using the methods of Section 3 and the numerical implementation technique of Section 4. The incident waves are taken with directions d uniformly distributed over the unit circle (if only one wave is used the incoming wave is $e^{ikx\cdot d}$ with $d = (1, 0)$). We also denote by m the degree of the trigonometric polynomial used for the approximation of the boundary, i.e., the parametrization of the boundary

$$x(t) = (x_1(t), x_2(t)),\ t \in [0, 2\pi]$$

where $x_l(t) = b_0^l + \sum_{j=1}^{m}(b_j^l \cos jt + b_j^l \sin jt), l = 1, 2.$

In the following examples, we used $M = 64$ collocation points in (25) and the same number of equidistant quadrature points for the solution of the Tikhonov regularization for (6) and (16). To avoid an inverse crime, we use the far-field patterns obtained by approximately solving the direct scattering problem by the double-layer boundary integral equation method [6, 7]. The double-layer potential approach leads to an integral equation of the second kind which we solved numerically by Nyström's method [6] with 128 collocation points.

To illustrate the stability for solving the inverse problem, we also give the results for noisy data. For noisy data, random errors were added point-wise to the values of the far-field pattern at the 64 points with the percentage given in terms of the l^2-norm. In addition, the wave number k we use is 1.0, and the black arrows in the following figures represent the directions of incoming waves except for the case of twelve incoming waves.

Example 1. we consider the reconstruction of a peanut-shaped boundary curve given by the parametrization

$$\partial D = \{\sqrt{\cos^2(t) + 0.25\sin^2(t)}\ (\cos(t), \sin(t)),\ t \in [0, 2\pi]\}.$$

For parameters we choose $\alpha = 0.5 \times 10^{-10}$ for the exact far-field and $\alpha = 0.5 \times 10^{-5}$ for noise data with the 3% noise, and $m = 8$. The iterations were started with the auxiliary curve and the initial guess of radius 0.3 and 0.7 centered at origin, respectively. The numerical results after six iterations and with different directions are illustrated in figures 1 and 2.

Example 2. we consider the reconstruction of an acorn-shaped boundary curve given by the parametrization

$$\partial D = \{(2 + 0.5\cos(3t))\,(\cos(t), \sin(t)),\ t \in [0, 2\pi]\}.$$

For parameters we choose $\alpha = 0.5 \times 10^{-10}$ for the exact far-field and $\alpha = 0.5 \times 10^{-5}$ for noise data with the 3% noise, $m = 6$. The iterations were started with the auxiliary curve and the initial guess of radius 1.0 and 1.8 centered at origin, respectively. The numerical results after eight iterations and with different directions are illustrated in figures 3 and 4.

Example 3. we consider a domain that is contained in the approximation space and given by

$$\partial D = \{\frac{2 + 1.8\cos(t) + 0.2\sin(2t)}{1 + 0.75\cos(t)}\ (\cos(t), \sin(t)),\ t \in [0, 2\pi]\}.$$

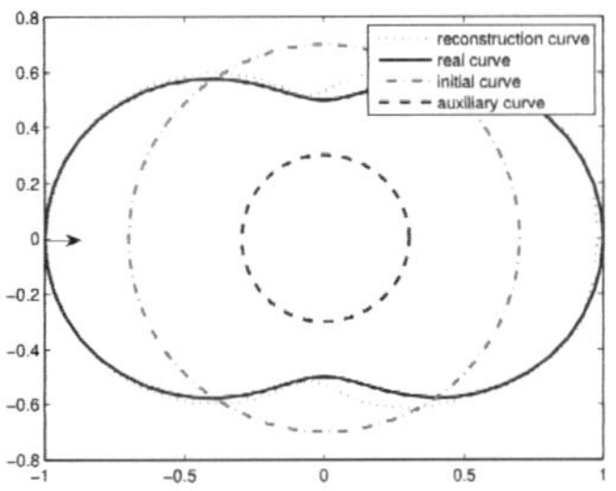

(a) One incoming wave

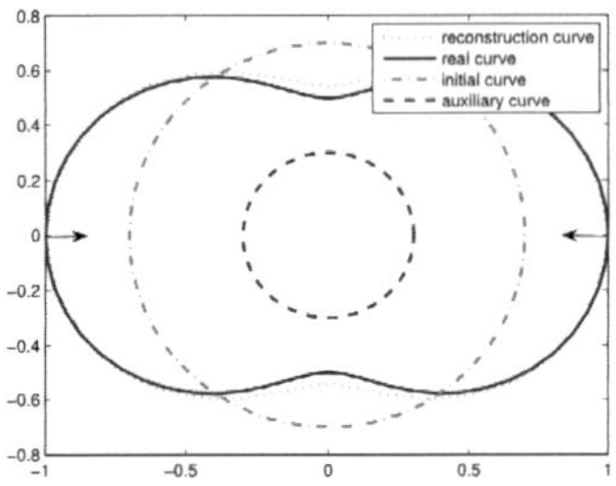

(b) Two horizontal incoming wave

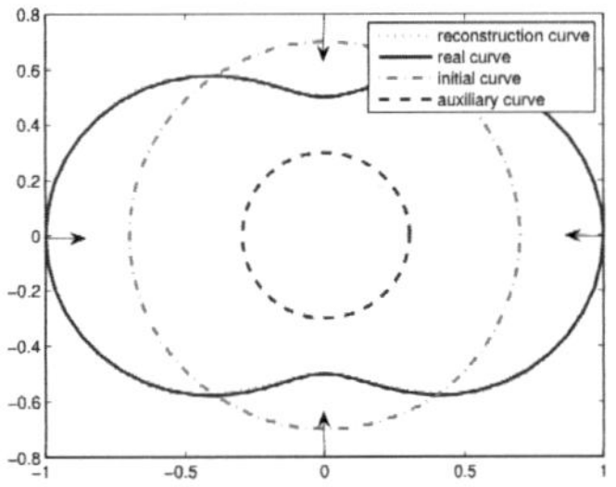

(c) Four incoming waves

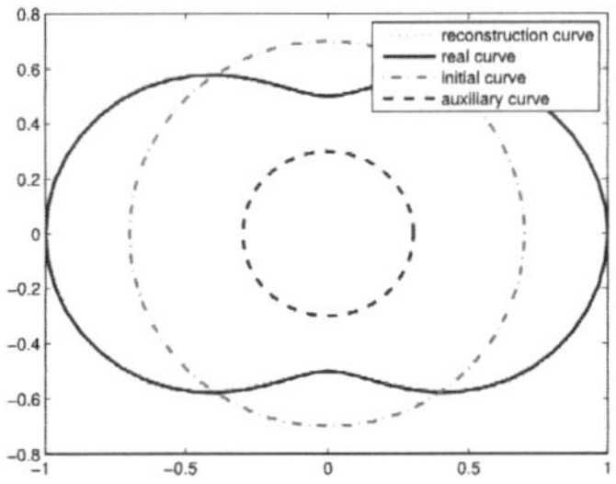

(d) Twelve incoming waves

Figure 1: Reconstructions without noise after six iterations.

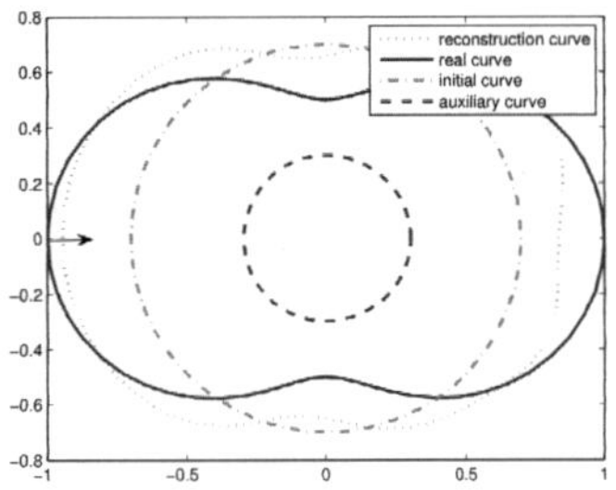

(a) One incoming wave

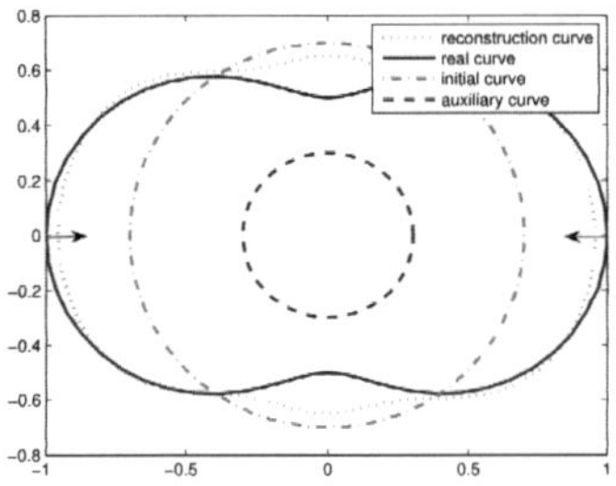

(b) Two horizontal incoming wave

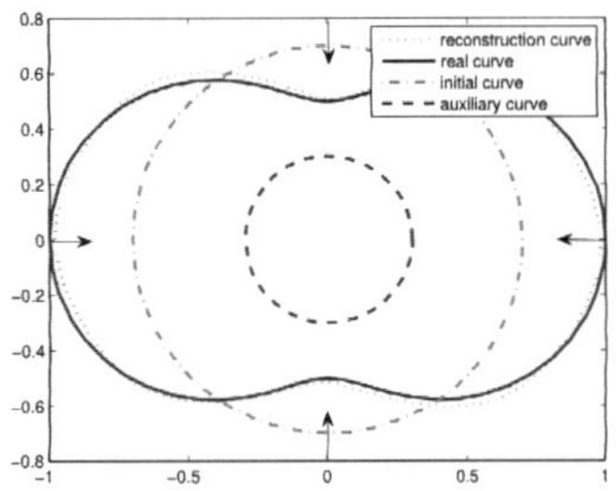

(c) Four incoming waves

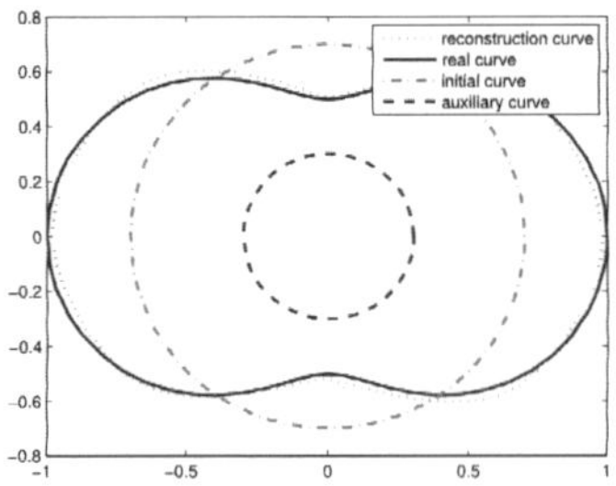

(d) Twelve incoming waves

Figure 2: Reconstructions with 3% noise after six iterations.

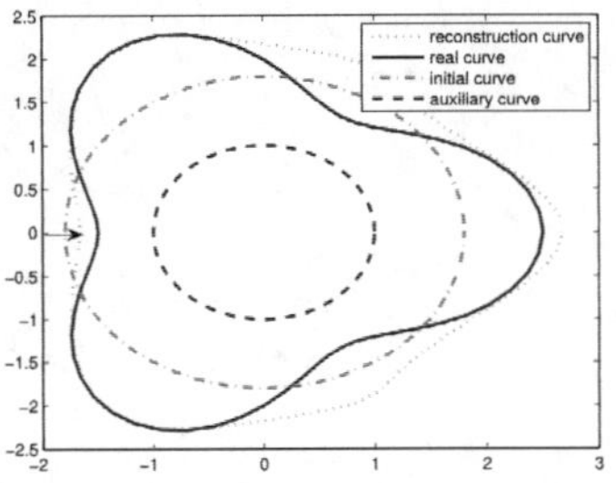

(a) One incoming wave

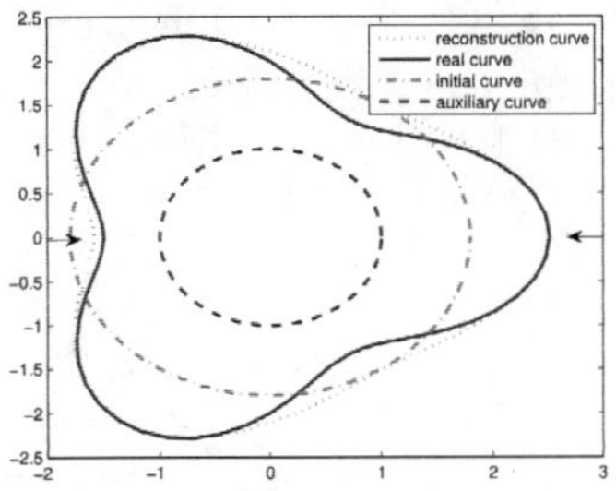

(b) Two horizontal incoming wave

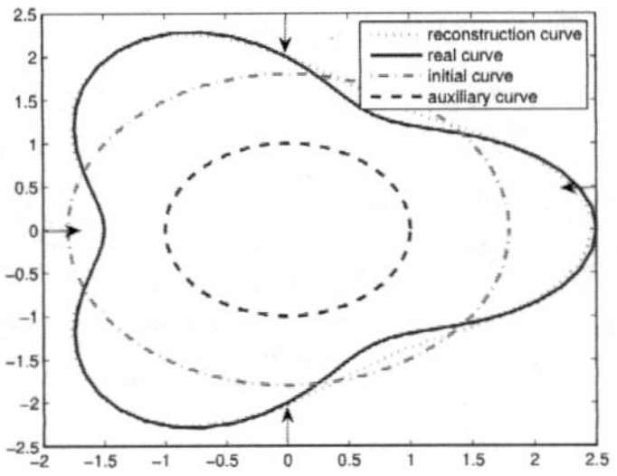

(c) Four incoming waves

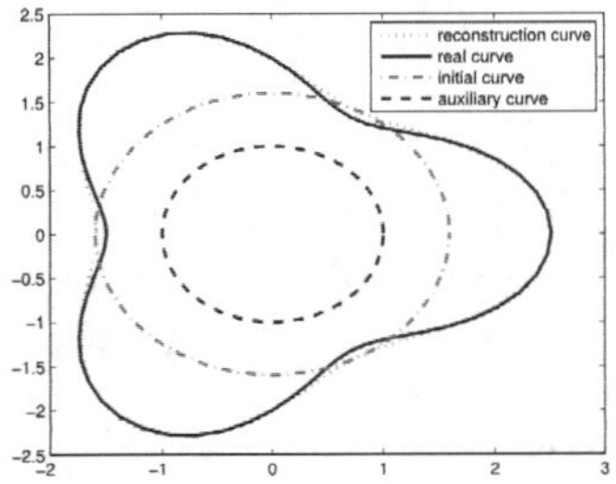

(d) Twelve incoming waves

Figure 3: Reconstructions without noise after six iterations.

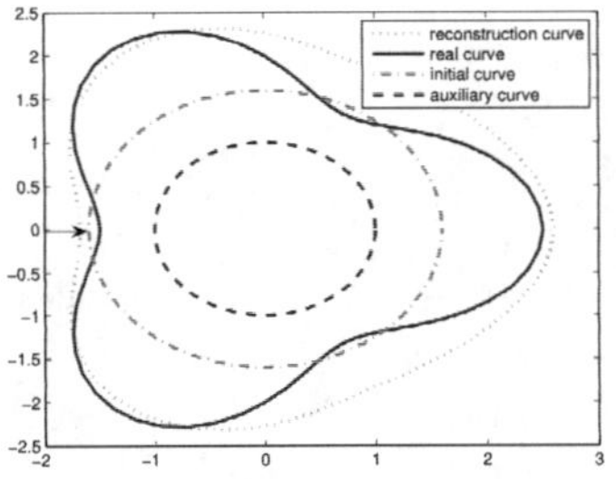

(a) One incoming wave

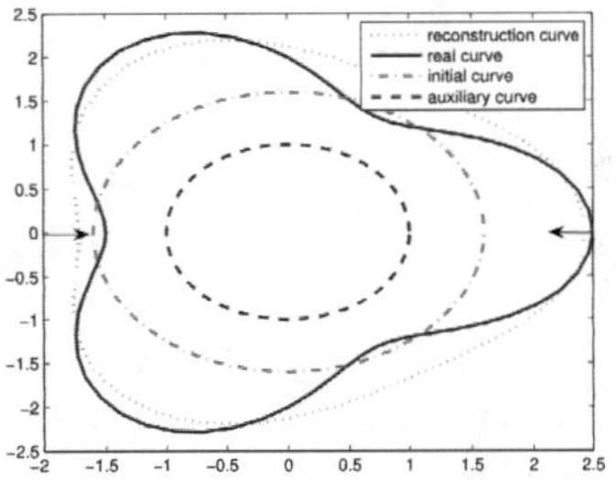

(b) Two horizontal incoming wave

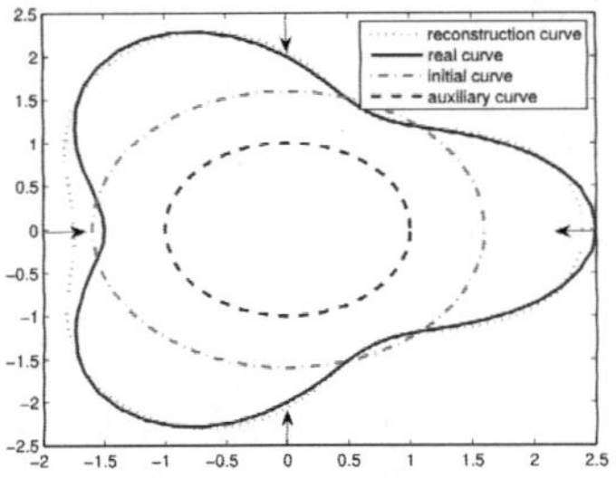

(c) Four incoming waves

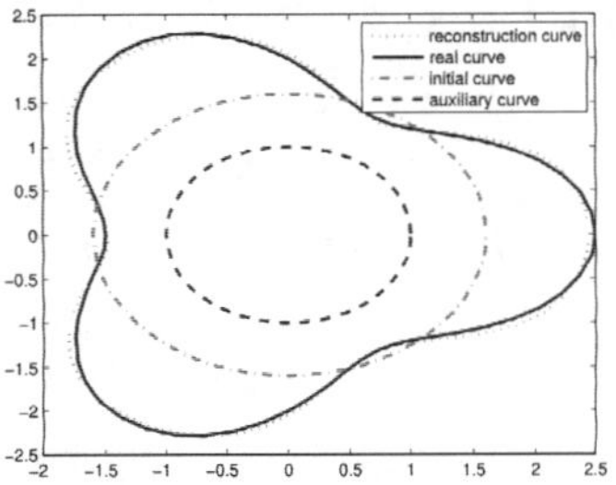

(d) Twelve incoming waves

Figure 4: Reconstructions with 3% noise after six iterations.

For parameters we choose $\alpha = 0.5 \times 10^{-12}$ for the exact far-field and $\alpha = 0.5 \times 10^{-6}$ for noise data with the 3% noise, $m = 16$. The iterations were started with the auxiliary curve and the initial guess of radius 0.5 and 1.6 centered at origin, respectively. The numerical results after ten iterations and with different directions are illustrated in figures 5 and 6.

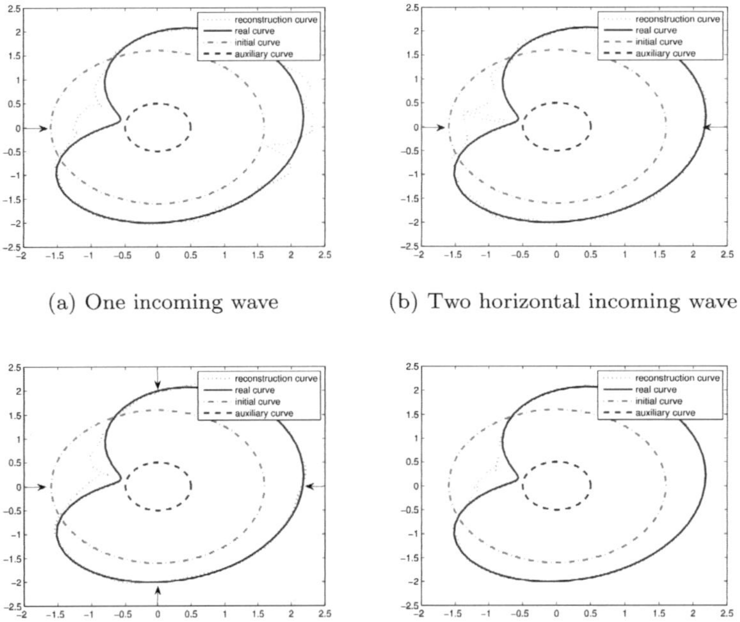

(a) One incoming wave (b) Two horizontal incoming wave

(c) Four incoming waves (d) Twelve incoming waves

Figure 5: Reconstructions without noise after six iterations.

Our numerical results show that the alternative method of Kirsch-Kress method does not need a forward solver and yields reasonable reconstructions of the obstacle, provided the initial guess is close enough to the exact boundary. Moreover, the accuracy of the reconstructions is satisfactory with 4 or more than 4 incoming waves. For a good initial guess, we can combine this method with the probe method [11, 12] or linear sampling method [13, 14]. On the other hand, we note that the regularization parameters were chosen by trial and error; and the number of Newton steps was also chosen by trial and error. Therefore, these are our work in progress. For example, we may choose the regularization parameters by the model function method [15, 16]. Moreover, we can extend this method to handle other boundary conditions such as the sound-hard case (Neumann boundary condition) and the impedance case (Robin boundary condition).

Acknowledgments

This work is supported by NSFC (10861001) and NSF of Jiangxi Province(2009GZS0001,2010GZS0010).

References

[1] Hohage T 1998 Convergence rates of a regularized Newton method in sound-hard inverse scattering. *SIAM J. Numer. Anal.*, **36** 125-42.

[2] Hettlich F 1995 Fréchet derivatives in inverse obstacle scattering. *Inverse Problems*, **11** 371-82.

[3] Potthast R 1996 Fréchet differentiability of the solution to the acoustic Neumann scattering problem with respect to the domain. *J. Inverse Ill-Posed Problems* **4** 67-84.

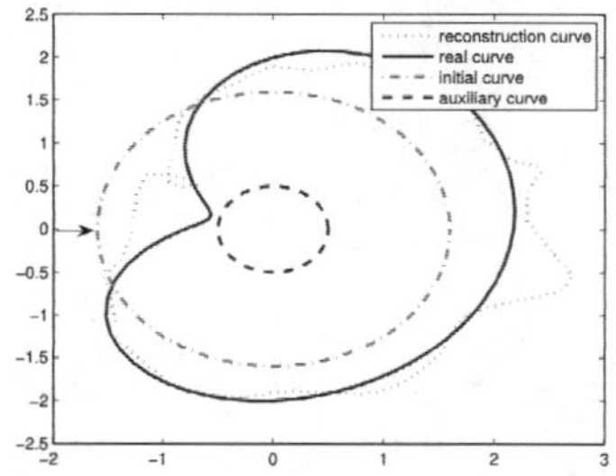

(a) One incoming wave

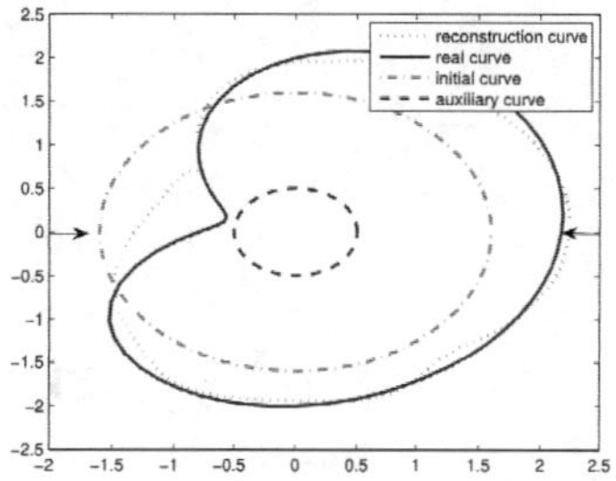

(b) Two horizontal incoming wave

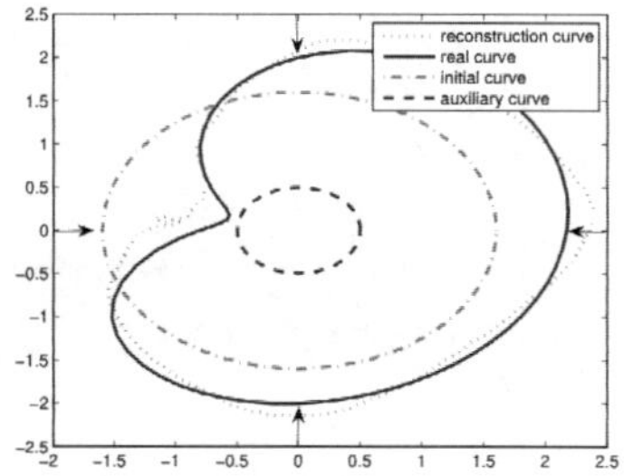

(c) Four incoming waves

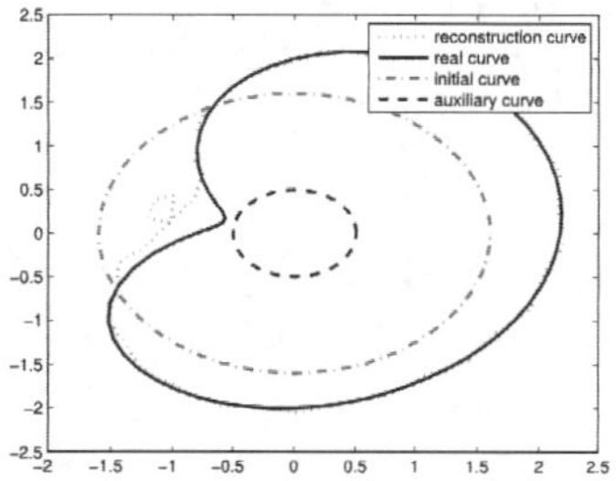

(d) Twelve incoming waves

Figure 6: Reconstructions with 3% noise after six iterations.

[4] Kirsch A and Kress R 1987 A numerical method for an inverse scattering problem Inverse Problems. Engl H, Groetsch C W, eds. *Inverse Problems* (New York: Academic Press Orlando) 279-90.

[5] Kirsch A and Kress R 1988 Two methods for solving the inverse acoustic scattering problem. *Inverse Problems*, **4** 749-70.

[6] Colton D and Kress R 1998 *Inverse Acoustic and Electromagnetic Scattering Theory(2nd ed)* (Berlin: Springer).

[7] Colton D and Kress R 1983 *Integral equation methods in scattering theory*(John Wiley & Sons).

[8] Kress R 2003 Newton's method for inverse obstacle scattering meets the method of least squares. *Inverse Problems* **19** 91-104.

[9] Serranho P 2006 A hybrid method for inverse scattering for shape and impedance. *Inverse Problems* **22** 663-80.

[10] Kress R 2007 Serranho P. A hybrid method for sound-hard obstacle reconstruction. *J. Comput. Appl. Math.* , **204** 418-27.

[11] Cheng J, Liu J J and Nakamura G 2005 The numerical realization of the probe method for the inverse scattering problems from the near-field data. *Inverse Problems* **21** 839-55.

[12] Yuan M and Liu J J 2006 Numerical realization of probe method for 2-D inverse scattering[J]. *Mathematica Numerica Sinica*, **28** 189-200.

[13] Cakoni F and Colton D 2005 *Qualitative Methods in Inverse Scattering Theory: An Introduction* (New York: Springer).

[14] Guan J Y and Wang L T 2008 Inverse of two dimensional penetrable objects based on linear sampling method. *Pure and Applied Mathematics*, **24** 114-20.

[15] Wang Z W and Liu J J 2009 New model function methods for determining regularization parameters in linear inverse problems. *Applied Numerical Mathematics* **59** 2489-506.

[16] Liu J J and Ni M 2008 A model function method for determining the regularizing parameter in potential approach for the recovery of scattered wave. *Applied Numerical Mathematics* **58** 1113-28.

A direct numerical method for solving inverse heat source problems

Xiong Xiangtuan
Department of Mathematics, Northwest Normal University, China
E-mail: xiongxt@gmail.com

Yan Yaomei
Department of Mathematics, Northwest Normal University, China
E-mail: yanyaomei@gmail.com

Wang Junxia
Department of Mathematics, Northwest Normal University, China
E-mail: wangjx@gmail.com

Abstract. Inverse source problems is a classical ill-posed problem. There exist several literatures on inverse heat source problems. In this study we use some stable methods to deal with the inverse problems of determining space-wise and time-wise dependent heat source in the heat diffusion equation. The problems of inverse heat source sources are transferred to the problems of numerical differentiation. By simple and effective methods which are used to regularize the problems of numerical differentiation before, we can obtain stable solutions. The numerical results show that our proposed procedure yields stable and accurate approximation.

1. Introduction

In this paper, we consider the problems of identifying the heat source from measured data for heat equation. This model has many important applications in practice, e.g., in finding a pollution source intensity and also for designing the final state in melting and freezing processes. The problem is ill-posed and any small change in the input data may result in a dramatic change in the solution. To obtain a stable numerical solution for these kinds of ill-posed problems, some regularization strategies should be applied.

The problems of identification of heat source have been discussed by many authors. For theoretical papers, we can refer to the references [3, 4, 5, 13, 17, 18, 19, 22]. For computational papers, we can refer to the references [2, 8, 9, 10, 14, 15, 23] and the references therein.

Recently, In the case of space-wise dependent heat source, Johansson and Lesnic proposed an iterative method [11] and a variational method [12] for solving the problem of this kind numerically. However, these methods need to solve a direct problem at each iterative step. The computational cost is very large. In the case of time-wise dependent heat source, Yan et al in [23] provided a fundamental solution method for solving the problem. The method has one drawback: it needs source points outside the domain which are not easy to place properly.

In this paper, we present an alternative method for solving these problems. The method is simple and effective. The main aim of this paper is to provide a new method for solving the problems of unknown heat source. Furthermore, our method can be extended to the case of higher dimensions easily.

2. The inverse problem of space-wise dependent heat source

We consider the following inverse problem: find the temperature u and the heat source f which satisfy the heat conduction equation, namely

$$u_t(x,t) = u_{xx}(x,t) + f(x), 0 < x < 1, 0 < t < T, \tag{1}$$

with initial data and boundary conditions

$$u(x,0) = u_0(x), 0 < x < 1 \tag{2}$$

$$u(0,t) = s(t),\ u(1,t) = l(t), 0 < t < T, \tag{3}$$

where we need to reconstruct $f(x)$ from the over-specified data $u(x,T) = g(x)$. Assume that the given functions satisfy the compatibility conditions

$$\begin{aligned} u_0(0) = s(0),\ u_0(1) = l(0), \\ g(0) = s(T),\ g(1) = l(T). \end{aligned} \tag{4}$$

One may ask a question: How much is the degree of ill-poseness of this problem? In order to answer this question, for convenience of analysis, let $s(t) = l(t) = 0$. Here we use finite difference method to give the answer in Subsection 2.1.

2.1. Analysis on degree of ill-posedness

First we divide the domain $(x,t) \in [0,1] \times [0,T]$ into several sub-domains, i.e. $x_1 = 0 < x_2 < \cdots < x_i = (i-1)h < \cdots < x_{n+1} = 1$, $t_1 = 0 < \cdots < t_i = (i-1)\Delta t < \cdots < t_{m+1} = T$, where $h = 1/n, \Delta t = T/m$. Let $u_t(x_i,t_j) \approx \frac{1}{\Delta t}(u_i^{j+1} - u_i^j)$, $u_{xx}(x_i,t_j) \approx \frac{1}{h}(u_{i+1}^j - 2u_i^j + u_{i-1}^j)$, where $u_i^j = u(x_i,t_j)$, let $f(x_i) = f_i$, $g(x_i) = g_i$, the equation (1)-(3) can be discretized as

$$\frac{1}{\Delta t}(u_i^{j+1} - u_i^j) - 1/h^2(u_{i+1}^j - 2u_i^j + u_{i-1}^j) = f_i, \tag{5}$$

Let $r = \Delta t/h^2$, then

$$u_i^{j+1} = ru_{i+1}^j + (1-2r)u_i^j + ru_{i-1}^j + \Delta t f_i, \tag{6}$$

we can rewrite the above equation as

$$U^{j+1} = AU^j + \Delta t F, \tag{7}$$

where $U^j = [u(x_2,t_j),\cdots,u(x_n,t_j)]^{tr}$, $G = [g_2,\cdots,g_n]^{tr}$, the superscript tr denotes the transposition of a matrix,

$$A = \begin{pmatrix} 1-2r & r & & & \\ r & 1-2r & r & & \\ & r & 1-2r & r & \\ & & \ddots & \ddots & \ddots \\ & & & r & 1-2r \end{pmatrix}_{(n-1)\times(n-1)}.$$

By induction, we have

$$
\begin{aligned}
&U^2 = AU^1 + \Delta t G \\
&U^3 = AU^2 + \Delta t G = A[AU^1 + \Delta t G] + \Delta t G \\
&= A^2 U^1 + \Delta t A G + \Delta t G \\
&\vdots \\
&U^{m+1} = A^m U^1 + (A^{m-1} + A^{m-2} + \cdots + A + I)\Delta t F.
\end{aligned}
$$

If $r < 1/2$, then $\|A\|_2 < 1$, $A^{m-1} + A^{m-2} + \cdots + A + I = (I-A)^{-1}(I-A^m)$, Noting that $U^{m+1} = G$, we have

$$G = A^m U^1 + (I-A)^{-1}(I-A^n)\Delta t F. \tag{8}$$

Denote $K = (I-A)^{-1}(I-A^m)\Delta t$, we can get

$$KF = G - A^m U^1. \tag{9}$$

As A is a symmetric matrix, we can easily obtain its eigenvalues

$$\lambda_j = 1 - 4r\sin^2(j\frac{\pi}{2}h),\ 1 \le j \le n-1, \tag{10}$$

then the eigenvalues of K are

$$\mu_j = [(1-\lambda_j^m)/(1-\lambda_j)]\Delta t,\ 1 \le j \le n-1.$$

By a simple calculation,

$$\mu_j = \frac{1 - [1 - 4r\sin^2(j\frac{\pi}{2}h)]^m}{4r\sin^2(j\frac{\pi}{2}h)}\Delta t,\ 1 \le j \le n-1.$$

Recall that $r = \Delta t/h^2 = n^2/m$, and r is fixed and satisfies $r \le 1/2$. If $c > 0$, n is large sufficiently, we have $(1-c/n)^n \approx \exp(-c)$, $\sin(j\frac{\pi}{2}h) \approx j\frac{\pi}{2}h$, then

$$\mu_j \approx \frac{1 - [1 - 4rj^2\frac{\pi^2}{4n^2}]^{n^2/r}}{4rj^2\frac{\pi^2}{4n^2}}\frac{r}{n^2},\ 1 \le j \le n-1.$$

We get

$$\mu_j \approx \frac{1 - e^{-\pi^2 j^2}}{j^2\pi^2},\ 1 \le j \le n-1. \tag{11}$$

If j is large enough, then

$$\mu_j \approx \frac{1}{j^2\pi^2},\ 1 \le j \le n-1. \tag{12}$$

From above formula, we can conclude that the degree of the problem (1)-(3) is equivalent to that of second-order numerical differentiation.

2.2. A simple method for the inverse problem

In this subsection, we assume that $s(t)$ and $l(t)$ in (1)-(3) belong to $C^1(0,T)$, and $g(x)$, $u_0(x)$ belong to $C^2(0,1)$.

Differentiating Equation (1) with respect to t, it yields

$$u_{tt} = u_{xxt}, \tag{13}$$

let $w = u_t$, and

$$u(x,t) = \int_0^t w(x,\tau)d\tau + u_0(x). \tag{14}$$

Thus, (13) becomes

$$w_t = w_{xx}, \tag{15}$$

with boundary conditions

$$w(0,t) = s'(t), \;\; w(1,t) = l'(t). \tag{16}$$

Integrating Equation (13) with respect to t in the interval $[0,T]$, it is obtained

$$\int_0^T u_{tt}dt = \int_0^T u_{xxt}dt, \tag{17}$$

i.e.

$$w(x,T) - w(x,0) = g''(x) - u_0''(x). \tag{18}$$

Integrating Equation (13) with respect to t in the interval $[t,T]$, we obtain

$$u_t(x,t) = u_{xx}(x,t) + u_t(x,T) - g''(x). \tag{19}$$

Therefore, we can get

$$f(x) = w(x,T) - g''(x). \tag{20}$$

Since $g(x)$ is measured data and there exist measurement errors, i.e. only the noisy data $g^\delta(x)$ is available. This involves an ill-posed problem of second-order numerical differentiation. Hence we should deal with the second-order numerical differentiation by a stable approximation method firstly. The problem of numerical differentiation is discussed by several authors, we can consult the references [1, 16, 20, 21]. In the sequent section, we use the Tikhonov regularization method to approximate the second-order derivative of noisy data g^δ. Minimize the following functional

$$H_\alpha(g_\alpha^\delta) = \int_0^1 (g_\alpha^\delta(x) - g^\delta(x))^2dx + \alpha\int_0^1 (\frac{d^2g_\alpha^\delta}{dx^2}(x))^2dx, \tag{21}$$

and we can get the minimizer $g_\alpha^\delta(x)$. Thus we have a stable approximation $(g_\alpha^\delta)''(x)$ to the second-order derivative $g''(x)$.

Now let us summarize what we have:

- Step 1. Use the Tikhonov method (or other methods such as central difference method) to stably approximate the second-order derivative g''.
- Step 2. Numerically solve the following problem:

$$\begin{cases} w_t(x,t) = w_{xx}(x,t), & 0 < x < 1, 0 < t < T, \\ w(0,t) = s'(t), w(1,t) = l'(t), & 0 < t < T, \end{cases} \tag{22}$$

and the following condition

$$w(x,T) - w(x,0) = g''(x) - u_0''(x). \tag{23}$$

- Step 3. Compute the heat source $f(x)$ according to (20).

2.3. Generalization to two-dimensional case
We consider the following two-dimensional problem

$$u_t(x,y,t) = u_{xx}(x,y,t) + u_{yy}(x,y,t) + f(x,y), 0 < x < 1, 0 < y < 1, 0 < t < T, \quad (24)$$

with initial data and boundary conditions

$$u(x,y,0) = \varphi(x,y), 0 < x < 1, 0 < y < 1, \quad (25)$$

$$u(0,y,t) = f_1(y,t),\ u(x,0,t) = f_2(x,t),\ 0 < t < T, \quad (26)$$

$$u(1,y,t) = f_3(y,t),\ u(x,1,t) = f_4(x,t),\ 0 < t < T, \quad (27)$$

here we need to reconstruct $f(x,y)$ from the over-specified data $u(x,y,T) = g(x,y)$. Similarly, we have the following algorithm:

- Step 1. Use the Tikhonov method to stably approximate the second-order derivative $g_{xx}(x,y)$ and $g_{yy}(x,y)$.
- Step 2. Numerically solve the following problem:

$$\begin{cases} w_t(x,y,t) = w_{xx}(x,y,t) + w_{yy}(x,y,t), & 0 < x < 1, 0 < y < 1, 0 < t < T, \\ u(0,y,t) = f_1'(y,t),\ u(x,0,t) = f_2'(x,t), & 0 < t < T \\ u(1,y,t) = f_3'(y,t),\ u(x,1,t) = f_4'(x,t), & 0 < t < T, \end{cases} \quad (28)$$

  where the symbol a' denotes differentiating the function a with respect to t and the following condition

$$w(x,y,T) - w(x,y,0) = g_{xx}(x,y) + g_{yy}(x,y) - \varphi_{xx}(x,y) - \varphi_{yy}(x,y). \quad (29)$$

- Step 3. Compute the heat source $f(x,y)$ according to

$$f(x,y) = w(x,y,T) - g_{xx}(x,y) - g_{yy}(x,y). \quad (30)$$

3. The inverse problem of time-wise dependent heat source

3.1. A direct method
Similar to Section 2, we consider the following inverse problem: find the temperature u and the heat source f which satisfy the heat conduction equation, namely

$$u_t(x,t) = u_{xx}(x,t) + f(t), 0 < x < 1, 0 < t < T, \quad (31)$$

with initial data and boundary conditions

$$u(x,0) = u_0(x), 0 < x < 1, \quad (32)$$

$$u_x(0,t) = s(t),\ u_x(1,t) = l(t), 0 < t < T. \quad (33)$$

We want to reconstruct $f(t)$ from the over-specified data $u(x_f,t) = g(t)$ where x_f is a fixed point inside $[0,1]$. Assume that the given functions satisfy the compatibility conditions

$$\begin{aligned} &(u_0)_x(0) = s(0),\ (u_0)_x(1) = l(0), \\ &g(0) = u_0(x_f). \end{aligned} \quad (34)$$

Furthermore, we assume that $s(t)$ and $l(t)$ in (33) belong to $C(0,T)$ and $g(t)$ belongs to $C^1(0,T)$, and $u_0(x)$ belongs to $C^1(0,1)$.

Differentiating Equation (31) with respect to x, it yields

$$u_{tx} = u_{xxx}, \tag{35}$$

let $w = u_x$, and

$$u(x,t) = \int_{x_f}^{x} w(y,t)dy + g(t). \tag{36}$$

Thus, (35) becomes

$$w_t = w_{xx}, \tag{37}$$

with initial and boundary conditions

$$w(x,0) = (u_0)_x(x), \ w(0,t) = s(t), \ w(1,t) = l(t). \tag{38}$$

Integrating Equation (35) with respect to x in the interval $[x_f, x]$, it is obtained

$$\int_{x_f}^{x} u_{tx}dx = \int_{x_f}^{x} u_{xxx}dx, \tag{39}$$

i.e.

$$u_t(x,t) = u_{xx}(x,t) + g'(t) - u_{xx}(x_f,t). \tag{40}$$

Therefore, we can get

$$f(t) = g'(t) - u_{xx}(x_f,t). \tag{41}$$

Since $g(t)$ is measured data and there exist measurement errors, i.e. only the noisy data $g^\delta(t)$ is available. This involves an ill-posed problem of first-order numerical differentiation. It is well-known that the step lengths in the standard numerical method such as finite difference method have regularization effect. Hence we can choose appropriate step lengths to compute the first-order derivative stably.

Remark. *Alternatively the first-order numerical differentiation can be solved by other stable approximation methods. For example, we can use the Tikhonov regularization method to approximate the first-order derivative of noisy data g^δ. Minimize the following functional*

$$H_\alpha(g_\alpha^\delta) = \int_0^T (g_\alpha^\delta(t) - g^\delta(t))^2 dt + \alpha \int_0^T (\frac{dg_\alpha^\delta}{dt}(t))^2 dt \tag{42}$$

and we can get the minimizer $g_\alpha^\delta(t)$. Thus we have a stable approximation $(g_\alpha^\delta)'(t)$ to the first-order derivative $g'(t)$.

Now let us summarize what we have:

- Step 1. Use the central difference formula to stably approximate the first-order derivative g' with appropriate step lengths.
- Step 2. Numerically solve the following problem to get $w_x(x_f,t)$:

$$\begin{cases} w_t(x,t) = w_{xx}(x,t), & 0 < x < 1, 0 < t < T, \\ w(0,t) = (u_0)_x(x), & 0 < x < 1, \\ w(0,t) = s(t), \ w(1,t) = l(t), & 0 < t < T. \end{cases} \tag{43}$$

- Step 3. Compute the heat source $f(t)$ according to (41) by noting that $w = u_x$.

3.2. Generalization to two-dimensional case

Consider the following two-dimensional problem

$$u_t(x,y,t) = u_{xx}(x,y,t) + u_{yy}(x,y,t) + f(t), 0 < x < 1, 0 < y < 1, 0 < t < T, \tag{44}$$

with initial data and boundary conditions

$$u(x,y,0) = \varphi(x,y), 0 < x < 1, 0 < y < 1, \tag{45}$$

$$u_x(0,y,t) = f_1(y,t),\ u_y(x,0,t) = f_2(x,t),\ 0 < t < T, \tag{46}$$

$$u_x(1,y,t) = f_3(y,t),\ u_y(x,1,t) = f_4(x,t),\ 0 < t < T, \tag{47}$$

here we need to reconstruct $f(t)$ from the over-specified data $u(\overline{x},\overline{y},t) = g(t)$ where $(\overline{x},\overline{y}) \in (0,1) \times (0,1)$. Similarly, from the transform $v(x,y,t) = u(x,y,t) - \int_0^t f(\tau)d\tau$, we have the following algorithm:

- Step 1. Use the central difference formula to stably approximate the first-order derivative g' with appropriate step sizes..
- Step 2. Numerically solve the following problem to get $v(\overline{x},\overline{y},t)$:

$$\begin{cases} v_t(x,y,t) = v_{xx}(x,y,t) + v_{yy}(x,y,t), & 0 < x < 1, 0 < y < 1, 0 < t < T, \\ v(x,y,0) = \varphi(x,y), & 0 < x < 1, 0 < y < 1, \\ v_x(0,y,t) = f_1(y,t),\ v_x(1,y,t) = f_3(y,t), & 0 < y < 1,\ 0 < t < T, \\ v_y(x,0,t) = f_2(x,t),\ v_y(x,1,t) = f_4(x,t), & 0 < x < 1,\ 0 < t < T. \end{cases} \tag{48}$$

- Step 3. Compute the heat source $f(t)$ according to

$$f(t) = g'(t) - v_t(\overline{x},\overline{y},t). \tag{49}$$

4. Numerical examples

In this section, we implement the numerical algorithm by using Matlab in IEEE double precision with unit round off $1.1 \cdot 10^{-16}$ and use uniform grids in finite difference methods. Due to the similarity, we only give two numerical examples for the case of time-wise dependent heat source.

In the following numerical experiment, the problem (43) in Step 2 is solved by the explicit forward Euler method which is the simplest method and requires that $\frac{\Delta t}{(\Delta x)^2} \leq \frac{1}{2}$ for the reason of stability, where Δt and Δx are the time step length and space step length, respectively.

In this experiment, in order to investigate the stability of the numerical solution, the data $g(t)$ is perturbed as

$$G^\delta = G + \epsilon * randn(size(G)), \tag{50}$$

where the randn function generates arrays of random numbers whose elements are normally distributed with mean 0, variance $\sigma^2 = 1$, and standard deviation $\sigma = 1$, G is the column vector discretized by $g(t)$ at grid points, ϵ is the noise level.

To test the accuracy of the approximate solution, we use the root mean square error (RMSE) and the relative root mean square error (RSE), which are defined as follows:

$$RMSE(f) = \sqrt{\frac{1}{N}\sum_{i=1}^{N}(f_c(t_i) - f_e(t_i))^2}; \tag{51}$$

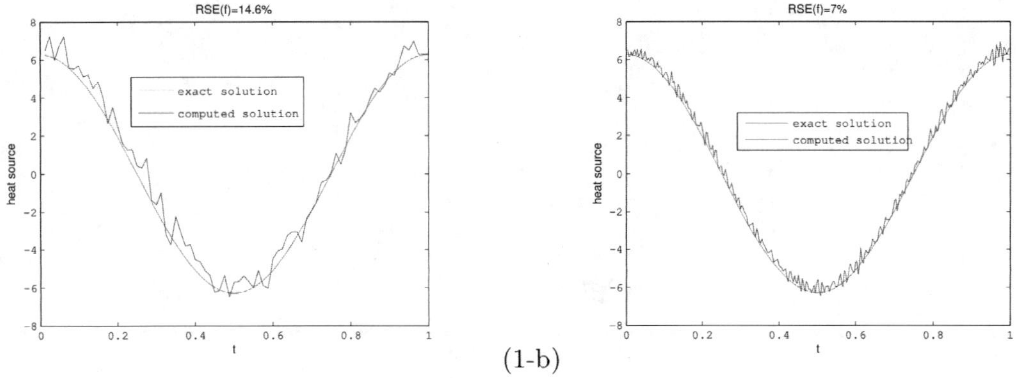

(1-a) (1-b)

Figure 1. Reconstruction of heat source for Example 1.

$$RSE(f) = \frac{\sqrt{\sum_{i=1}^{N}(f_c(t_i) - f_e(t_i))^2}}{\sqrt{\sum_{i=1}^{N}(f_e(t_i))^2}}, \tag{52}$$

where N is the total number of test points in $[0, T]$, f_c and f_e represent the computed heat source and the exact heat source, respectively.

Example 1. We try to reconstruct the heat source defined by

$$f(t) = 2\pi\cos(2\pi t), \ t \in (0, 1). \tag{53}$$

If we consider boundary conditions

$$u_x(0, t) = 0 = s(t) \text{ and } u_x(1, t) = 2 = l(t) \text{ for } t \in [0, 1]. \tag{54}$$

The initial condition is given by

$$u(x, 0) = x^2 \text{ for } x \in [0, 1]. \tag{55}$$

Then in this case the forward problem given by (31)-(33) with f given by (53) has the exact solution

$$u(x, t) = x^2 + 2t + \sin 2\pi t, \ \text{ for } (x, t) \in [0, 1] \times [0, 1]. \tag{56}$$

Based on (56), we can obtain the data at $x_f = \frac{1}{2}$ given by

$$g(t) = \frac{1}{4} + 2t + \sin 2\pi t, \ \text{ for } t \in [0, 1]. \tag{57}$$

Figure 1-a shows the reconstruction effect with $M = 6$, $N = 80$ and $\epsilon = 0.01$, where M is the total number of grid points at x-axis and N is the total number of grid points at t-axis. Figure 1-b shows the approximation effect of $f(t)$ with $M = 10$, $N = 300$ and $\epsilon = 0.001$. In the experiment, we find that the approximation effect varies with the time step length $1/N$ if ϵ is fixed. This shows that the time step length has the effect of regularization.

Example 2. In this example, we want to reconstruct the heat source given by

$$f(t) = \begin{cases} 0, & 0 \le t < \frac{1}{4}, \\ \frac{1}{2}, & \frac{1}{4} \le t < \frac{1}{2}, \\ 1, & \frac{1}{2} \le t \le 1, \end{cases} \tag{58}$$

(2-a) 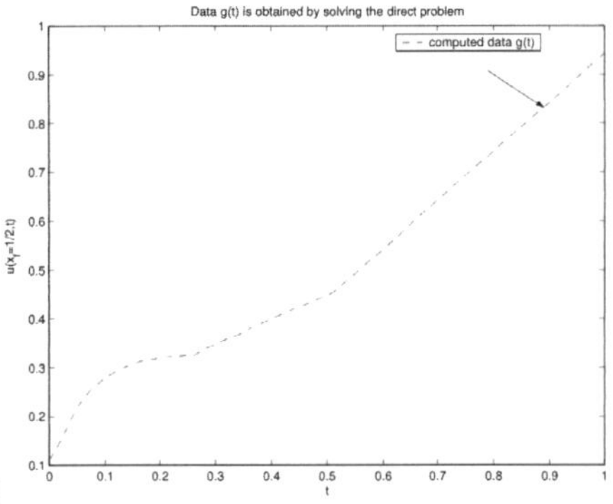

(2-b) 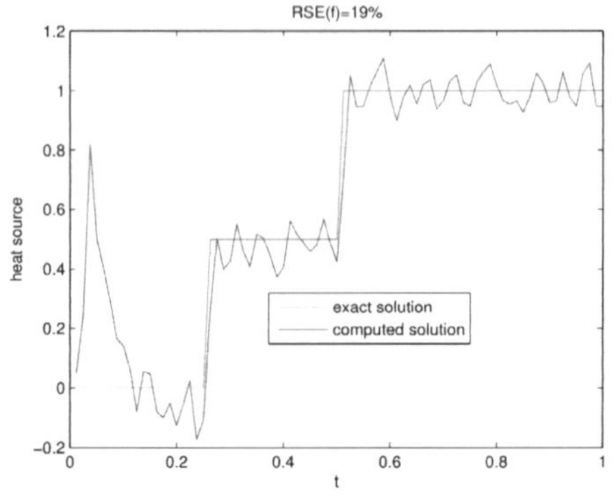

Figure 2. Reconstruction of heat source for Example 2.

with the initial and boundary conditions

$$\begin{aligned} &u_0(x) = x^2, && 0 \le x \le 1, \\ &u_x(1,t) = u_x(0,t) = 0, && 0 \le t \le 1. \end{aligned} \tag{59}$$

In this case, we don't have the exact solution $u(x,t)$. We solve the direct problem numerically to get $u(x_f,t)$, then we add random noise to the data $g(t) := u(x_f,t)$, finally we reconstruct the heat source from the noisy data. The parameters in this example are given as follows: $M = 6$, $N = 80$ and $\epsilon = 0.001$, $x_f = 1/2$. The numerical results are shown as Figure 2. The large oscillation near zero in Fig. 2-b is caused by the incompatibility of the data $g(0) = u_0(x_f)$ in the Eq. (34).

5. Conclusions

In this paper we can see that degree of ill-posedness of the problems of unknown time-wise and space-wise dependent heat source are as the same as that of the problem of first-order and second-order numerical differentiation, respectively. For sideways heat equation, as Eldén pointed in the References [6, 7], the ill-posedness is caused by time derivative and the time step length has regularization effect. Therefore, by choosing appropriate step lengths, we can make the direct numerical method work well with moderate noise levels. Furthermore, we can conclude that our method can be generalized to higher dimensional case. Further work should be considered, for example, how to reconstruct the style of heat source such as $f(x,t)$ in (1)-(3)? This problem will be discussed in the forthcoming paper.

Acknowledgement. This work is partially supported by the National Natural Science Foundation of China (nos 11001223 11061030), the Research Fund for the Doctoral Program of Higher Education of China (no 20106203120001) and the Doctoral Foundation of Northwest Normal University, China (no 5002-577).

The first author would like to thank Prof. Y.C Hon and Dr. Ming Li for their hospitality during staying at City University of Hong Kong. And the authors also thank Prof. Ruyun Ma(Department of Mathematics, Northwest Normal University, China) for his help.

References

[1] Ahn S, Cho U J, Ramm A G 2006 *J. Comput. Appl. Math.* **186** 325
[2] Badia A E and Duong T H 2002 *J. Inverse Ill-Posed Probl.* **10** 585
[3] Cannon J R 1986 *SIAM J. Numer. Anal.* **5** 275
[4] Choulli M and Yamamoto M 2004 *J. Inverse Ill-Posed Probl.* **12** 233
[5] Dou F F, Fu C L and Yang F L 2009 *J. Comput. Appl. Math.* **230** 728

[6] Elén L 1995 *Inverse Problems* **11** 913
[7] Elén L 1997 *J. Heat Transfer, Trans. ASME* **119** 406
[8] Farcas A and Lesnic D 2006 *J. Eng. Math.* **54** 375
[9] Fatullayev A G 2002 *Math. Comput. Simul.* **58** 247
[10] Ikehata M 2007 *Inverse Problems* **23** 183
[11] Johansson T and Lesnic D 2007 *J. Comput. Appl. Math.* **209** 66
[12] Johansson T and Lesnic D 2007 *IMA J. Appl. Math.* **72** 748
[13] Li G 2006 *Appl.Math.Comput.* **173** 566
[14] Li G, Liu J, Fan X and Ma Y 2008 *Appl. Math. Comput.* **196** 646
[15] Ling L, Hon Y C, Yamamoto M and Takeuchi T 2006 *Inverse Problems* **22** 1289
[16] Ramm A G and Smirnova A B 2001 *Math. Comput.* **70** 1131
[17] Savateev E G 1995 *J. Inverse Ill-Posed Probl.* **3** 83
[18] Solov'ev V V 1990 *Diff. Equ.* **25** 1114
[19] Trong D, Quan P and Alain P 2005 *J. Comput. Appl. Math.* **191** 50
[20] Wang Y B, Jia X Z and Cheng J 2002 *Inverse Problems* **18** 1461
[21] Wei T, Hon Y C and Wang Y B 2005 *Inverse Problems* **21** 657
[22] Yamamoto M 1993 *Math. Comput. Modelling.* **18** 79
[23] Yan L, Fu C L and Yang F L 2008 *Eng. Anal. Boundary. Elem.* **32** 216

An inverse problem of thickness design for bilayer textile materials under low temperature

Dinghua Xu[1]*, Jianxin Cheng[1], Yuanbo Chen[1] and Meibao Ge[2]

[1]Department of Mathematics, College of Sciences, Zhejiang Sci-Tech University, Hangzhou 310018, Zhejiang, P.R.China

[2]College of Sciences and Arts, Zhejiang Sci-Tech University, Hangzhou 311121, Zhejiang, P.R.China

E-mail: `dhxu6708@zstu.edu.cn(*Corresponding author)`

Abstract. The human heat-moisture-comfort level is mainly determined by heat and moisture transfer characteristics in clothing. With respect to the model of steady-state heat and moisture transfer through parallel pore textiles, we propose an inverse problem of thickness design for bilayer textile material under low temperature in this paper. Adopting the idea of regularization method, we formulate the inverse problem solving into a function minimization problem. Combining the finite difference method for ordinary differential equations with direct search method of one-dimensional minimization problems, we derive three kinds of iteration algorithms of regularized solution for the inverse problem of thickness design. Numerical simulation is achieved to verify the efficiency of proposed methods.

1. Introduction

With the rapid developments in sciences and technologies, people have increasing requirements on functional clothing, such as strong, durable clothing in the past, meanwhile both fashionable and comfortable clothing in the recent years. As for the humanbody-comfort textiles, it is hoped that the textiles are of fast decalescence, fast heat radiation, soft or stand-up apparel. Of course, clothing should be light and keeps body warm under low temperature, meanwhile sweat vaporizes fast and body feels cool under high temperature. Hence people need take full account of the functional material design based mainly on heat and moisture transfer characteristics.

We should consider the physical characteristics (e.g. sorption, condensation), structure characteristics (e.g. porous media, parallel pore structure, pellets accumulation pore structure; single-layered or multi-layered clothing) in textile materials and moreover various heat and mass transfer, momentum transfer (e.g. conduction, convection, radiation and molecular diffusion, etc.) [1-2]. From 1930s, researchers discussed a few models of heat and moisture transfer and corresponding numerical simulations of thermal and water vapor concentration in porous media [3-14]. Henry first proposed the establishment of coupled heat and moisture transfer model base on differential element in 1939 [3] and further analyzed the model in 1948 [4], where two parabolic partial differential equations were used to describe the process of heat and moisture transfer and coupling term was used to describe the fiber absorption and desorption of moisture and the latent heat.

Published under licence by IOP Publishing Ltd

Different textile fibers have different heat capacity and thermal conductivity, consequently, heat and moisture transfer in fabric will be different. In addition, the process of moisture and humidity changes have latent heat release and absorption, which also affect heat and moisture transfer in the fabric. Although considerable previous work has been carried out on the diverse aspects of simultaneous heat and moisture transfer in the literature both mathematically(in mathematical modeling way) and experimentally, little has been done on the coupling of heat and moisture transfer with phase change and condensation before 2000. From 2000, some researchers (e.g. Fan and Li [5-13], Huang [14]) have already put forward a few mathematical models of coupled heat and moisture transfer through porous clothing assemblies and fibrous insulation. Based on these models, they have designed different numerical methods to solve these problems [11-14], such as finite difference method, finite volume method, finite element method and controllability volume-time-domain recursive method, and the numerical results are well matched with experimental results. On the basis of present research results and the analysis of the basic features of various models, we can see that coupled models of heat and moisture transfer are the most promising theoretical model, so it is necessary that we should study well-posed conditions and numerical algorithms for such problems.

Recently, a kind of coupled model of heat and moisture transfer was established on the basis of ordinary differential equations [15]. Textile is certainly of a complex multi-pore structure, which can be described as the parallel pore structure and pellets accumulation pore structure. The heat and moisture transfer through textile is affected by the textile structure. In [15], Based on the parallel pore structure textile and a system of human-textile-environment, a model of heat and moisture transfer through the parallel pore textiles is established. It is a system of coupled ordinary differential equations on temperature, water vapor pressure and water vapor mass flux through textile and condensation on the surface of textile. According to this model, we propose an inverse problem of thickness design for single layer textile material [16], and propose numerical algorithms of the inverse problem, numerical simulation is achieved in order to verify the validity of proposed methods.

In this paper, we propose an inverse problem of thickness design for bilayer textile material with parallel pore structure. In this humanbody-clothing-environment system, thickness of the outer fabric is known but thickness of the inner fabric is unknown and need determining. We design numerical algorithm for the inverse problem of thickness design. The numerical results indicate the correctness of mathematical models and validity of numerical algorithms.

2. Mathematical formulation

In this section, we consider an inverse problem of thickness design for bilayer textile material based on the steady-state model presented in [15]. A clothing assembly consisting of a layer of inner fabric, a layer of outer fabric was considered in a cold climate in this paper. Such clothing assembly, as well as human body and environment, constitutes a system of Humanbody-Clothing-Environment.

2.1. Assumptions and restrictions

The following assumptions are made:

(A_1). Each fabric material is isotropic.

(A_2). Volume changes of the fibers due to changing moisture and water content are neglected.

(A_3). Local thermal equilibrium exits among all phases.

(A_4). Diffusion within the fiber is considered to be so rapid that moisture content at the fiber surface is in sorptive equilibrium with that of the surrounding air. We only consider the condensation of water on the fibre surface.

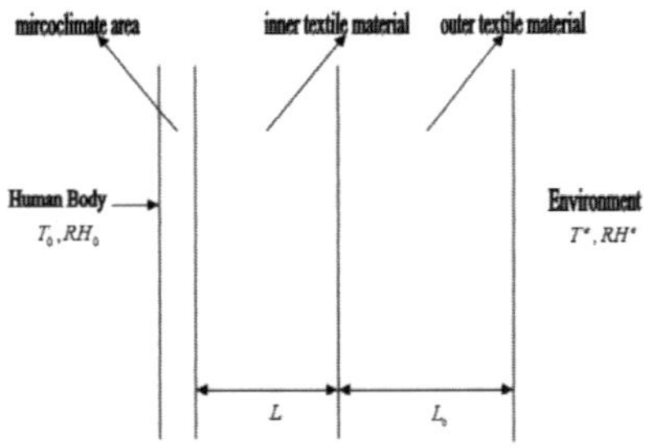

Figure 1. Schematic diagram of the body-clothing-environment system.

(A_5). Free convection in the inner fabric is negligible.

(A_6). Temperature, water vapor pressure , mass flux of water vapor and heat fluxes on the boundary of inner fabric and outer fabric are all continuous.

2.2. Mathematical formulation of the inverse problem of thickness design

Under the above assumptions, we can establish the following mathematical equations for the coupled heat and mass transfer in the parallel pore fabric according to the conservation of heat energy and mass balance.

The model of the inner fabric is given as follows

$$\begin{cases} \frac{k_1\varepsilon_1 r_1}{\tau_1}\cdot\frac{p_v}{T^{3/2}}\cdot\frac{dp_v}{dx}+m_v(x)=0 \\ \frac{dm_v}{dx}+\Gamma(x)=0 \\ \kappa_1\frac{d^2T}{dx^2}+\lambda\Gamma(x)=0 \\ \Gamma(x)=\frac{-k_2\varepsilon_1 r_1}{\tau_1}\cdot(p_{sat}-p_v)\cdot\frac{1}{\sqrt{T}}, \end{cases} \tag{1}$$

where $0 \le x \le L$, and the boundary conditions are prescribed as follows

$$\begin{cases} T(0)=T_L \\ m_v(0)=m_{v,0}. \end{cases} \tag{2}$$

Meanwhile the model of the outer fabric is given as follows

$$\begin{cases} \frac{k_1\varepsilon_2 r_2}{\tau_2}\cdot\frac{p_v}{T^{3/2}}\cdot\frac{dp_v}{dx}+m_v(x)=0 \\ \frac{dm_v}{dx}+\Gamma(x)=0 \\ \kappa_2\frac{d^2T}{dx^2}+\lambda\Gamma(x)=0 \\ \Gamma(x)=\frac{-k_2\varepsilon_2 r_2}{\tau_2}\cdot(p_{sat}-p_v)\cdot\frac{1}{\sqrt{T}}, \end{cases} \tag{3}$$

where $L \le x \le L + L_0$, and boundary value conditions are prescribed as follows

$$\begin{cases} T(L+L_0)=T_R \\ \kappa_2 T'(L+L_0)=\frac{T^e-T(L+L_0)}{R_{t0}+(1/h_t)} \\ m_v(L+L_0)=m_{v,R} \\ p_v(L+L_0)=p_{v,R}. \end{cases} \tag{4}$$

Nomenclatures and Symbols:

$T(x)$ is temperature(K);

$p_v(x)$ is water vapor pressure(Pa);

$m_v(x)$ is mass flux of water vapor $(kg/(m^2 \cdot s))$;

$\Gamma(x)$ is the rate of condensation $(kg/(m^3 \cdot s))$.

ε_i, r_i and τ_i (i=1,2) denote the porosity of textile surface(%), the radius of cylindrical pore(m) and the effective tortuosity of the textile, respectively.

κ_1 and κ_2 are thermal conductivities of inner and outer fabric respectively.$(W/(m \cdot K))$;

R_{t0} is thermal resistance of outer textile material$(K \cdot m^2/W)$;

h_t is the convective heat transfer coefficient between the outer surface of the textile material and the environment$(W/(K \cdot m^2))$;

λ is latent heat of sorption and condensation of water vapor(J/kg); T_R is temperature of the out boundary of outer textile material;

T^e is the environmental temperature;

$m_{v,R}$ is mass flux of water vapor of the out boundary of outer textile material;

$p_{v,R}$ is water vapor pressure of the out boundary of outer textile material;

k_1 and k_2 are constants which are related with molecular weight and gas constant.

The saturation vapor pressure within the parallel pore is given as follows [8]:

$$p_{sat}(T) = 100 \cdot e^{18.956 - \frac{4030}{(T-273.16)+235}}$$

Let $k_{3,i} = \frac{\kappa_i}{\lambda}$, $A_i = \frac{\varepsilon_i r_i}{\tau_i}$ $(i = 1, 2)$.

The model(3) combined with boundary conditions(4) constitute the terminal value problem of coupled ordinary differential equations, it is usually called a direct problem (**DP** in abbreviation).

In above heat and moisture transfer process, heat conductivity, condensation and diffusion are involved in the double layer parallel cylindrical pore fabric.

Next, we consider an inverse problem of thickness design for double-layer textile material with parallel cylindrical pore structure.

Suppose that the environmental temperature and relative humidity are given as follows:

$$(T^e, RH^e) \in [T_{min}, T_{max}] \times [H_{min}, H_{max}],$$

where T_{min} and T_{max} are minimum average temperature and maximum average temperature at a specific place during a specific time period respectively; Similarly H_{min} and H_{max} are minimum average relative humidity and maximum average relative humidity respectively.

Suppose that the structure and type of both layer fabric are known. The structure of textile includes the radius of pore, porosity of textile surface and effective tortuosity of the fabric.

The literatures on clothing thermal comfort have indicated that the comfort indexes in the clothing microclimate, which is located between the skin surface and the inner surface of fabric, are given as follows (refer to [2]): temperature $(32 \pm 1)^o$C, relative humidity $(50\% \pm 10\%)$, wind speed (0.25 ± 0.15)m/s.

According to the requirements of clothing thermal comfort, we intend to determine the thickness L of inner fabric. Thus, the inverse problem of thickness design can be formulated as follows:

Inverse Problem of Thickness Design (IPTD in abbreviation): Given the environmental temperature and relative humidity and the above comfort indexes, according to the boundary conditions (2) and right boundary conditions (4), we intend to determine the thickness L of inner fabric, where T_R and $p_{v,R}$ are both related with environmental temperature and humidity.

3. Numerical Algorithms of the IPTD

With regard to the inverse problem of thickness design for bilayer fabric, we construct numerical algorithm by adopting idea of regularization method. The similar algorithm was constructed to solve the inverse problem of thickness design for single-layer fabric in [16].

3.1. Regularized Solution of the IPTD

In order to obtain the regularized solution, we discretize the combination of environmental temperature and humidity as (T_i^e, RH_j^e) $(i = 1, 2, \cdots, k;\ j = 1, 2, \cdots, m)$. Under this combination, we can obtain the numerical solution of the temperature at inner side of outer fabric, $T_b^{i,j}$, and the water vapor pressure at inner side of outer fabric, $p_{v,b}^{i,j}$. As temperature and water vapor pressure are both continuous on the boundary, we can use $T_b^{i,j}$ and $p_{v,b}^{i,j}$ as the right boundary value of inner fabric. Let $RH_{i,j,0}(x)$ be relative humidity of inner fabric, thus we can easily obtain its numerical solution through the model(1) and following conditions:

$$\begin{cases} T(0) = T_L \\ T(L) = T_b^{i,j} \\ m_v(0) = m_{v,0} \\ p_v(L) = p_{v,b}^{i,j} \end{cases} \tag{5}$$

Suppose that RH_0^* is experience value of relative humidity in comfortable state.

We can attribute the inverse problem to the following least squared problem:

$$min \sum_{i=1}^{k} \sum_{j=1}^{m} (RH_{i,j,0}(x) - RH_0^*)^2$$

Since above least squared problem doesn't exist unique solution or the solutions are unstable, we should adopt regularization idea to improve the least squared method.

In this respect, we define the following function:

$$J(x) = \alpha \cdot x^2 + \sum_{i=1}^{k} \sum_{j=1}^{m} (RH_{i,j,0}(x) - RH_0^*)^2$$

This function is different from the least squared function, as it is added a penalty term on the least squared function, where $\alpha > 0$ is a regularization parameter. Set $M = [0, L]$, which is called the permissible solution set. If x_{reg} satisfies

$$J(x_{reg}) = min_{x \in M} J(x)$$

then it is called the regularized solution of the IPTD, or the generalized solution.

3.2. Iteration Algorithms of the Regularized Solution

According to the idea that direct problem solving is combined with search iteration of one-dimension minimization problem, we construct the iteration algorithm:

$$x_{n+1} = x_n + \Delta_n \cdot d_n, n = 1, 2, \cdots$$

to solve the minimization problem such that $J(x_{n+1}) < J(x_n)$, where x_1 is arbitrarily given, and x_n satisfies:

$$\begin{cases} \frac{k_1 \varepsilon_1 r_1}{\tau_1} \cdot \frac{p_v}{T^{3/2}} \cdot \frac{dp_v}{dx} + m_v(x) = 0, 0 < x < x_n \\ \frac{dm_v}{dx} + \Gamma(x) = 0, 0 < x < x_n \\ \kappa_1 \frac{d^2T}{dx^2} + \lambda \Gamma(x) = 0, 0 < x < x_n \\ \Gamma(x) = \frac{-k_2 \varepsilon_1 r_1}{\tau_1} \cdot (p_{sat} - p_v) \cdot \frac{1}{\sqrt{T}}, 0 < x < x_n \end{cases} \tag{6}$$

$$\begin{cases} T(0) = T_L \\ T(x_n) = T_b^{i,j} \\ m_v(0) = m_{v,0} \\ p_v(x_n) = p_{v,b}^{i,j}, \end{cases} \tag{7}$$

where the search step Δ_n, and the search direction d_n can be determined by search method of one-dimensional minimization problems.

3.3. Search Method of The One-Dimensional Minimization Problem

The optimization problem involved in this paper is a single variable problem. As we know, $RH_{i,j,0}(x)$ is relative humidity of inner fabric which is a numerical solution calculated by coupled ordinary differential equations, and it is difficult to obtain the derivative of $RH_{i,j,0}(x)$, hence we have to use the direct search method. Taking this actual situation into account, we use Hooke-Jeeves pattern search algorithm[17-18], direct search algorithm by Cai[19] and Golden Section Method [20] respectively to solve the above optimization problem.

(1) **Hooke-Jeeves' Pattern Search Algorithm**

Step 1. x_1 is given. Set initial step $\Delta_1 > 0$, acceleration factor $\gamma \geq 1$, reduced rate $\beta \in (0,1)$, permissible error $\varepsilon > 0$, search direction $e_1 = 1, e_2 = -1$. set $y_1 = x_1, i = 1$.

Step 2. If $J(y_1 + \Delta_1 \cdot e_1) < J(y_1)$, then

$$y_2 = y_1 + \Delta_1 \cdot e_1$$

carry out the step 4; otherwise, carry out the step 3.

Step3. If $J(y_1 + \Delta_1 \cdot e_2) < J(y_1)$, then

$$y_2 = y_1 + \Delta_1 \cdot e_2$$

carry out the step 4; otherwise, if $J(y_1 + \Delta_1 \cdot e_2) \geq J(y_1)$, then

$$y_2 = y_1,$$

carry out the step 4.

Step 4. If $J(y_2) < J(x_i)$, carry out the step 5; otherwise, if $J(y_2) \geq J(x_i)$, carry out the step 6.

Step 5. $x_{i+1} = x_i; y_1 = x_{i+1} + \gamma(x_{i+1} - x_i); i = i + 1$; go to the step 2.

Step 6. If $\Delta_1 \leq \varepsilon$, then stop,$x^* = x_i$; otherwise, $\Delta_1 = \beta \cdot \Delta_1; y_1 = x_i; x_{i+1} = x_i; i = i + 1$; go to the step 2.

(2) **Direct search algorithm proposed by Cai**

Step 1. $\eta \in [\frac{1}{2}, 1)$ is given. Set initial values $x_1^0, x_2^0 \in [0, L], x_1^0 \neq x_2^0$. Suppose $J(x_1^0) < J(x_2^0)$ (otherwise, exchange x_1^0 and x_2^0), $k = 0$.

Step 2. If $\frac{|x_1^k - x_2^k|}{1-\eta} \geq L$, continue; otherwise, choose the initial values again.

Step 3. If $|x_1^k - x_2^k| \leq \varepsilon$, go to the step 9.

Step 4. $x_*^k = x_1^k - \eta(x_2^k - x_1^k)$

Step 5. If $J(x_*^k) \leq J(x_1^k)$ and $x_*^k \in [0, L]$, then $x_1^{k+1} = x_*^k; x_2^{k+1} = x_1^k; k = k + 1$; go to the step 3.

Step 6. If $J(x_*^k) > J(x_1^k)$ or $x_*^k \notin [0, L]$, then $x_{**}^k = x_1^k + \eta(x_2^k - x_1^k)$.

Step 7. If $J(x_{**}^k) < J(x_1^k)$, then $x_1^{k+1} = x_{**}^k; x_2^{k+1} = x_1^k; k = k + 1$, go to the step 3.

Step 8. If $J(x_1^k) \leq J(x_{**}^k) < J(x_2^k)$, then $x_1^{k+1} = x_1^k; x_2^{k+1} = x_{**}^k; k = k + 1$, go to the step 3.

Step 9. $x^* = x_1^k$, stop.

(3) **Golden Section Method**

Step 1. An initial interval $[a_1, b_1]$ is given. Set permissible error $\varepsilon > 0$. Choose the explosive point λ_1 and μ_1, calculate the function values $J(\lambda_1)$ and $J(\mu_1)$:

$$\lambda_1 = a_1 + (1-\beta)(b_1 - a_1), \mu_1 = a_1 + \beta(b_1 - a_1).$$

where $\beta = (\sqrt{5}-1)/2$. $k = 1$

Step 2. If $b_k - a_k < \varepsilon$, then stop,$x^* = \frac{1}{2}(a_k + b_k)$; otherwise, if $J(\lambda_k) > J(\mu_k)$, go to the step 3; if $J(\lambda_k) \leq J(\mu_k)$, go to the step 4.

Step 3.$a_{k+1} = \lambda_k, b_{k+1} = b_k, \lambda_{k+1} = \mu_k,$

$$\mu_{k+1} = a_{k+1} + \beta(b_{k+1} - a_{k+1}),$$

calculate $J(\mu_{k+1})$ go to the step 5.

Step 4. $a_{k+1} = a_k, b_{k+1} = \mu_k, \mu_{k+1} = \lambda_k,$

$$\lambda_{k+1} = a_{k+1} + \beta(b_{k+1} - a_{k+1}),$$

calculate $J(\lambda_{k+1})$, go to the step 5.

Step 5. $k = k + 1$, go to the step 2.

4. Numerical simulations

In this section, numerical simulation is carried out to verify the validity of above numerical method and the efficiency of the numerical algorithms.

4.1. Micro-structure and Type of Textile Materials

We choose down material as inner fabric, and nylon as outside fabric. Structure and type parameters of these two textile materials are listed in Table 1.

Table 1. Geometric and physical parameters of textile materials

Material	Down (inner fabric)	Nylon (outer fabric)
Thickness(m)	Unknown and need determining	2.73×10^{-4} (fixed in advance)
Radius(m)	1.0×10^{-5}	3.71×10^{-7}
Porosity	0.915	0.85
Heat Conductivity	0.024	0.25
Effective tortuosity	1.2	1.1

4.2. The parameters in simulation process

We consider numerical solution for the heat and mass transfer model consisting of down material and nylon batting under the given boundary conditions of temperature and water vapor concentration in order to verify the correctness of heat and moisture model and efficiency of the proposed algorithm.

Suppose that the initial mass flux of water vapor is $m_v(0) = 3.3084 \times 10^{-5} kg/(m^2 \cdot s)$; The temperature of the inner side of fabric is assumed to be 32°C to guarantee that temperature in microclimate is in the comfort index interval, that is $T(0) = 305.16K$; The convective heat transfer coefficient between the outer surface of the textile material and the environment is $h_t = 8W/(m^2 \cdot K)$; Thermal resistance of outer textile material is $R_{t0} = 0.0315K \cdot m^2/w$; The mass flux of water vapor between the outside fabric and environment is assumed to be $m_v(L + L_0) = 1.5 \times 10^{-5}kg/(m^2 \cdot s)$; In the model, $k_1 = 0.00006, k_2 = 0.00007$ are constants which are related with molecular weight and gas constant; T^e is the environmental temperature; T_R and $p_{v,R}$ are related to the temperature and moisture of environment.

The latent heat of sorption or condensation of water vapor is determined by moisture level in the fabrics, and it has nothing to do with the type of material, here we choose $\lambda = 2260\times10^3 J/kg$.

We assume that heat conductivity of textiles is a constant due to small changes in the environment:

$$T^e \in [-10^\circ\mathrm{C}, 10^\circ\mathrm{C}], \mathrm{RH}^e \in [30\%, 90\%]$$

4.3. Measurements on Environmental Temperature and relative Humidity

Two different environmental conditions under low temperature are chosen for simulation.

Case 1. Environmental conditions: $T^e \in [-10^\circ\mathrm{C}, 0^\circ\mathrm{C}], \mathrm{RH}^e \in [40\%, 90\%]$

Case 2. Environmental conditions: $T^e \in [0^\circ\mathrm{C}, 10^\circ\mathrm{C}], \mathrm{RH}^e \in [30\%, 85\%]$

We make isometry subdivision on the interval $[T_{min}, T_{max}]$ and $[H_{min}, H_{max}]$ respectively, where $[T_{min}, T_{max}]$ is decomposed into k equal portions, and $[H_{min}, H_{max}]$ is decomposed into m equal portions. In simulation, we choose $k = 10$, $m = 10$.

4.4. The uniqueness of regularized solution from the Graphs of $J(x)$

Let's study the property of function $J(x)$ for different materials and environmental conditions through the following function graphs.

The graphs of $J(x)$ for down material in case 1 and case 2 are shown in Figure 2 and Figure 3 respectively.

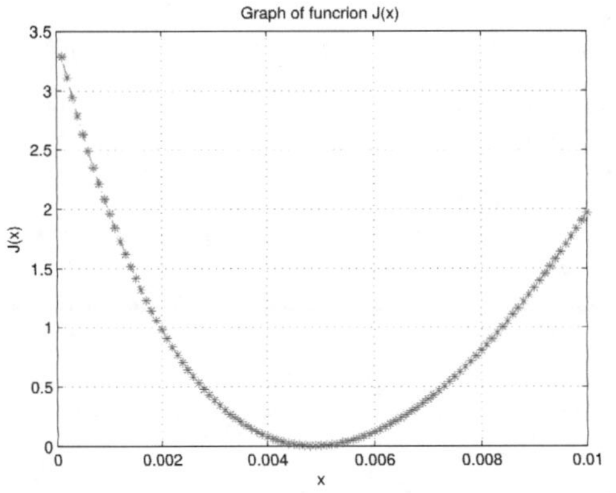

Figure 2. The Graph of Function J(x) in case 1.

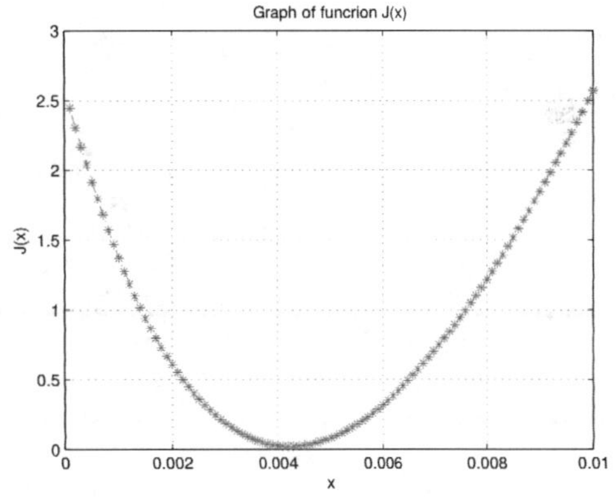

Figure 3. The Graph of Function J(x) in case 2.

From above two graphs, it's easy to see that the minimum point of function $J(x)$ is unique in the interval [0, 0.01]. This property illustrates that the regularized solution of the inverse problem of thickness design for double-layer textile materials is uniquely determined.

4.5. *Realization of thickness determination of the inner textile material*

Using above iteration algorithm for forward and inversion problems in section 3 and 4, we can obtain approximate values of thickness of the inner textile material respectively in Case 1 and Case 2.

Some parameters in above three kind of search methods of the one-dimensional minimization problems are given as follows:

(1) Hooke-Jeeves' Pattern Search Algorithm

Acceleration factor $\gamma = 3$, reduced rate $\beta = 0.5$, permissible error $\varepsilon = 10^{-4}$.

(2) Cai's Method

$x_1^0 = 0.001$, $x_2^0 = 0.008$, $\eta = 0.5$, $\varepsilon = 10^{-5}$.

(3) Golden Section Method

Initial interval $[a_1, b_1] = [0.001, 0.008]$, $\varepsilon = 10^{-5}$.

In Case 1, the numerical solutions of the thickness of down material are shown in table 2.

Table 2. Approximate values of down material's thickness determined by three algorithms in Case 1

Algorithms	Hooke-Jeeves' Method	Cai's Method	Golden Section Method
Initial Values(m)	Thickness(mm)	Thickness(mm)	Thickness(mm)
0.0005	4.8625		
0.001	5.025	4.869	4.867
0.005	4.875		
0.008	4.875		

In Case 2, the numerical solutions of the thickness of down material are shown in the table 3.

Table 3. Approximate values of down material's thickness determined by three algorithms in Case 2

Algorithms	Hooke-Jeeves' Method	Cai's Method	Golden Section Method
Initial Values(m)	Thickness(mm)	Thickness(mm)	Thickness(mm)
0.0005	4.45		
0.001	4.2125	4.5	4.209
0.005	4.2125		
0.008	4.075		

4.6. *The Validation of the Numerical Results*

In order to verify correctness and rationality of the above numerical results, we take the inversion results of thickness of down material as the thickness of inner textile material, and obtain the numerical results of relative humidity $RH_{i,j,0}(x)$ $(i = 1, 2, \cdots, k;\ j = 1, 2, \cdots, m)$ in the microclimate area under the combination of environmental temperature and humidity. Then we will try to verify whether $RH_{i,j,0}(x)$ belong to the comfort index interval. If $RH_{i,j,0}(x)$ belong to the comfort index interval generally, then it indicates that the numerical results are reasonable.

Under the environmental condition 1 (Case 1), we show in figure 4 that the relative humidity $RH_{i,j,0}(x)$ $(i = 1, 2, \cdots, k;\ j = 1, 2, \cdots, m)$ in the microclimate area vary as the environmental temperature and humidity $T_i^e \in [-10^\circ\mathrm{C}, 0^\circ\mathrm{C}], \mathrm{RH}_j^e \in [40\%, 90\%]$ change. The relative humidity in the microclimate area $RH_{i,j,0}(x) \in [49\%, 52\%]$, which belong to the comfort index interval.

Under the environmental condition 2 (Case 2), we show in figure 5 that the relative humidity $RH_{i,j,0}(x)$ $(i = 1, 2, \cdots, k;\ j = 1, 2, \cdots, m)$ in the microclimate area vary as the environmental temperature and humidity $T_i^e \in [0^\circ\mathrm{C}, 10^\circ\mathrm{C}], \mathrm{RH}_j^e \in [30\%, 85\%]$ change. The relative humidity in the microclimate area $RH_{i,j,0}(x) \in [48\%, 56\%]$, which belong to the comfort index interval.

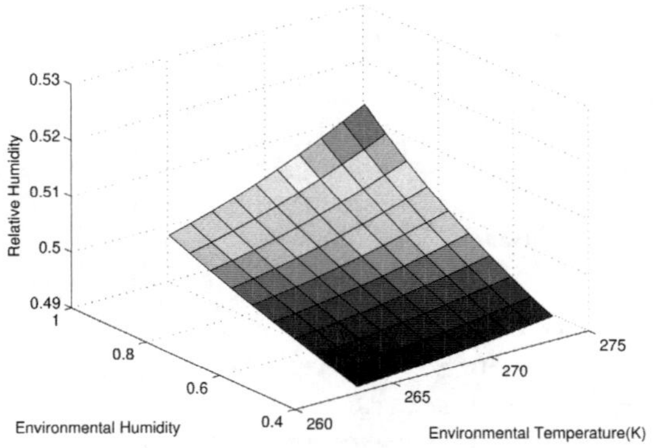

Figure 4.

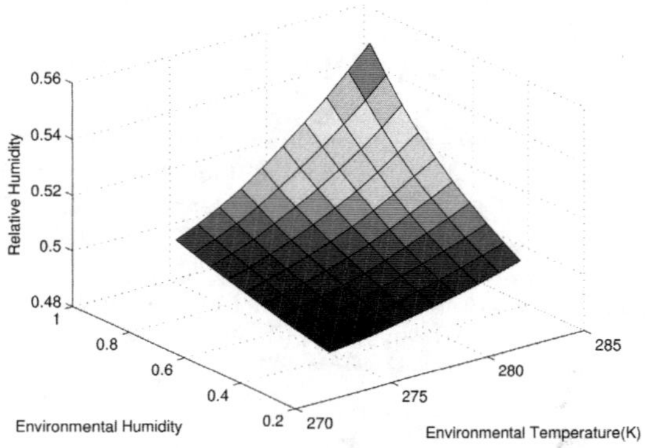

Figure 5.

According to the figure 4 and figure 5, we prove that the determined thickness results of down material are correct and reasonable.

5. Conclusions and Prospects

On the basis of the stationary model of heat and moisture transfer through textiles, we present numerical algorithms for the inverse problems of thickness design for bilayer textile materials under low temperature. The associated direct problems are solved by FDM. Meanwhile, by means of three kinds of direct search methods of minimization of a regularized function, the IPTD are effectively solved.

The numerical results prove that the formulation of the IPTD is correct and the numerical algorithms are efficient in solving the IPTD. Therefore we are sure that the research results are necessary fundamentals to theoretical support and scientific explanations for laminated textile materials design experiments.

Hooke-Jevees' method, Cai's method and Golden Section Method can applied to solve the IPTD for bilayer textile materials. The Golden Section Method is much simpler than the other two methods. As for Hooke-Jeeves's pattern search algorithm, different initial values lead to different iterations and computational time, so its convergence is much sensitive to the initial value.

In this paper, we study the stationary model of heat and moisture transfer through textiles by formulation of the IPTD and corresponding numerical algorithms. It is necessary that we further study well-posedness of the IPTD and mathematical analysis of proposed numerical methods, which will be discussed in forthcoming papers.

6. Acknowledgments

The research is partially supported by National Natural Science Foundation of China (Contract Grant No. 10561001 and 11071221). The first-named author would like to extend his sincere thanks to the organizing committee of International Conference on Inverse Problems, held in Hong Kong City University on December 13-17 2010, for the financial support and the Outstanding Presentation IOP Award which I was luckily granted. Meanwhile he appreciates the inspiring discussion with Professor Y. Li in Donghua University and Hong Kong Polytechnic University, Professor J.T. Fan in Zhejiang Sci-Tech University and Hong Kong Polytechnic University and Prof W Sun in Hong Kong City University.

References

[1] Yao M 1996 *Science of Textile Materials(2nd ed)* (Beijing: China Textile Press).
[2] Huang J H 2008 *Clothing Comfort* (Beijing: Scientific Press).
[3] Henry P S H 1939 Diffusion in absorbing media *Pro.R.Soc.* vol 17, 215-41.
[4] Henry P S H 1948 The diffusion of moisture and heat through textiles *Discuss. Faraday. Soc.* **3** 243-57.
[5] Li Y and Zhu Q Y 2003 A model of coupled liquid moisture and heat transfer in porous textiles with consideration of gravity *Numer Heat Transfer, Part A.* **43** 501-23.
[6] Li Y and Luo Z X 2000 Physical mechanisms of moisture transfer in hygroscopic fabrics under humidity transients *J. Textile Res.* **91** 302-16.
[7] Fan J T, Luo Z X and Li Y 2000 Heat and moisture transfer with sorption and condensation in porous clothing assemblies and numerical simulation *Int. J. Heat Mass Transfer.* **43** 2989-3000.
[8] Wu H J and Fan J T 2008 Study of Heat and Moisture Transfer Within Multi-layer Clothing Assemblies Consisting of Different Types of Battings *Int. J. Thermal Sciences* **47** 641-647.
[9] Fan J T, Cheng X and Sun W 2004 An improved model of heat and moisture transfer with phase change and mobile condensates in fibrous insulation and comparison with experimental results *Int. J. Heat Mass Transfer* **47** 2343-52.
[10] Fan J T and Wen X 2002 Modeling heat and moisture transfer *Int. J. Heat Mass Transfer* **45** 4045-55.
[11] Li Y and Luo Z X 1999 An improved mathematical simulation of the coupled diffusion of moisture and heat in wool fabrics *J. Textile Res.* **69** 760-68.
[12] Li F Z, Li Y and Liu Y X 2004 Numerical simulation of coupled heat and mass transfer in hygroscopic fabrics considering the influence of atmospheric pressure *Numer Heat Transfer, Part B* **45** 249-62.
[13] Li F Z, Li Y and Cao Y L 2009 The control volume time recursive expansion algorithm for solving coupled heat and moisture transfer through fabrics *J. Nanjing University of Aeronautics and Astronautics* **41** 319-23.
[14] Huang H X, Ye C H and Sun W W 2008 Moisture transport in fibrous clothing assemblies *J. Engineering Math.* **61** 35-54.
[15] Xu D H, Cheng J X and Zhou X H 2010 A model of heat and moisture transfer through the parallel pore textiles *Proc. Textile Bioengineering and Informatics Symposium (TBIS 2010, Shanghai)* 1151-56.
[16] Xu D H, Cheng J X and Zhou X H 2010 An inverse problem of thickness design for single layer textile material under low temperature *J.Math-for-Industry* **2** 139-146.
[17] Hooke R and Jeeves T A 1961 Direct search solution of numerical and statistical problems *J. Assoc. Comput. Mach.* **8** 212-29.
[18] Torczon V 1997 On the convergence of pattern search algorithms *SIAM. J. Optim.* **7** 1-25.
[19] Cai Z J and Chen D Q 2006 The direct search algorithm of nonlinear optimization and its convergence proof *Journal of Fudan University(Natural Science Edition)* **45** 396-403.
[20] Chen B L 2005 *Optimization Problems and Algorithms* (Beijing: Tsinghua University Press)

Journal of Physics: Conference Series **290** (2011) 012019 doi:10.1088/1742-6596/290/1/012019

Upward radiation condition for wave in essentially stratified medium

Yongzhi Xu

Department of Mathematics, University of Louisville, Louisville, KY 40292, USA

E-mail: ysxu0001@louisville.edu

Abstract. Waves from a point source in a homogeneous medium will propagate spherically. This can be described by the Sommerfeld radiation condition. However, it is hard to use simple formula to describe radiation condition for wave propagation in a stratified medium. A generalized Sommerfeld radiation was proposed by Xu in 1990, applying contour integral and asymptotic approximation. The study in this area becomes more active recently. Papers in 2008 and 2009 from different aspects presented rigorous analysis for results similar to Xu's condition. On the other hand, Chandler-Wolde, Bo Zhang and their coworkers introduced the upward propagating radiation condition.

In this paper we introduce an upward propagating radiation condition for waves in essentially stratified medium with unbounded inhomogeneity or unbounded rough boundary.

The upward propagating radiation conditions can be used to formulate direct and inverse scattering problems in nonlocal perturbed medium.

1. Introduction

Waves from a point source in a homogeneous medium will propagate spherically. This can be described by the Sommerfeld radiation condition. However, it is hard to use simple formula to describe radiation conditions for wave propagation in a stratified medium. A generalized Sommerfeld radiation was proposed by Xu in 1990, applying contour integral and asymptotic approximation. In [13], Xu studied the scattering of a harmonic acoustic wave by an obstacle in a stratified medium. In a stratified medium, sound waves may be trapped by acoustic ducts and caused to propagate horizontally. (See [3], [12], [11]) Therefore, the waves scattered by a compact obstacle or local inhomogeneity in a stratified medium may not satisfy the well-known Sommerfeld radiation condition which is the fundamental assumption of scattering theory in a homogeneous medium. In [13][14], a new radiation condition, which is a generalization of the Sommerfeld radiation condition, was found based on the analysis of the behavior of point source acoustic wave in a stratified medium. Using the generalized Sommerfeld condition, we proved the uniqueness and existence of some exterior problems.

The study in this area becomes more active recently. Recently some papers from different aspects presented rigorous analysis for results similar to Xu's condition. In [7], [8], Ciraolo and Magnanini presented a similar radiation condition based on the same wave splitting using a weaker form of Sommerfeld type radiation conditions for the various components. In [9], Bonnet-Ben Dhia, Dakhia, Hazard, and Chorfi presented a modal radiation condition using a generalized Fourier transform. The generalized Fourier transform appears as the operator of decomposition on the family of eigenfunctions and generalized eigenfunctions, which diagonalizes the transverse part of the unperturbed Helmholtz operator. Based on their result, a further asymptotic analysis for continuous spectrum and discrete spectrum should lead to a similar generalized Sommerfeld radiation condition presented in [13] [14].

Nonlocal rough surface or rough medium scattering problems are seem in many applications, including acoustic and electromagnetic wave propagation over outdoor ground and sea surfaces, and have been studied in many papers. It is known that the Sommerfeld radiation condition is not appropriate for such problems. A mathematically rigorous, approach is the so-called upward propagating radiation condition (UPRC) introduced by Chandler-Wilde [4], and further studied by Chandler-Wilde and Zhang [5][6][20].

More references on radiation conditions may be found, for examples, in [1] [2] [7] [8] [9].

In this paper, we study the scattering of time-harmonic acoustic waves in an essentially stratified medium with nonlocal rough surface or nonlocal inhomogeneity. (We will clarify the meaning of essentially stratified medium in the later discussion.) we introduce an upward propagating radiation condition for waves in essentially stratified medium with unbounded inhomogeneity or unbounded rough boundary. This condition plays an important role for formulating direct and inverse scattering problems for unbounded object in a stratified medium.

2. Scattering problem in an essentially stratified medium with nonlocal rough surface or nonlocal inhomogeneity

We study the scattering problems in a nonlocally perturbed layered half-space. Let $R^2_+ = \{\mathbf{x} \in R^2 | x_2 \geq 0\}$, where $\mathbf{x} = (x_1, x_2)$. Let

$$n_0(x_2) = \begin{cases} n_0, & for\ 0 < x_2 < h, \\ 1, & for\ h < x_2 < \infty, \end{cases} \tag{1}$$

where $n_0 > 1$ is a constant. We assume that the inhomogeneity is contained in a slab domain $R^2_d = \{\mathbf{x} \in R^2 | 0 < x_2 < d < \infty\}$.

The nonlocal perturbed medium scattering problem (Problem M) is defined as following: given incident wave u^i satisfying

$$\triangle u^i + k^2 n_0^2(x_2) u^i = f(\mathbf{x}) \ , \ in\ R^2_+, \tag{2}$$

find the total field $u = u^s + u^i$ such that

$$\triangle u^s + k^2 n^2(\mathbf{x}) u^s = 0, \ , \ in\ R^2_+, \tag{3}$$

$$u(\mathbf{x}) = u^s(\mathbf{x}) + u^i(\mathbf{x}) = 0 \ , \ when\ x_2 = 0, \tag{4}$$

The nonlocal rough surface scattering problem (Problem S) is defined as following: given incident wave u^i satisfying

$$\triangle u^i + k^2 n_0^2(x_2) u^i = f(\mathbf{x}) \ , \ in\ R^2_+, \tag{5}$$

find the total field $u = u^s + u^i$ such that

$$\triangle u^s + k^2 n_0^2(x_2) u^s = 0, \ , \ in\ R^2_h, \tag{6}$$

$$u(\mathbf{x}) = u^s(\mathbf{x}) + u^i(\mathbf{x}) = 0 \ , \ on\ \Gamma = \{(x_1, x_2) | x_2 = h(x_1), -\infty < x_1 < \infty\}, \tag{7}$$

where $h \in C^1(R)$.

Here $\triangle = \frac{\partial^2}{\partial x_1^2} + \frac{\partial^2}{\partial x_2^2}$, u^i, u^s and u are the incident, scattered and total field of the time harmonic waves. f is a function with compact support.

We need to assign a suitable radiation condition. If the inhomogeneity is in a bounded region, a generalized Sommerfeld radiation condition can be applied [10]. However, if the inhomogeneity is unbounded, a kind of upward propagating radiation condition may be more appropriate.

First we recall the Green's function in stratified medium and its far-field behavior. We outline the construction of the Green's function for the time-harmonic waves in a layered half-space and obtain its free-wave and guided-wave far-field behavior. More details can be seen in [10], also in [13][14].

A function $G(\cdot\,; x_1^0, x_2^0)$ is the outgoing Green's function from the source at $\mathbf{x^0} = (x_1^0, x_2^0)$ for the time-harmonic wave in a layered half-space, if $G(x_1, x_2; x_1^0, x_2^0)$ satisfies

$$\triangle G + k^2 n_0^2(x_2) G = -\delta(|\mathbf{x} - \mathbf{x^0}|)\ , \ in\ R_+^2 \tag{8}$$

in the generalized function sense;

$$G(x_1, 0; x_1^0, x_2^0) = 0, \tag{9}$$

and satisfies the out-going radiation condition.

Using the Fourier Transform, and considering the outgoing property in horizontal direction, we obtain a representation of $G(x_1, x_2; x_1^0, x_2^0)$

$$G(x_1, x_2; x_1^0, x_2^0) = \frac{1}{2\pi}\int_{-\infty}^{\infty} \tilde{G}(\xi, x_2; x_1^0, x_2^0) e^{i\xi|x_1 - x_1^0|} d\xi, \tag{10}$$

$\tilde{G}(\xi, x_2; x_1^0, x_2^0)$ has representation

$$\tilde{G}(\xi, x_2; x_1^0, x_2^0) = \frac{\phi^+(\xi, x_{2>})\phi^-(\xi, x_{2<})}{W(\phi^+, \phi^-)}, \tag{11}$$

where

$$\phi^+(\xi, x_2) = \begin{cases} e^{i\tau x_2}, & h < x_2 < \infty \\ B_1 e^{i\tau_0 x_2} + B_2 e^{-i\tau_0 x_2}, & 0 < x_2 < h, \end{cases} \tag{12}$$

$$\phi^-(\xi, x_2) = \begin{cases} \sin(\tau_0 x_2), & 0 < x_2 < h \\ C_1 \cos(\tau x_2) + C_2 \sin(\tau x_2), & h < x_2 < \infty, \end{cases} \tag{13}$$

where

$$B_1 = \frac{1}{2}\left(1 + \frac{\tau}{\tau_0}\right) e^{i(\tau - \tau_0)h},\ B_2 = \frac{1}{2}\left(1 - \frac{\tau}{\tau_0}\right) e^{i(\tau + \tau_0)h},$$

$$C_1 = \cos(\tau h)\sin(\tau_0 h) - \frac{\tau_0}{\tau}\sin(\tau h)\cos(\tau_0 h),$$

$$C_2 = \sin(\tau h)\sin(\tau_0 h) + \frac{\tau_0}{\tau}\cos(\tau h)\cos(\tau_0 h),$$

$\tau = \sqrt{k^2 - \xi^2}$ and $\tau_0 = \sqrt{k^2 n_0^2 - \xi^2}$. $W(\phi^+, \phi^-)$ is the Wronskian

$$W(\phi^+, \phi^-) = e^{i(\tau h)}\left[\tau_0 \cos(\tau_0 h) - i\tau \sin(\tau_0 h)\right]. \tag{14}$$

We obtain from (10)and (11):

$$G(x_1, x_2; x_1^0, x_2^0) = \frac{1}{2\pi}\int_{-\infty}^{\infty} \frac{\phi^+(\xi, x_{2>})\phi^-(\xi, x_{2<})}{W(p^+, p^-)} e^{i\xi|x_1 - x_1^0|} d\xi. \tag{15}$$

Setting the Wronskian be 0, we have $\tan(\tau_0 h) = i\frac{\tau}{\tau_0}$., which has at most a finite number of solutions for $\xi^2 > k^2$. The out-going radiation condition as $|x_1| \to \infty$ implies that the positive zeros should be used. Denote them by $k < \xi_1 < \xi_2 < \cdots < \xi_N$. They are poles of the integrant.

There are branch points at $\pm k$ and $\pm kn_0$. Choosing branch cuts Γ and Γ_0 [10] for $\tau = \sqrt{k^2 - \xi^2}$ and $\tau_0 = \sqrt{k^2 n_0^2 - \xi^2}$, and denoting $\tau_n = \sqrt{k^2 - \xi_n^2}$ and $\tau_{0n} = \sqrt{k^2 n_0^2 - \xi_n^2}$, we can represent the Green's function as:

$$G(x_1, x_2; x_1^0, x_2^0) = G_f(x_1, x_2; x_1^0, x_2^0) + G_g(x_1, x_2; x_1^0, x_2^0)$$

$$= \frac{1}{2\pi} \int_\Gamma \frac{\phi^+(\xi, x_{2>})\phi^-(\xi, x_{2<})e^{i\xi|x_1 - x_1^0|}}{e^{i\tau h}[\tau_0 \cos(\tau_0 h) - i\tau \sin(\tau_0 h)]} d\xi + \sum_{n=1}^{N} \frac{\phi^+(\xi_n, x_{2>})\phi^-(\xi_n, x_{2<})e^{i\xi_n|x_1 - x_1^0|}}{\frac{d}{d\xi} W(\phi^+, \phi^-)\Big|_{\xi = \xi_n}}. \tag{16}$$

For $x_2 > h$,

$$G_g(x_1, x_2; x_1^0, x_2^0) = \sum_{n=1}^{N} \frac{e^{-|\tau_n|x_2} \sin(\xi_n, x_2^0) e^{i\xi_n|x_1 - x_1^0|}}{\frac{\xi(i + h\tau_n)e^{i\tau_n h}}{\tau_{0n}^2 \tau_n}(k^2 n_0^2 - k^2)\sin(\tau_{0n} h)} = O\left(e^{-|\tau_1|x_2}\right). \tag{17}$$

$G_g(x_1, x_2; x_1^0, x_2^0)$ corresponds to the guided wave that is trapped in the layer $0 < x_2 < h$.

$$G_f(x_1, x_2; x_1^0, x_2^0) = \frac{1}{2\pi} \int_\Gamma \frac{e^{i\xi|x_1 - x_1^0| + i\tau x_2} \sin(\tau_0 x_2^0)}{e^{i\tau h}[\tau_0 \cos(\tau_0 h) - i\tau \sin(\tau_0 h)]} d\xi$$

$$= \frac{1}{2\pi} \int_{-\infty}^{\infty} e^{i\zeta|x_1 - x_1^0| + i\sqrt{\zeta^2 + k^2(n_0^2 - 1)}x_2} \Phi_f(\zeta) d\zeta = \frac{1}{2\pi} \int_{-\infty}^{\infty} e^{ir\Theta_f(\zeta)} \Phi_f(\zeta) d\zeta \tag{18}$$

Using the Method of Stationary Phase, we obtain an approximation of $G_f(x_1, x_2; x_1^0, x_2^0)$ for $x_2 > h$ and $r = |\mathbf{x}|$ large:

$$G_f(x_1, x_2; x_1^0, x_2^0) = \frac{e^{ikr}}{\sqrt{2\pi i r}} \left[G_f^\infty(\theta; x_1^0, x_2^0)\right] + O\left(\frac{1}{r^{3/2}}\right) \tag{19}$$

$G_f(x_1, x_2; x_1^0, x_2^0)$ corresponds to the free-wave that propagates to the half plane. $G_f^\infty(\theta; x_1^0, x_2^0)$ is the free-wave far-field pattern.

In summery, a point source wave in a layered medium contains two components, the free-wave component decays at the rate of $O\left(\frac{1}{r^{1/2}}\right)$ in all upward angles, while the guided-wave component decays exponentially in the upward direction, but will not decay in the horizontal direction.

For related discussion of free-wave and guided-wave in stratified media, see [15, 16, 14, 17, 18, 19].

3. Upward propagating radiation condition in essentially stratified medium

Now we consider the scattering problems (M) and (S) in stratified medium. Let $\Phi(x, y)$ denote the free-space Green's function for $\triangle + k_+^2$, that is, $\Phi(x, y) = \frac{i}{4} H_0^{(1)}(k_+|x - y|)$, for $x, y \in R^2$, $x \neq y$, where $H_0^{(1)}$ is the Hankel function of the first kind of order zero. The upward radiation condition in a medium without guided wave is defined as:[20]

Given a domain $G \in R^2$, say that v satisfies the UPRC in G if, for some d,

$$v(x) = 2 \int_{\Gamma_d} \frac{\partial \Phi(x, y)}{\partial y_2} \phi(y) ds(y), \quad x \in U_d, \tag{20}$$

where $U_d = \{x | x_2 > d\}$, and $\Gamma_d = \{x | x_2 = d\}$.

In some of the papers of Chandler-Wilde and Zhang, conditions are given to prevent any guided wave. For example, in [20], the condition (A3) (absorbing medium), or conditions (A4) and (A5) (a counter-waveguide medium) are given explicitly.

Now we consider a perturbed Holmholtz equation that does not satisfy the aforementioned (A3) nor (A4) and (A5):

$$\triangle G + k(x)G = \delta(x - x^0), \tag{21}$$

where

$$k(x) = \begin{cases} k_-, & for\ x \in E_R = \{-R < x_1 < R, 0 < x_2 < h\}, \\ k_+, & otherwise. \end{cases} \tag{22}$$

Then

$$\triangle G + k_+ G = (k_+ - k(x))G + \delta(x - x^0)$$

has solution

$$G(x, x_0) = \int_{E_R} \Phi(\xi, x)\, \mathrm{m}(\xi)\mathrm{G}(\xi, \mathrm{x_0})\, \mathrm{d}\xi + \Phi\left(\mathrm{x}^0, \mathrm{x}\right), \tag{23}$$

where $m(\xi) = k_+ - k(x)$. Recall

$$\Phi(\xi, x) = O\left(\frac{1}{\sqrt{|\xi|}}\right), \text{ and } G(\xi, x_0) = O\left(\frac{1}{\sqrt{|\xi|}}\right), \ as|\xi| \to \infty,$$

$$\int_{E_R} \Phi(\xi, x)\, \mathrm{m}(\xi)\mathrm{G}(\xi \mathrm{x_0})\, \mathrm{d}\xi = \int_{\mathrm{E_R}} \Phi(\xi, \mathrm{x})\, (\mathrm{k_+} - \mathrm{k_-})\mathrm{G}(\xi, \mathrm{x_0})\, \mathrm{d}\xi$$

$$= (k_+ - k_-) \int_{-R}^{R} \int_0^h \Phi(\xi, x)\, \mathrm{G}(\xi, \mathrm{x_0})\, \mathrm{d}\xi_2 \mathrm{d}\xi_1 = \mathrm{O}(\log \mathrm{R}).$$

$$G(x, x_0) = \int_{E_R} \Phi(\xi, x)\, \mathrm{m}(\xi)\mathrm{G}(\xi, \mathrm{x_0})\, \mathrm{d}\xi + \Phi\left(\mathrm{x}^0, \mathrm{x}\right) \text{ diverges as } \mathrm{R} \to \infty.$$

Actually, as discussed above, the Green's function for layered medium $G(x, y)$ has representation

$$G(x, y) = G_f(x, y) + \sum_{l=1}^{N} G_l(x, y), \tag{24}$$

with

$$G_l(x, y) = \frac{\sin(\xi_n, y_2)}{\frac{\xi(i + h\tau_n)e^{i\tau_n h}}{\tau_{0n}^2 \tau_n}(k^2 n_0^2 - k^2)\sin(\tau_{0n} h)} e^{-|\tau_n| x_2} e^{i\xi_n |x_1 - y_1|}$$

which does not decay in the x_1 direction. This shows that the waves in stratified medium may not satisfy the UPRC (20). That is, a different upward propagating radiation condition is needed for stratified medium.

We introduce an Upward Propagating Radiation Condition for Essentially Stratified Medium (UPRCESM): the scattered wave in an essentially stratified medium satisfies the representation formula

$$u(x) = \int_{\Gamma_d} \frac{\partial G(x, y)}{\partial y_2} \phi(y) ds(y), \ x \in U_d, \tag{25}$$

where $U_d = \{x | x_2 > d > h\}$,

$$G(x, y) = G_f(x, y) + \sum_{l=1}^{N} G_l(x, y),$$

with

$$G_l(x, y) = \frac{\sin(\xi_n, y_2)}{\frac{\xi(i + h\tau_n)e^{i\tau_n h}}{\tau_{0n}^2 \tau_n}(k^2 n_0^2 - k^2)\sin(\tau_{0n} h)} e^{-|\tau_n| x_2} e^{i\xi_n |x_1 - y_1|}.$$

The volume potential u with density $\phi \in L_\infty(E_B)$, where E_B is a box or a slab containing the inhomogeneity, by

$$u(x) = \int_{E_B} G(x,y)\phi(y)dy$$
$$= \int_{E_B} G_f(x,y)\phi(y)dy + \sum_{l=1}^{N} \int_{E_B} G_l(x,y)\phi(y)dy \qquad (26)$$

satisfies the perturbed stratified Holmholtz equation, the related boundary equation, and the UPRCESM.

Since

$$G_f = O\left(R^{-\frac{1}{2}}\right),\ \frac{\partial G_f}{\partial \nu} - ikn_\infty G_f = O\left(R^{-\frac{3}{2}}\right),$$

uniformly as $R \to +\infty$ on the set of $C_R = \{x||x| = R, x_2 > 0\}$, we have

$$\int_{\partial\Omega_R} |G_f|^2 d\sigma = O(1),\ \int_{\partial\Omega_R} \left|\frac{\partial G_f}{\partial \nu} - ikn_\infty G_f\right|^2 d\sigma = O(R^{-1})$$

$$\left|G_f(x_1,x_2;x_1^0,x_2^0) - \frac{1}{2\pi}\log r\right| \le C,\ r = |(x_1,x_2) - (x_1^0,x_2^0)|.$$

$|G_f(x_1,x_2;x_1^0,x_2^0) - \Phi(x_1,x_2;x_1^0,x_2^0)|$ is uniformly bounded.

$$\int_{\partial\Omega_R^l} |G_l|^2 d\sigma = O(R),\ l = 1,2,...,N,\ \text{as } R \to +\infty,$$

$$\int_{\partial\Omega_R^l} \left|\frac{\partial G_l}{\partial \nu} - ikn_l G_l\right|^2 d\sigma \text{ vanishes exponentially as } R \to +\infty.$$

Therefore,

$$u_f(x) = \int_{\Gamma_h} \frac{\partial G_f(x,y)}{\partial y_2}\phi(y)ds(y),\ x \in U_d,$$

has similar convergent properties as the UPRC in perturbed homogeneous medium studied by Chandler-Wilde, Zhang, et al. [5] [6] [20].

However,

$$u_g(x) = \sum_{l=1}^{N} \int_{\Gamma_h} \frac{\partial G_l(x,y)}{\partial y_2}\phi(y)ds(y) = \sum_{l=1}^{N} \int_{\Gamma_h} \Psi_l(x_2,h)\phi(y_1,h)e^{i\xi_n|x_1-y_1|}dy_1,\ x \in U_d,$$

behaves in the same way as a Fourier transform. Therefore, UPRCESM must be understood in the sense of distribution.

Alternatively, we can represent UPRCESM as

$$u_f(x) = u(x) - \sum_{l=1}^{N} u_l(x) = \int_{\Gamma_h} \frac{\partial G_f(x,y)}{\partial y_2}\phi(y)ds(y),\ x \in U_d, \qquad (27)$$

or

$$u(x) = \sum_{l=1}^{N} u_l(x) + \int_{\Gamma_h} \frac{\partial G_f(x,y)}{\partial y_2}\phi(y)ds(y),\ x \in U_d, \qquad (28)$$

where $u_l(x)$ may be represented as

$$u_l(x) = \int_{U_B} G_l(x,y)\psi_l(y)dy, \text{ for } l = 1, 2, ..., N.$$

Using the Green's function for stratified medium, the scattered wave for a point source incident wave satisfies the integral equation

$$u(x) + \int_\Omega G(\xi, x)\, \mathrm{m}(\xi)\mathrm{u}(\xi)\, d\xi = \mathrm{G}\left(\mathrm{x}^0, \mathrm{x}\right). \tag{29}$$

Since $|u(x)|$ is uniformly bounded, we have:

(1) $u(x)$ satisfies radiation condition in stratified medium, if

$$\sup_{x \in R_+^2} \int_\Omega |G(\xi, x)\, \mathrm{m}(\xi)|\, d\xi \le C < \infty. \tag{30}$$

Now we can define the **essentially stratified medium**: A perturbed stratified medium is said **essentially stratified**, if the perturbation $m(\xi) = k_+ - k(x)$ satisfies the condition (30).

(2) If

$$\sup_{x \in R_+^2} \int_\Omega |G(\xi, x)\, \mathrm{m}(\xi)|\, d\xi < 1, \tag{31}$$

then the scattering problem has unique solution.

Following are some examples of essentially stratified media:

(a) There is a function $x_2 = f(x_1) > 0$ satisfying $f(x_1) \le C|x|^{-\delta}$ for $\delta > 1$ as $|x| \to \infty$, such that the support of inhomogeneity $supp(m) \subset \{x|0 \le x_2 \le f(x_1)\}$.

(b) $|m(x)| \le C|x_1|^{-\delta}$ for $\delta > 1$ as $|x| \to \infty$, and the support of inhomogeneity $supp(m) \subset \{x|0 \le x_2 \le H\}$.

Similarly we can study rough surface in essentially stratified medium:

Let $\Gamma := \{x|x_2 = h(x_1) > 0, \|h\|_1 \le C\}$ be the sound-soft surface. If there exists $x_2 = f(x_1)$ satisfying $f(x_1) \le C|x|^{-\delta}$ for $\delta > 1$ as $|x| \to \infty$, such that $0 \le h(x_1) \le f(x_1)$ for $x_1 \in R$, then a solution of the Dirichlet problem $u|_\Gamma = g(x)$ has a representation

$$u(x) = \int_\Gamma G(x,y)\psi(y)ds(y), \; x \in \{x \in R_+^2 | x_2 > h(x_1)\},$$

where $|\psi| \le C_2$ is bounded and satisfies the boundary integral equation

$$\int_\Gamma G(x,y)\psi(y)ds(y) = g(x), \; x \in \Gamma.$$

Since

$$|G(x,y)| \le C_1|x_2|, \text{ for } x_2 \text{ small}.$$

$$\left|\int_\Gamma G(x,y)\psi(y)ds(y)\right| \le C_2 \int_\Gamma |G(x,y)|\, ds(y)$$

$$\le C \int_\Gamma |y_2|\, ds(y) = C \int_{-\infty}^{\infty} h(y_1)\sqrt{1 + [h'(x_1)]^2} dy_1 < \infty.$$

We can use the UPRCESM to establish uniqueness of solution for rough surface scattering problems with a Dirichlet or a Robin type boundary condition on the scattering surface provided the medium is essentially stratified.

Acknowledgments
This research was supported in part by an University of Louisville IRIG grant.

References
[1] T. Arens, Uniqueness for elastic wave scattering by rough surfaces, SIAM J. Math. Anal., 33, 461-476 (2001).
[2] T. Arens, T. Hohage, On radiation conditions for rough surface scattering problems, IMA J. Appl. Math. 70 (2005) 839847.
[3] L. Brekhovskikh, *Waves in Layered Media*, Second Edition, Academic Press, New York, (1980).
[4] Chandler-Wilde, Boundary value problems for the Helmholtz equation in a half-plane, in Proceedings of the 3rd International Conference on Mathematical and Numerical Aspects of Wave Propagation, 188-197, SIAM, 1995.
[5] S. N. Chandler-Wilde and B. Zhang, Electromagnetic scattering by an inhomogenous conducting or dielectric layer on a perfectly conducting plate, Proc. R. Soc. Lon. A, 454, 519-542 (1998).
[6] S. N. Chandler-Wilde and B. Zhang, A uniqueness result for scattering by infinite rough surfaces, SIAM J. Appl. Math., 58, 1774-1790 (1998).
[7] Giulio Ciraolo, A radiation condition for the 2-D Helmholtz equation in stratified medium, Communications in Partial Differential Equations, Volume 34, Issue 12, December 2009 , pages 1592 -1606
[8] Giulio Ciraolo and Rolando Magnanini, A radiation condition for uniqueness in a wave propagation problem for 2-D open waveguides, Mathematical Methods in the Applied Sciences, Volume 32 Issue 10, (2008), Pages 1183 1206
[9] Bonnet-Ben Dhia, Dakhia, Hazard, and Chorfi, Diffraction by a defect in an open waveguide: a mathematical analysis based on a modal radiation condition, SIAM J. Appl. Math. Volume 70, Issue 3, pp. 677-693 (2009) 79.
[10] R. Gilbert, Klaus Hackl and Y. Xu, Inverse problem for wave propagation in a perturbed layered half-space, Mathematical and Computer Modeling, 45 (1-2), 2007, 21-33.
[11] R. Weder, *Spectral and scattering theory for wave propagation in perturbed stratified media*, Springer-Verlag, New York, (1991).
[12] C. H. Wilcox, *Sound Propagation in Stratified Fluids*, Springer-Verlag, New York, (1984).
[13] Y. Xu, *Scattering of acoustic wave by obstacle in stratified medium*, in Partial Differential Equations with Real Analysis, H. Begehr and A. Jeffrey ed., Pitman Research Notes in Mathematics series 263, Longman Sc & Tech., Essex, UK (1992).
[14] Y. Xu, Radiation condition and scattering problem for time-harmonic acoustic waves in a stratified medium with a nonstratified inhomogeneity, IMA J. Appl. Math. 54 (1995), no. 1, pp. 9-29.
[15] Xu, Yongzhi; "Scattering of acoustic wave by obstacle in stratified medium," *Partial Differential Equations with Real Analysis* Ed. H. Begehr, A. Jeffrey, Pitman Research Notes in Mathematics Series 263, (1992) 147-168.
[16] Xu, Yongzhi; "Reciprocity relations and completeness of far-field pattern vectors for obstacle scattering of acoustic wave in a stratified medium," *Math. Meth. in Appl. Sciences*, Vol. 18, 51-66 (1995).
[17] Xu, Yongzhi; "Object shape determination using incomplete data in a stratified medium", *Automatic Object Recognition*, Firooz A. Sadjadi, Editor, Proc. SPIE 2756, (1996), 164-173.
[18] Xu, Yongzhi; "Some Inverse Problems in Stratified Media", *Generalized Analytic Functions*, ed. H. florian et al., Kluwer Academic Publishers, 255-264, (1998).
[19] Xu, Yongzhi; "An inverse obstacle scattering problem in a stratified medium", *Applicable Analysis*, Vol. 68(3-4), 261-280, (1998).
[20] B. ZHANG AND S. N. CHANDLER-WILDE, Acoustic scattering by an inhomogeneous layer on a rigid plate, SIAM J. APPL. MATH. Vol. 58, No. 6, pp. 19311950, 1998.

AUTHOR INDEX

Institute of Physics
Dirac House, Temple Back
Bristol BS1 6BE UK

CURRAN ASSOCIATES INC
proceedings.com